GREAT IDEAS OF

MODERN
MATHEMATICS

Their Nature and Use

By

JAGJIT SINGH

DOVER PUBLICATIONS, INC.
NEW YORK

Published in Canada by General Publishing Company, Ltd., 30 Lesmill Road, Don Mills, Toronto,
Ontario.

*Great Ideas of Modern Mathematics: Their Na-
ture and Use* is a new work, first published by Dover
Publications, Inc., in 1959. It is published simulta-
neously in the United Kingdom by Hutchinson &
Company, Ltd., under the title *Mathematical Ideas
—Their Nature and Use*.

International Standard Book Number: 0-486-20587-8
Library of Congress Catalog Card Number: 60-1473

Manufactured in the United States of America
Dover Publications, Inc.
180 Varick Street
New York, N.Y. 10014

CONTENTS

ACKNOWLEDGMENTS

IN writing a book of this kind one is inevitably indebted to more people than one knows. The authors that one has to read to provide the background material are too numerous to be listed. Nevertheless, I cannot forgo the pleasure of acknowledging with gratitude my debt to Professor Kenneth O. May, Sir K. S. Krishnan, F.R.S., Prof. Mahalanobis, F.R.S., Prof. R. Vaidyanathanswamy, Dr. C. R. Rao and Dr. Martin Davidson, who read the book and made several valuable suggestions for improvement.

Thanks are also due to Mr. M. J. Moroney, whose frank and forthright criticism of some earlier drafts of the book helped to remove a number of faults, and to Sardar Khushwant Singh, who read the book to spotlight the difficulties of the enquiring layman and thus acted as a vigilant censor on his behalf.

FOREWORD

THIS is neither a text-book nor a general exposition of mathematics. It is an explanation of certain extremely useful branches of mathematics, some of which are little known to non-specialists. Though originally only of academic interest, they have been developed to tackle problems which proved unsolvable by the old mathematics. Some general knowledge of them is essential to the understanding of modern science, almost every department of which has its own special mathematical technique. This mathematisation of science is quite a recent phenomenon, for, until about the close of the nineteenth century, most branches of science were largely descriptive. It is true that the physical sciences, such as physics and astronomy, did use a good deal of mathematics, but even in these sciences one could get along and often make useful contributions without it. In fact, some of the most significant contributions to these sciences were made by non-mathematicians. Nowadays, even descriptive sciences, *e.g.* biology, zoology, genetics, psychology, neurology, medicine, economics, philology, *etc.*, have begun to employ elaborate mathematical techniques. The mathematics used is not always difficult, but it is often unfamiliar even to people who have had some mathematical training at a university. This is inevitable because the mathematics used is of recent origin and has not yet found its way into the school and college curricula.

One consequence of this development is that science is becoming the exclusive preserve of specialists. This is particularly unfortunate as science is the agency whereby our society has been changed in the past and will be changed still more rapidly and profoundly—perhaps within our own lifetime. It is dangerous to let this knowledge remain in the hands of a small group of specialists, however gifted, for the potentialities of science for good and evil are infinite. If this danger is to be averted, everyone, including you and I, must make a serious effort to understand what contemporary scientists are doing; but this understanding is impossible without some insight into the ideas of modern mathematics.

Fortunately, the need for popularising mathematical ideas is now universally recognised. More than twenty years ago an excellent popularisation of mathematics appeared in Professor Lancelot Hogben's *Mathematics for the Million*. This book deals with mathematics as it developed till about the middle of the eighteenth century. There is need for a similar popularisation of mathematical ideas that have come into being during the last two hundred years and are now proving so fruitful in genetics, economics, psychology, evolution, *etc.*, as well as in physics, astronomy and chemistry. The task will probably require many hands, if only for the reason that no single person can hope to master more than a few branches of the subject.

In this book I have tried to give a popular and, I hope, a not-too-inexact exposition of some major mathematical ideas that have been invented during the past two centuries or so. In explaining these topics, I have assumed the reader to have some knowledge of elementary mathematics such as could be obtained from any text-book on school algebra and geometry. I have tried to show how the most fruitful of these newer ideas arose as a result of man's impulse to mould his environment according to his heart's desire, and that where this impulse has been lacking mathematical progress has been stunted.

Although some of the theories explained in this book are still controversial, I have often taken 'sides' in presenting the different points of view, preferring to present things as I see them rather than as they might appear to an imaginary observer. However, the reader should find the treatment on the whole sufficiently unbiased for him to judge the issues for himself.

Calcutta   JAGJIT SINGH

1

THE NATURE OF MATHEMATICS

W E learn about the universe around us by experience and observation on the one hand, and by thought and deductive reasoning on the other. Although in practice we get most of our knowledge by continually combining observation with deduction, it is possible subsequently to formulate certain types of knowledge by pure deduction starting from a set of 'axioms'—that is, statements accepted as true without proof because we feel that their truth is self-evident. The classic example of deductive method is school geometry, where we postulate certain definitions and axioms concerning points, lines, *etc.*, and deduce a logical chain of theorems concerning lines, angles, triangles, and so forth. The great advantage of the deductive method is the certainty of its conclusions. If there is no fallacy in our reasoning the conclusions must be correct—unless there is something wrong with our axioms. But the question whether the axioms chosen at the outset are valid is a difficult one. Certain axioms, which appear obvious to some, may seem clearly false or at least very doubtful to others. A familiar example of this kind is Euclid's axiom that parallel lines never meet. Mathematicians now recognise that it cannot be accepted as self-evident even though schoolboys are still taught geometry as if it were true. Indeed, the examination of the validity of any given axiom system is such a vexed question that some people propose to cut the Gordian knot by claiming that 'pure' mathematics is merely concerned with working out the consequences of stated axioms with no reference whatever to whether there is anything in the real world that satisfies these axioms.

Further support to this view was lent by the profound and penetrating studies of the foundations of Euclidean geometry towards the close of the nineteenth century. They revealed that geometrical proofs depend mainly on diagrams embodying properties which we accept as part of our equipment without including them in the axioms. Consequently an attempt was made to prove geometrical theorems without using the meanings of geometrical terms—like points and lines—as understood in everyday speech, but only their *relations and properties* as explicitly stated in the initial axioms. It was declared that 'if geometry is to be deductive, the deduction must everywhere be independent of the *meaning* of geometrical concepts, just as it must be independent of diagrams; only the relations specified in the axioms employed may legitimately be taken into account.'

But the insistence that the proof be independent of the meanings of the terms used, and employ only their mutual relations as explicitly stated in the basic axioms, did not mean that these terms were to be 'meaningless'* —only that they should have no specific reference to any particular thing. The terms of the axiom system should remain deliberately undefined— that is, free from association with any specific thing so that they become pure symbols formally related to one another in certain ways embodied in the axioms of the system.

There are two advantages in adopting such a procedure. First, by cutting out all associations that otherwise cling to terms used in the ordinary meanings of everyday speech, we eliminate the tendency to use meanings and relations other than those expressly stipulated in our axiom system— a tendency so prominent in Euclidean proofs. Second, the axiom system acquires a generality otherwise impossible. For example, it becomes possible to encompass within the single framework of an axiom system concepts, such as *group*† and *abstract space*, appearing in seemingly unrelated branches of mathematics. These very valuable gains are not to be despised. But the systematic draining of all meaning and content from the terms of our discourse, and thus turning them into pure symbols, means that mathematical proof becomes a sort of game with symbols. In a somewhat over-simplified form this is the view advocated by some mathematicians. Instead of playing-cards, dice, or pawns, bishops, rooks and knights, we may start with a collection of symbols such as $\sim$, $-$, $\times$, $+$, $=$, *etc.*, and a set of 'rules' or 'axioms' according to which they may be combined. We then proceed to play a game which consists of arranging these symbols to form 'meaningless' expressions according to the given 'rules'. We could change the symbols or the rules or both in any arbitrary manner we liked; the result would always be 'pure' mathematics.

Like all games this game of manipulating paper marks, that is, 'pure' mathematics, has, according to its present-day exponents, no ostensible object in view except the fun of playing it and playing it well. If men, nevertheless, find its results of great practical utility in their daily lives, that is not its *raison d'être*. Its sole function has been and should be merely to divert the human mind by the 'elegance' and 'beauty' of its expressions, irrespective of their utility. It is, no doubt, possible to argue in favour of this view. In fact, G. H. Hardy has done so in his charming little book, *A Mathematician's Apology*. His argument seems to rest on a distinction between what he calls 'school' or 'Hogben' mathematics, that is both

---

* This usage of the word must be distinguished from that of everyday speech as a term of disparagement. The word 'meaning-free' comes closer to the sense in which it is used here.

† See Chapter 7.

'trivial' and 'dull' but has considerable practical utility, and 'serious' mathematics that alone is the 'real' mathematician's delight despite its remoteness from everyday life.

It is evident that even those who hold this view cannot boast too seriously about the uselessness of their work and the 'meaninglessness' of the symbols and rules of their game. They do not go so far as to express dismay when their work turns out to be useful! Moreover, they are sufficiently sane to allow that a 'certain sense of fitness of things' should at any rate forbid a 'wild overturning of the law and order established in the development of mathematics'.

Mathematics is too intimately associated with science to be explained away as a mere game. Science is serious work serving social ends. To isolate mathematics from the social conditions which bring mathematicians (even of the Game Theory school) into existence is to do violence to history. Hogben and other writers have shown how great mathematical discoveries and inventions have throughout history been rooted in the social and economic needs of the times. Most books take us little beyond the eighteenth century in tracing this connection. A mistaken view has grown up in certain quarters that modern mathematics, particularly during the last 150 years, is a 'free creation' of the human mind, having little or nothing to do with the technological and social demands of the time. If science and technology have been able to make use of such mathematical inventions as, for example, the tensor calculus in Relativity and the matrix theory in Quantum physics, that in no way influenced their creation. Nevertheless, as will appear in the sequel, there is a close tie-up between the practice and theory (which has largely remained mathematical even up to the present time) of science and technology. Thus, while the empirical practices of the eighteenth-century mechanical engineers from Savery to Watt led to Thermodynamics, the 'pure' theory of Faraday and Maxwell paved the way for the practical inventions of Edison and Marconi. As Leonardo da Vinci remarked, science is the captain and practice the soldiers: both must march together. One reason why Einstein's Relativity theory has not yet advanced very much beyond the stage it reached during the second decade of the twentieth century may be the fact that it has made no practical difference in the calculation of the astronomical tables in the *Nautical Almanac*. If, at some future date, we are able to undertake interplanetary voyages, relativity might find a field of applications for want of which it has languished.

In spite of the 'game' theory, mathematics is still largely inspired by contemporary social, technological and scientific demands, as in the new 'pure' mathematics that has been created to deal with the problems of cosmic rays, stellar dynamics, stochastic processes and cybernetics. The present-day needs of science and technology for speedier methods of calculation by means of electronic machines may well inaugurate as great a mathematical revolution as the Hindu invention of zero and the positional system of writing numbers.

Far from despising utility and practical applications, the early pioneers of modern science and mathematics—men like Huygens, Newton and Leibnitz, to name only a few—cultivated them mainly for the advancement of technics. Is it any wonder then that the founders, patrons and some of the first scientists of such scientific academies as the Italian *Academie del Simento*, the English Royal Society or the French *Academie des Sciences*, were kings, nobles, courtiers, magnates and city merchants? The idea of a machine as the demiurge of a new heaven on earth was so uppermost in their minds that the very first standing committee of the Royal Society had for its terms of reference the 'consideration and improvement of all mechanical inventions'.

What, then, has given rise to the recent idea that mathematics is a game, a *jeu d'esprit* or 'free creation' of the mind divorced from the practical problems of daily life? It is the fact that the intimate connection between mathematics and reality is lost sight of in the abstract logical schemes which a mathematician constructs, though these always embody certain essential features abstracted from some sphere of reality. These logical schemes created by mathematicians do often *look* like games of manipulating symbols according to certain rules. This, however, does not mean that arithmetic, geometry, algebra, the calculus, *etc.*, arose by someone constructing these theories as games played according to some rules. Quite the contrary. They arose as abstractions from concrete applications, though their *subsequent logical* formulation may appear like games played with symbols. The authors of the game theory are, of course, aware of this distinction between the historical genesis of mathematical knowledge and its subsequent logical formulation. But when they claim that mathematics is a game they seem to confuse the means of expressing mathematical truths and the mathematics itself. (We shall deal with this subtle question more fully in the last chapter.)

2

NUMBER AND NUMBERS

WHEN Ulysses had left the land of the Cyclops, after blinding Polyphemus, the poor old giant used to sit every morning near the entrance to his cave with a heap of pebbles and pick up one for every ewe that he let pass. In the evening when the ewes returned, he would drop one pebble for every ewe that he admitted to the cave. In this way, by exhausting the stock of pebbles that he had picked up in the morning he ensured that all his flock had returned.

The story is apocryphal, but this is precisely what the primitive shepherd did with his sheep before he learnt to count them. This also is not very far from what a modern mathematician does when he wants to compare two infinite collections, which *cannot* be counted in the ordinary way. However, the important difference between the two is that while the former used this tallying process without knowing what he was doing, like M. Jourdain speaking prose, the latter uses it with knowledge and insight. He thus acquires certain powers, otherwise unattainable, such as the power to count the uncountable. We shall see later (in Chapter 5) how the mathematician, by refining the primitive shepherd's practice, has succeeded in accomplishing this and other seemingly paradoxical feats. Meanwhile, we may note the theory behind the shepherd's practice. This theory is based on the fact that if the individuals of a flock can be matched, one by one, with those of a heap of pebbles so that both are exhausted together, then the two groups are equal. If they are not, the one that gets exhausted earlier is the lesser.

What gives this matching process its great power is that it can be applied universally to all kinds of aggregates—from collections of ewes and pebbles to those of belles and braces, apples and angels, or virtues and vipers. Any two aggregates whatever can be matched so long as the mind is able to distinguish their constituent members from one another.

Gradually men formed the notion of having a series of standard collections for matching the members of any given group or aggregate. One such series consisted of the ten different collections formed by including one or more fingers of their two hands. All collections, which could, for example, be matched on the fingers of one hand were 'similar' in at least one respect, however they might otherwise differ among themselves. They were, as we now say, all equal. These standard collections were then given names—

One, Two, Three . . . *etc.* This is the social origin of the practice of counting. Thus, when we now say that the number of petals in a rose is five, all that we mean is that if we start matching the petals one by one with the fingers of one hand, the members of both the collections are exhausted simultaneously. By long practice in handling the abstract symbols 1, 2, 3 . . . we are liable to forget that they are only a shorthand way of describing the result of an operation, *viz.*, that of matching the items of an aggregate with those of some set of standard collections that are presumed to be known. The process is so habitual that it usually escapes notice. This has caused endless confusion in the past, when, for long centuries, even learned men failed to understand the nature of number, particularly when they began to handle negative and imaginary numbers. If we keep in view the fact that whole numbers or integers are a mere shorthand for describing the result of a matching process, in which one of the collections is presumed to be known, we shall avoid a lot of trouble in understanding the nature of more sophisticated types of numbers in mathematical literature.

We have seen that originally man formed his standard collections for counting with the fingers of his hands. In the beginning this sufficed, there being no occasion to budget for atomic piles, armament races, refugee reliefs, or Marshall Aid. But presently, even in the days of the river-valley civilisations of antiquity, the needs of armies, taxation and trade gave rise to collections which could hardly be matched on the fingers of the two hands. What could man do about it? He could use the marks on his fingers instead of the fingers themselves for the purpose; but even so he would not have enough of them. But as the matching process was independent of the nature of the members constituting the collections, it did not matter whether he formed them by means of fingers or finger-marks or anything else. So he conceived the idea of generating a new standard collection from one already known by mentally adding just one more item to it. And as the process could be repeated indefinitely, he produced an unending succession of standard collections some one of which sufficed to match any given collection, however large. Thus to the original idea of matching or tallying was grafted another—that of order—in virtue of which relative rank is given to each object in the collection. Out of the union of the two arose the idea of integral number—an unending, ordered sequence of integers.

In matching two collections we have hitherto considered them as mere crowds of individuals without any internal order between themselves. The concept of number as the characteristic of a class of similar collections evolved from the practice of matching unordered aggregates (*e.g.* using any pebble in Cyclops' hand for tallying any ewe in the fold) is known as the *cardinal* number. However, we can also conceive of ordered aggregates, such as soldiers in a battle-array or ewes arranged in a straight file, in which every element has a rank or place. When we match such ordered aggregates and conceive of number as the characteristic of a class of similar ordered aggregates, which exhaust themselves together in a matching process, it is known as the *ordinal* number. Primitive man used both these concepts without making any distinction between them. When he used pebbles like Cyclops, he was using *cardinal* numbers; but when he used his fingers he probably used them in a definite order, possibly first his right-hand thumb, then the index, middle, ring and little fingers. In the latter case, he was using *ordinal* numbers. The distinction between the two types of numbers is somewhat subtle and was not noticed by mathematicians themselves till about the end of the nineteenth century. Fortunately, it is of no great importance for all ordinary purposes and may be ignored. The sole justification for introducing what may appear to some a pedantic distinction is the importance it assumes in the theory of transfinite numbers.*

Now, quite early in life we are taught the technique of adding and multiplying the integers. Underlying this technique are certain general laws of addition and multiplication. Though known by high-sounding names they merely express in symbolic language just one commonplace fact of everyday experience. That is, that it makes no difference in what order you add the various sets of objects. Thus whether you buy two books on the first occasion and three on the second or vice versa, your total purchase remains the same. We express facts like this by the formula $2 + 3 = 3 + 2$. This formula is a particular case of a more general law, the *commutative law of addition*, which requires that the sum of any two integers such as a and b is the same in whatever order we may choose to add them. In symbols,

$$a + b = b + a. \qquad . \qquad . \qquad . \qquad . \quad (1)$$

Next, there is the *associative addition law*, expressed by the equality

$$(a + b) + c = a + (b + c)^\dagger. \qquad . \qquad . \qquad . \quad (2)$$

Since multiplication is only reiterated addition, there are naturally also corresponding *commutative* and *associative multiplication laws*, the counterparts of the addition laws. Thus:

$$a.b = b.a \qquad . \qquad (3), \textit{corresponding to } (1)$$

and $\qquad\qquad (ab)c = a(bc) \qquad . \qquad (4), \textit{corresponding to } (2)$

Finally, we have the *distributive law*

$$a(b + c) = ab + ac \qquad . \qquad . \qquad . \qquad (5)$$

* See Chapter 5.
† The brackets here simply mean that the numbers within them are to be added first.

In the following sections we shall see how the number system of integers can be extended to include other types of numbers like negative, fractional and irrational numbers. It will also be shown that the above five fundamental laws hold equally well for these more complicated types of numbers. Further, they remain valid even for sets of symbols which are not numbers but behave *like* numbers in some respects. We shall cite several instances of such sets later.* But the fact that such sets of symbols are possible has enabled modern mathematicians to make truly amazing feats of abstraction. We can, for instance, picture a set of 'elements' a, b, c, d . . . about which we assume nothing except that they obey the five fundamental laws of arithmetic. Starting from this assumption we can prove a number of theorems about them. These theorems will hold, not only for numbers—whether integers, fractions or irrationals—but also for a much wider class of symbols which includes these numbers as a special case. In other words, we are able to subsume the properties of a vast variety of elements under one generic form that applies to them all. This is the method of abstraction which is the very life-breath of modern mathematics.

Indeed, the main difference between ancient and modern mathematics is just this. In its relentless drive towards greater abstraction it refuses to tie up the symbols of its discourse to anything concrete, so it cannot make much use of the meanings of these symbols. How could it, indeed—since it refuses to give them any meanings, or at any rate keeps them as meaning-free as possible? Mathematics thus has to base itself more on the mutual relations of abstract symbols, as embodied in laws like the five fundamental laws of arithmetic, than on the meanings of these symbols. That is why some people call ancient mathematics '*thing-mathematics*', meaning concrete, and modern mathematics '*relation-mathematics*', meaning abstract.

This difference does not imply that modern mathematicians are cleverer or more imaginative than their forefathers. It is largely the outcome of changes that have since occurred in the mode of civilized living. In the ancient world, where material production was largely for use and barter, things had not yet become completely metamorphosed into abstract embodiments of sale or money value. This could occur only with the change from a productive system for use or barter to a money economy producing commodities. In such a social system all qualitative differences between commodities are effaced in money and one begins to think in terms of their money values as—four talents, four livres or four guineas. From this it is but a step to think of the number 'four' as dropping its material crutches and coming into its own as an abstract conceptual symbol for the common quadruplicity of them all. But this is by no means

* See pages 21 and 146.

the end. A modern mathematician is able to carry on the process of abstraction much farther. All that he needs is a system of elements obeying some scheme of abstract laws like the five laws of arithmetic cited above.

You may wonder whether these abstract schemes about phantom entities obeying phantom laws are of any use at all. They are indeed. As science advances, sooner or later a stage is reached when it has to reckon with what is sometimes called interphenomena, that is, phenomena beyond the limit of direct observation. Thus no one can see what actually happens inside a star, an atom, a gene, a virus, an amacrine cell or an ultramicroscopic speck of nerve fibre. And yet a scientist must somehow figure it out if he is to give an intelligible account of perceptible phenomena. One way of doing it is to adopt some abstract scheme of a mathematician's fancy and see where it leads us. It may happen that it enables us to predict some observable phenomena capable of direct observation. If we do succeed in observing the predicted phenomena, we may be sure that the abstract scheme does embody at least some features of the interphenomena under study. Surprising as it may seem, the method actually works. It is by the use of such abstract mathematical schemes that scientists have been able to fathom what happens in the interiors of stars and atoms.

Stars and atoms may seem very remote, but they have now begun to influence our everyday lives very directly, for their study has revealed new sources of power which, in the case of atoms, we may use for war or peace as we may choose. For instance, thanks to these studies we can now imitate, though in a rudimentary manner, the cosmic processes at work in the solar interior and construct here on earth those miniature suns, the H-bombs, which threaten to wipe the human race out of existence. And yet the same theory which has produced the H-bomb is also potentially capable of putting, as it were, sunshine on tap for the advancement of technics, civilisation and life.

* * * *

As we have seen, early man's need to compare discrete groups, such as flocks of sheep, herds of cattle, fleets of vessels and quivers of arrows, gave rise to the concept of the integer. His other needs had even more fruitful consequences. For instance, he wanted to know whether the milk-yield of his cattle was rising or falling, or to make sure that no one encroached on his field at harvest time. Here he was faced with *continuously* varying quantities that could not be counted like the discrete objects of a group, such as eggs in a basket. Nevertheless, he found a way of reducing the problem of quantizing these continuous magnitudes to that of counting a discrete group. Thus, he took a standard yard-stick and counted the number of such sticks, which, placed end to end, covered the entire length of the field from one extremity to the other. The continuous length thereby became a discrete group of equal yard-sticks. Or he took a standard vessel and poured out of his milk-yield as many vessel-fuls as he possibly could. The continuous quantity—*i.e.*, milk-yield—was thus changed into a discrete collection of standard-sized vessels. In this way continuous quantity became discrete and could be counted. But the solution had its awkwardness. For, in reducing the length of a field to a discrete group of yard-sticks or unit-measures, it might happen that a residue was left. For instance, the yard-stick might cover a given length 200 times and then leave a residue smaller than the yard-stick. What was he to do with it?

There were two ways of handling it, namely to find a unit that leaves no residue or to ignore it altogether. The first method was used in measuring time when men divided the duration of daylight into twelve equal hours. The hour was thus a variable unit, for a duration of one hour at the time of summer solstice, for instance, was not the same as that at the time of vernal equinox. At a time when there were no Hours of Employment Regulations, overtime wages, payment by the hour, *etc.*, this meant no social inconvenience. It was otherwise with measuring the lengths of fields. For it would have been necessary to discover by trial and error a unit that would cover each length exactly without leaving a residue. Even if he found one it would most probably be of no use for measuring another length, for which it would be necessary to seek another unit. If, therefore, residues were to be eliminated by this method, almost every length would have needed a special unit of its own—very much like Chinese writing, which has a separate ideograph for every word in its vocabulary. With such a medley of different units, not only the calculus of lengths would have been more complicated than Chinese writing but few lengths could have been compared with each other. A fixed standard of length, as also of weight and volume, was therefore a prime necessity of social intercourse. Consequently, while all the early civilisations adopted variable units of time, the units of lengths, weights and measures were fixed. As early as 2300 B.C. the Sumerian Hammurabi, for example, issued edicts fixing these standards.

But if these units were to remain invariable, the problem of the residue had to be solved. Now it would be a mistake to imagine that early man became 'residue-conscious' overnight. Far from it. Actually this residue-consciousness took whole millennia acoming. Too often and too long men tried to ignore the residues and were content to make do with approximations, even ludicrous approximations, such as the value of $\pi = 3$ adopted in the *Book of Kings*, in *Chronicles** and by the Babylonians. Neverthe-

* 1 Kings vii, 23; 2 Chronicles iv, 2.

less, the problems of trade and the administration of the vast revenues of temples, city states and empires, kept in the forefront the mathematical problem of the residues. Even then our present insight into the nature of fractional numbers, which were created to solve it, was gained with the greatest difficulty. The Egyptians, who treated only fractions with the numerator 1, never understood that fractions were amenable to the same rules as integers. The Babylonians, who had begun to deal with fractions as early as 5000 B.C., did not acquire complete mastery of fractional numbers until 2000 B.C. We can well appreciate their difficulty: they faced problems unprecedented in human history. To solve them they had to find their way about in an utterly strange and uncharted domain. No wonder that, like the early navigators, they often reached their goal by the longest and most devious route. And yet it is easy, retrospectively, to state the residue problem and its solution. The problem of the residue is simply this:

If the fixed unit yardstick does not happen to go into the length to be measured an exact number of times, how are we to measure the residue? The solution is equally simple. Divide the unit yardstick itself into a number of aliquot parts and then measure the residue with it. For instance, the yard may be divided into 3 sub-units, each one foot long, and we may try to measure the residue with the sub-unit, the foot. It may be that the new sub-unit will lie along the residue from end to end exactly twice. We thus have a sub-unit which covers the residue twice and our original unit—the yardstick itself—thrice. This process gives us a number couple, *viz.* 2 and 3, which can be used as a measure of the residue. We may write it as $\frac{2}{3}$ as our Hindu ancestors did, or $\frac{2}{3}$ or 2/3, as in modern text books, or as (2,3) as some learned people might advocate; it matters little. The important point is that it is a shorthand for an operation just as the single integer by itself is a symbolical way of describing the result of a matching process. A number pair like $\frac{2}{3}$ or (2,3) says what in ordinary language would have to be expressed somewhat as follows:

'If you take a sub-unit that divides the fixed unit, say a yard, exactly thrice, and use it to span the residue in question, it will go into the residue exactly twice'.

The problem of measuring residues is, therefore, merely the problem of finding *some* sub-unit that will cover exactly both the residue and the fixed unit. Although, as we shall see later, strictly speaking it is not soluble in all cases, a solution that is good enough for all practical purposes can always be found. For, if we take a sufficiently small sub-unit of the original unit, it will either cover the residue exactly or, at most, leave a remainder less than the chosen sub-unit. But as the sub-unit can be made as small as we please, the remainder, if any, will always be smaller still.

Having found a way to measure residues or fractions, as they are generally called, it was necessary to devise methods of adding and multiplying
them. In time, general rules for adding and multiplying fractions were
framed. They are the well-known rules we all learnt at school, *viz.*:

Addition rule: $(a, b) + (c, d) = (ad + bc, bd)$,
Multiplication rule: $(a, b)(c, d) = (ac, bd)$,*

You could, if you were in doubt, even 'prove' these rules by strict logic.
All that you need do is to recall the meaning of number pairs (a, b) and
(c, d) and the way to add or multiply them would be clear. But the point
that is of greater interest is that these two rules for adding and multiplying
fractions show that fractions too obey the same five fundamental laws of
arithmetic as the integers. This too can be proved logically.

*   *   *   *

If you operate with the first three integers 1, 2, 3, you may easily verify
that you can combine them in pairs in $3 \times 3 = 9$ ways. These nine ways
lead to the nine fractional numbers:

$$(1, 1), (1, 2), (1, 3)$$
$$(2, 1), (2, 2), (2, 3)$$
$$(3, 1), (3, 2), (3, 3)$$

Likewise, four integers produce $4 \times 4 = 16$ fractions and five integers
$5 \times 5 = 25$ fractions. From this you may easily infer that with N integers
you can manufacture $N \times N = N^2$ fractions. This would seem to show
that the set of fractional numbers is vastly more numerous than that of
positive integers. But the set of positive integers is a never-ending or
infinite set. If we denote this infinity by the usual symbol ∞, the infinity of
fractional numbers would appear to be the much bigger infinity $\infty \times \infty =$
$(\infty)^2$. And yet if, in the manner of the Cyclops, you started matching the
infinite set of fractional numbers with the infinite set of integers, both the
sets would be exhausted together—provided one could speak of exhausting
inexhaustible or infinite sets! In other words, the two infinite sets of integers
and rational fractions are exactly 'equal'. This is, no doubt, paradoxical.
We shall explain this paradox in Chapter 5.

*   *   *   *

If early man noticed that one herd of cattle could be 'added' to another
and he thus formed the notion of 'addition', he also performed the reverse

* (a, b), (c, d) are here used to denote the fractions a/b, c/d.

operation—that of taking some cattle out of his herd as, for instance, for the purpose of bartering them for other goods. This is the origin of 'subtraction', the inverse of addition. Similarly, multiplication, which is only a reiterated addition, gave rise to its inverse, 'division', a reiterated subtraction. At first, these new operations caused him some confusion. For, while he could always add and multiply *any* two integers, he could not always perform the inverse operations. Thus, he could add *any* two herds of cattle, but he could not take out, say, fifty cows from a herd of only forty. Division, too, must have worried him at times, and he must have often wondered whether a division of, say, seven by two is possible at all. Like most children beginning to learn arithmetic, he, too, must have felt that there isn't a '*real*' half of seven.

Nevertheless, for two reasons early man had less trouble with division than with subtraction. First, he could always divide one integer, say 7, by another, say 2, and supplement the result by adding that the division is not 'exact' and leaves a *remainder*. Second, even if he had to divide a smaller number, say 5, by a larger, say 7, he could interpret the result as a number pair (5, 7)—the fractional number that he had already devised for measuring continuous magnitudes. But if he was asked to subtract, say 7 from 5, he was quite befogged. To make this magic possible, he had to wait for the rise of a banking system with an international credit structure, such as came into being in the towns of Northern Italy (particularly Florence and Venice) during the fourteenth century. The seemingly absurd subtraction of 7 from 5 now became possible when the new bankers began to allow their clients to draw seven gold ducats while their deposit stood at five. All that was necessary for the purpose was to write the difference, 2, on another side of the ledger—the debit side.

Although the attempt to resolve the difficulties of awkward divisions and subtractions did lead to the recognition of fractional and negative numbers, the realisation that they arose from the limitations of the integral number system itself, and could only be overcome by suitably extending that system, came much later. Thus, suppose we are given only the unending sequence of positive integers 1, 2, 3, ... We can clearly add any two of these integers, their sum being itself a positive integer. But we cannot always perform the inverse operation of subtraction. For instance, while we can subtract five from seven we cannot subtract seven from five. If we want to ensure that subtraction of one integer from another be as freely possible as addition, we must extend the number system of positive integers to include negative integers, so as to form a doubly unending set of positive and negative integers:

$$\ldots\; -4,\; -3,\; -2,\; -1,\; 0,\; 1,\; 2,\; 3,\; 4,\; \ldots$$

Only when we operate with such an extended set can we perform the operation of addition and its inverse subtraction on *any* two numbers without any restriction whatsoever. In other words, to make subtraction universally possible the system of positive integers must be extended to include negative integers too. This has as a consequence that to every positive integer, such as a, there corresponds its negative or inverse, $-a$, which belongs to the same set and is such that the sum $a + (-a)$ is zero. This fact may also be expressed by the statement that the equation $a + x = 0$ has always a solution $x = -a$ (also belonging to the set). If we consider the particular case of this equation when $a = 0$, we find that x too is zero. It therefore follows that the number zero of the set is its own inverse. It is called the '*identity element*' of the set for addition. We call it the *identity* element as its addition to any integer leaves the latter unaltered. If we imagine that the relation between an integer and its inverse is like that of an object and its mirror image, then the identity element zero is like the reflecting surface, which is its own image.

Our first extension of the number system is thus the doubly unending set of positive and negative integers complete with the identity element zero:

$$\ldots -4, -3, -2, -1, 0, 1, 2, 3, 4, \ldots$$

Such a system of positive and negative integers, including zero, is known as an *integral domain*. It is not possible to extend the integral domain any farther by means of addition and subtraction alone. No matter which two numbers of the domain we may add or subtract, we shall always end up with a positive or negative integer belonging to the integral domain. We may say that the integral domain is *closed* under addition and subtraction because the way to further extension of the domain by performing these two operations is blocked.

Just as we had to extend the system of positive integers to include negative integers in order to make subtraction universally possible, a similar consideration with regard to division leads to a further extension of integers to include rational fractions. For, so long as we work with integers, we can always multiply any two of them, but the inverse operation—division—is not always possible. To make division between any two integers as universally possible as multiplication, we have to extend the number system to include rational fractions, that is, number pairs devised to measure residues. In other words, the set of all integers must be extended to include all rational fractions so that each number a (other than zero) of the set has an inverse $1/a$ belonging to the set with respect to multiplication. We may express the same thing by saying that the equation $ax = 1$ has a solution $x = 1/a$ belonging to the set for all a's not equal to zero. If we consider the particular case of this equation when $a = 1$, we find that x is also 1. It therefore follows that the number 1 of the set is its own inverse. It is called the "identity element"* for multiplication, since multiplying it by any number leaves the latter unaltered. Such a system which includes all positive and negative integers as well as fractions is called a *field*.

With the construction of the field of rational numbers our second extension of the number system is in a way complete. It permits us to perform on *any* two numbers of the system not only addition and its inverse, subtraction, but also multiplication and its inverse, division, and express the result by a number belonging to the rational field. No further extension of the field is possible by performing any of these four arithmetical operations on any two numbers of the rational field. No matter which two numbers of the rational field we may add, subtract, multiply or divide, we shall always end up with a number belonging to the rational field. In other words, the field of rational numbers is *closed* under all the four arithmetical operations. But even so the number system still remains incomplete in some ways. It can be shown that certain magnitudes like the diagonal of a unit square cannot be measured by rational fractions. To measure the diagonal we have to find a sub-unit that goes an exact number of times into the side as well as the diagonal. Suppose, if possible, there is such a sub-unit which divides the side m times and the diagonal n times. Then the length of the diagonal is given by the number pair or fractional number n/m. We may assume that m and n are *not* both even, for if they were, we could cancel out the common factor 2 from the numerator and denominator till one of them became odd. Now, the lengths of the two sides of the square AB, BC, are both equal to unity, and that of the diagonal is n/m (see Fig. 1). But in a right-angled triangle like ABC,

$$AB^2 + BC^2 = AC^2.$$

In other words,
$$2 = n^2/m^2$$
or
$$n^2 = 2m^2.$$

But it can be proved that no two integers can satisfy this equation unless they are both even—a possibility which has been expressly excluded by our hypothesis. It is therefore impossible to find a rational number to measure the diagonal length AC.

The discovery of magnitudes which, like the diagonal of a unit square, cannot be measured by any whole number or rational fraction, that is, by means of integers, singly or in couples, was first made by Pythagoras some 2500 years ago. This discovery was a great shock to him. For he was a

---

* Note carefully that the number 1 plays the same role with regard to multiplication as the number zero with regard to addition. (See page 14.)

number mystic who looked upon integers in much the same reverential spirit as some present-day physicists choose to regard Dirac's *p*, *q* numbers, *viz.* as the essence and principle of all things in the universe. When, therefore, he found that the integers did not suffice to measure even the length of the diagonal of a unit square, he must have felt like a Titan cheated

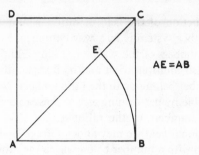

Fig. 1—The diagonal and the side of the square are incommensurable.

by the gods. He swore his followers to a secret vow never to divulge the awful discovery to the world at large and turned the Greek mind once for all away from the idea of applying numbers to measure geometrical lengths. He thus created an impassable chasm between algebra and geometry that was not bridged till the time of Descartes nearly 2000 years later.

Nevertheless, there is no great difficulty in measuring the length of the diagonal by an extension of the process that gave rise to fractions. Suppose we use the unit stick to measure the diagonal *AC* of the square. It goes once up to *E* and leaves a residue *EC* (see Fig. 1). Its length is therefore 1 + *EC*. Now suppose we use a sub-unit which goes into the original unit ten times, in order to measure *EC*. We shall find that it will cover *EC* four times, leaving again a small residue. This gives us a closer measure of the diagonal length, *viz.*, 1 + 4/10 *plus* a second residue. To obtain a still closer estimate, we try to measure the second residue with a still smaller sub-unit which divides our original unit into, say, 100 equal parts. We shall find that it goes into the second residue just once but again leaves a third residue. Our new estimate is thus 1 + 4/10 + 1/100 *plus* a small residue. To measure this last residue, we can again use a still smaller sub-unit, say, one-thousandth part of the unit, and observe how many times it covers it. We shall find that it covers it four times and still leaves a residue. The diagonal length then is 1 + 4/10 + 1/100 + 4/1000 *plus* a residue.

Now, the important point is that no matter how far we go, the residue always remains. For suppose, if possible, we were left with no residue when we began to use a scale equal to, say, a thousandth part of our unit. This would mean that a scale that covered the unit 1000 times would cover the diagonal 1414 times exactly. In other words, it would be measured by the fraction 1414/1000, but this, as we saw (page 15), is impossible. Thus, while we may be able to reduce the residue to as small a length as we please by prolonging this process far enough, we can never hope to abolish it altogether. How shall we express, then, the length of our diagonal if we want to do it exactly? We can do so by the never-ending sum:

$$1 + 4/10 + 1/100 + 4/1000 + 2/10,000 + \ldots$$

Or, if this is too cumbersome, we may use the non-terminating decimal expression, $1 \cdot 4142 \ldots$, which is only an abridged way of writing the same thing. A still shorter way of writing it would be 'square root of 2' or $\sqrt{2}$, which means that it is *some* number whose square is two. Numbers like $\sqrt{2}$ were called 'irrationals', as they did not appear to be amenable to reason; they escaped the number mesh cast by man to trap them. For instance, take $\sqrt{2}$ itself: if we do not wish to go beyond the first decimal place, the fraction $1 \cdot 4$ is a little too small, while $1 \cdot 5$ too large. By going to the second place, we can get a closer mesh, *viz.*, $1 \cdot 41$ and $1 \cdot 42$. The third decimal place gives us $1 \cdot 414$ and $1 \cdot 415$, fourth decimal place $1 \cdot 4142$ and $1 \cdot 4143$, and so on. If we square the end numbers of each of these meshes of the number-net we have

$$(1 \cdot 4)^2 = 1 \cdot 96 \qquad < 2 \quad < \quad (1 \cdot 5)^2 = 2 \cdot 25;$$
$$(1 \cdot 41)^2 = 1 \cdot 9881 \qquad < 2 \quad < \quad (1 \cdot 42)^2 = 2 \cdot 0164;$$
$$(1 \cdot 414)^2 = 1 \cdot 999396 \quad < 2 \quad < (1 \cdot 415)^2 = 2 \cdot 002225;$$
$$(1 \cdot 4142)^2 = 1 \cdot 99996164 < 2 \quad <(1 \cdot 4143)^2 = 2 \cdot 00024449;$$

. . . .

We note two properties of the end numbers of this mesh system that we have created to trap the square root of 2. First, the left-hand end numbers, *viz.* $1 \cdot 4$, $1 \cdot 41$, $1 \cdot 414$, $1 \cdot 4142 \ldots \ldots \ldots$ continually increase or at least never decrease. Second, no matter how far we go, the square of any number in the sequence always falls short of 2, though the difference continually decreases. Similarly the right-hand end numbers, *viz.* $1 \cdot 5$, $1 \cdot 42$, $1 \cdot 415$, $1 \cdot 4143 \ldots \ldots \ldots$ continually decrease or at any rate never increase. Likewise, the square of any member always exceeds 2, though the difference continually diminishes the farther we go. We have here a process whereby we generate two sequences of fractional numbers which continually approach the square root of 2 from both below and above, although they never actually reach it. Whatever the degree of precision required in the estimate, we can always pick up two numbers, one from each sequence, which are sufficiently close together, and between which the desired square root lies. Can we, then, say that this hunt for the square root of 2 is not as 'perfect and absolute a blank' as that of Lewis Carroll's crew in the hunting of the Snark? In other words, can we assert that the square root of 2 'exists', in spite of the fact that it cannot be expressed as an ordinary fraction? The question is not entirely academic, for unless we admit its 'existence', our number system is not complete. Our number vocabulary is not rich enough to quantize certain magnitudes. Without the irrationals we should have no numbers for exactly measuring certain lengths, although we might have increasingly finer sets of approximations. Voltaire once remarked that if God did not exist, it would be necessary to invent Him. With still greater justification, the mathematician says that if the square root of 2 does not exist, it is necessary to invent it, and he invents it by writing $\sqrt{2}$. It is the unique number towards which the infinite ever-increasing sequence of fractional numbers 1·4, 1·41, 1·414, 1·4142 . . . continually tends without ever reaching it. In other words, it is the *ultima Thule* or *limit* of this sequence which we usually abbreviate as the non-terminating, non-recurring decimal, 1·4142 . . . .

Now it is no use inventing numbers unless we know how to combine them by addition, multiplication, *etc.* What do we mean by adding two irrational numbers like $\sqrt{2}$ and $\sqrt{3}$? We defined $\sqrt{2}$ as the *limit* of an infinite never-decreasing sequence of rational fractions, such as:

$$1·4, 1·41, 1·414, 1·4142, \ldots \ldots$$

Likewise $\sqrt{3}$ is the limit of another infinite never-decreasing sequence of fractions, namely:

$$1·7, 1·73, 1·732, 1·7321 \ldots \ldots$$

The sum of $\sqrt{2}$ and $\sqrt{3}$ is merely the limit of the new infinite never-decreasing sequence formed by adding the corresponding terms of these two sequences *viz.*:

$$(1·4 + 1·7), (1·41 + 1·73), (1·414 + 1·732), (1·4142 + 1·7321), \ldots \ldots$$

Likewise, the product of $\sqrt{2}$ and $\sqrt{3}$ is the limit of the infinite sequence:

$$(1·4)(1·7), (1·41)(1·73), (1·414)(1·732), (1·4142)(1·7321), \ldots$$

Since fractional numbers are subject to the commutative, associative and distributive laws of arithmetic, so are the irrational numbers like $\sqrt{2}$ and $\sqrt{3}$, as they are defined as mere limits of infinite sequences of rational fractions. For instance, $\sqrt{2} + \sqrt{3}$ is the limit of the sequence

$$(1·4 + 1·7), (1·41 + 1·73), (1·414 + 1·732) \ldots \ldots \qquad . \quad (1)$$

and $\sqrt{3} + \sqrt{2}$ is the limit of

$$(1·7 + 1·4), (1·73 + 1·41), (1·732 + 1·414) \ldots \ldots \qquad . \quad (2)$$

Sequences (1) and (2) are obviously identical and consequently so also are their respective limits. In other words,

$$\sqrt{2} + \sqrt{3} = \sqrt{3} + \sqrt{2}.$$

The addition of irrational numbers to the field of rational numbers makes what is known as the *real number field*. It is the aggregate of all integral, fractional and irrational numbers, whether positive or negative. It is obvious that we can perform any of the four arithmetical operations on any two of its numbers and express the result as a number belonging to itself. This means that the real number field is also closed under the arithmetical operations. It might thus appear that our third extension of the number system is at last complete. But, as we shall see later, the real number field too is incomplete in some ways and needs further extension.

*   *   *   *

As we have seen, starting with positive integers, the number domain was extended to cover the entire set of real numbers by the invention of negative numbers, fractions and irrationals. We shall see later how the idea of vectors and complex numbers grew out of real numbers, and that of quaternions and hypercomplex numbers out of vectors and complex numbers. With the invention of hypercomplex numbers the art of number-making seemed to have reached its acme, for any kind of number could be shown to be a particular case of some hypercomplex number. With the closing of the field of number-making, mathematicians returned to the integer from which they had started and opened another. In endeavouring to discover the essence of the integer they created a new subject—mathematical logic. By the first two decades of the twentieth century, they had succeeded in creating a mere mathematician's delight, and that to such a degree that it was in real danger of becoming what the Americans call 'gobbledygook'. Fortunately it was rescued from this disaster by the practice of electronic engineers, who applied it to produce new types of ultra-rapid automatic calculating machines employing all manner of electrical apparatus. With the invention of these new electronic devices it was possible to apply the abstract ideas of mathematical logic to advance the design of calculating machines far beyond the dreams of early pioneers like Pascal and Leibnitz or even Babbage.

The reason why mathematical logic has had such great influence on the art of numerical computation is that the calculus of reasoning is symbolically identical with the calculus of number. Since in logic we deal with statements or propositions which have some meaning, every such proposition is either true or false. Let us assign the truth value $T = I$ when the proposition is true, and T = O when it is false. Every proposition such as A will then have a truth value T which may be either zero or one. If we have another proposition B, we can form a compound proposition from these two in two ways. First, we may produce a compound proposition S which is considered true provided *either* A *or* B is true. In this case S is the logical sum of A and B and the process of obtaining it is the analogue of numerical addition. Second, we may obtain another compound proposition P which is considered true if, and only if, both A *and* B are true. P is known as the logical product of A and B and the process of obtaining it is the counterpart of arithmetical multiplication.* For example, let A be the proposition 'Socrates drank the hemlock' and B the proposition 'Voltaire wrote *Gulliver's Travels*'. S, the logical sum of A and B, will then be the compound proposition:

$$S \begin{cases} \textit{Either} \text{ 'Socrates drank the hemlock'} \\ \textit{or} \quad \text{'Voltaire wrote } \textit{Gulliver's Travels'}, \end{cases}$$

P, the logical product of A and B, will, on the other hand, be the compound proposition:

$$P \begin{cases} \text{'Socrates drank the hemlock'} \\ \quad \textit{and} \\ \text{'Voltaire wrote } \textit{Gulliver's Travels'}. \end{cases}$$

Since we know that in this case A is true and B false, then S will be true but P false. Consequently when the truth value of A is 1 and of B zero, that of S will be 1 and of P zero. In the same way we can easily work out the truth values of S and P, given those of A and B in any other case. In general, as mentioned earlier, for S to be true only *one* of the two constituents A and B need be true, whereas for P to be true *both* A and B have to be true. This rule suffices to evaluate the truth values of S and P as we shall now show.

Suppose both A and B are true so that the truth values of both are one. Since S is true when either A or B is true, the truth value of S is 1. This leads to the summation rule:

$$1 + 1 = 1.$$

If both A and B are false, then obviously their logical sum S too is equally false so that the summation rule now is:

$$O + O = O.$$

---

* *S* is also known as the disjunction of *A* and *B* and is written as $S = A \vee B$.
 P is also known as the conjunction of *A* and *B* and is written $P = A \cdot B$.

But if only one of the two, *viz.* A or B, is true, then S is also true, because S is true when either of them is true. This leads to the summation rules:

$$O + I = I, \quad I + O = I.$$

We may summarise these summation rules in the table of logical addition:

Logical Addition

+	O	I
O	O	I
I	I	I

To read the result of the addition of any two truth values, say O and I, take the row O and the column I; these are easily seen to intersect at I. The same rule applies in reading all other tables described in this section.

Consider now the product proposition P. Since P is true only when both A and B are true, its truth value is I only when that of both A and B is I. In every other case P is not true and therefore its truth value is zero. This leads to the product rules:

$$I \times I = I, \quad O \times I = O, \quad I \times O = O, \quad O \times O = O.$$

This may be summarised in the table of logical multiplication:

Logical Multiplication

×	O	I
O	O	O
I	O	I

We shall now show that these tables* of logical addition and multiplication are very similar to their counterparts of arithmetical addition and multiplication. Although we use the ten digits, 0, 1, 2, 3, 4, 5, 6, 7, 8, 9 to write numbers, this is merely due to the physiological accident that we have ten fingers. If we had had only eight, we might have worked with only eight digits *viz.* 0, 1, 2, 3, 4, 5, 6, 7. In that case what we now write as '8', '9' and '10' would be written '10', '11' and '12', respectively. A number like 123 written in this octonal notation would really be an abbreviation of $1(8)^2 + 2(8)^1 + 3(8)^0$ just as 123 in the decimal notation is a shorthand for $1(10)^2 + 2(10)^1 + 3(10)^0$. In the octonal notation, therefore, the number '123' would be $64 + 16 + 3 = 83$ in the usual decimal notation.

What notation we choose for writing numbers, whether decimal,

---

* With the help of these tables you may readily verify that the symbols O, I, though not numbers in the ordinary sense, yet obey the five fundamental laws of arithmetic. This is an instance of a set of 'elements' other than numbers satisfying the laws of arithmetic.

octonal or any other, is arbitrary. In principle we are free to make any choice we like. Of all the possible choices the simplest, though not the most familiar, is the binary notation in which we work with only two digits 0 and 1. It is remarkable that we can express any number whatever in the binary notation, using only these two digits. Thus, the number two would be written in the binary notation as 10, three as 11, four as 100, five as 101, six as 110 and so on. For 10 in the binary notation is $(2)^1 + 0(2)^0 = 2$ in the decimal notation. Likewise, 110 in the binary notation is $1(2)^2 + 1(2)^1 + 0(2)^0 = 4 + 2 + 0 = 6$ in the decimal notation. A binary 'millionaire' would be a very poor man indeed, for the figure 1,000,000 in the binary scale is a paltry:

$$1(2)^6 + 0(2)^5 + 0(2)^4 + 0(2)^3 + 0(2)^2 + 0(2)^1 + 0(2)^0 = (2)^6 = 64$$

in the decimal notation. Nevertheless, the binary notation is potentially as capable of expressing large numbers as the decimal or any other system. The only difference is that it is a bit lavish in the use of digits. Thus the very large number of grains of wheat which the poor King Shirman* was inveigled into promising his sly Grand Vizier as a reward for the latter's invention of chess could, in the binary notation, be expressed simply as a succession of sixty-four ones:

$$111, 111, 111, 111, 111, 111, \ldots \text{ sixty-four times.}$$

In the decimal notation of everyday use we should need twenty digits to write it.

The rules of ordinary addition and multiplication in the binary notation are

$$0 + 0 = 0; 0 + 1 = 1; 1 + 0 = 1; 1 + 1 = 10.$$

If we remember that while adding one to one, we get a 'one' which should be 'carried' to the next place, we can summarise the addition rules in the table of arithmetical addition:

Arithmetical Addition

+	0	1
0	0	1
1	1	1

* The allusion here is to the well-known legend of the Grand Vizier who asked for one grain of wheat in the first square of a chessboard, two in the second, four in the third, eight in the fourth, sixteen in the fifth and so on till the sixty-fourth square. The poor king never suspected till it was too late that the total number of grains required to fill the board in this manner would exceed the total world production of wheat during two millennia at its present rate of production!

Similarly, the rules of ordinary multiplication are:

$$0 \times 0 = 0; 0 \times 1 = 0; 1 \times 0 = 0; 1 \times 1 = 1.$$

They too can be summarised in a similar table of arithmetical multiplication:

Arithmetical Multiplication

$\times$	0	1
0	0	0
1	0	1

A glance at the tables of logical and arithmetical addition shows that they are identical. So also are those of logical and arithmetical multiplication.

Now the ideal calculating machine must be such that with an initial input of data it turns out the final answer with as little human interference as possible until the very end. This means that after the initial insertion of the numerical data the machine must not only be able to perform the computation but also be able to decide between the various contingencies that may arise during the course of the calculation in the light of *instructions* also inserted into it along with the numerical data at the beginning. In other words, a calculating machine must also be a logical machine capable of making a choice between 'yes' and 'no', the choice of adopting one or other of two alternative courses open to it at each contingency that may arise during the course of the computation. It is because of the formal identity of the rules of the logical and arithmetic calculi (in the binary notation) that the apparatus designed to mechanise calculation is also able to mechanise processes of thought. That is why the binary system is superior to other systems in both arithmetic and logic.

Another advantage of the binary system is this. A calculating machine can operate in only two ways. First, it may consist of a device which translates numbers into physical quantities measured on specified continuous scales—such as lengths, angular rotations, voltages, *etc*. After operating with these quantities it measures some physical magnitude which gives the result.* Second, it may consist of a device which operates with numbers directly in their digital form by counting discrete objects such as the teeth of a gear-wheel, or discrete events such as electrical pulses. Such, for instance, is the case with the ordinary desk calculating machines like the Brunsviga and Marchant. Naturally, the accuracy of the first type

---

* For example, a product xy may be evaluated by converting the logarithm of numbers x,y into lengths on a slide rule. We first read the length corresponding to the logarithm of the number x and add to it the length corresponding to that of number y. We then read the number corresponding to the combined lengths to obtain the product xy.

depends on the accuracy of the construction of the continuous scale, and that of the second on the sharpness with which the discrete set of events, such as wheel teeth or electrical pulses, can be distinguished from one another. Since it is easier to distinguish a set of discrete events than to construct a fine continuous scale, the latter type, *viz.* the digital machine, is preferable for highly accurate work. Further, since it is easier to distinguish between two discrete events than ten, digital machines constructed on the binary scale are superior to those on the decimal scale. In other words, the structure of the ideal machine should be a bank of relays each capable of two conditions—say, 'on' and 'off'; at each stage the relays must be able to assume positions determined by the position of some or all of the relays of the bank at a previous stage. This means that the machine must incorporate a clocking arrangement for progressing the various stages by means of one or more central clocks.

Now, as Norbert Wiener has remarked, the human and animal nervous systems, which too are capable of the work of a computation system, contain elements—the nerve cells or neurons—which are ideally suited to act as relays:

'While they show rather complicated properties under the influence of electrical currents, in their ordinary physiological action they conform very nearly to the "all-or-none" principle; that is, they are either at rest, or when they "fire" they go through a series of changes almost independent of the nature and intensity of the stimulus.' This fact provides the link between the art of calculation and the new science of Cybernetics, recently created by Norbert Wiener and his collaborators.

This science (cybernetics) is the study of the 'mechanism of control and communication in the animal and the machine', and bids fair to inaugurate a new social revolution likely to be quite as profound as the earlier Industrial Revolution inaugurated by the invention of the steam engine. While the steam engine devalued brawn, cybernetics may well devalue brain—at least, brain of a certain sort. For the new science is already creating machines that imitate certain processes of thought and do some kinds of mental work with a speed, skill and accuracy far beyond the capacity of any living human being.

The mechanism of control and communication between the brain and various parts of an animal is not yet clearly understood. We still do not know very much about the physical process of thinking in the animal brain, but we do know that the passage of some kind of physico-chemical impulse through the nerve-fibres between the nuclei of the nerve cells accompanies all thinking, feeling, seeing, *etc.* Can we reproduce these processes by artificial means? Not exactly, but it has been found possible to imitate them in a rudimentary manner by substituting wire for nerve-fibre, hardware for flesh, and electro-magnetic waves for the unknown impulse in the living nerve-fibre. For example, the process whereby flat-worms exhibit negative phototropism—that is, a tendency to avoid light—has been imitated by means of a combination of photocells, a Wheatstone bridge and certain devices to give an adequate phototropic control for a little boat. No doubt it is impossible to build this apparatus on the scale of the flatworm, but this is only a particular case of the general rule that the artificial imitations of living mechanisms tend to be much more lavish in the use of space than their prototypes. But they more than make up for this extravagance by being enormously faster. For this reason, rudimentary as these artificial reproductions of cerebral processes still are, the thinking machines already produced achieve their respective purposes for which they are designed incomparably better than any human brain.

As the study of cybernetics advances—and it must be remembered that this science is just an infant barely ten years old—there is hardly any limit to what these thinking-machines may do for man. Already the technical means exist for producing automatic typists, stenographers, multi-lingual interpreters, librarians, psychopaths, traffic regulators, factory-planners, logical truth calculators, *etc.* For instance, if you had to plan a production schedule for your factory, you would need only to put into a machine a description of the orders to be executed, and it would do the rest. It would know how much raw material is necessary and what equipment and labour are required to produce it. It would then turn out the best possible production schedule showing who should do what and when.

Or again, if you were a logician concerned with evaluating the logical truth of certain propositions deducible from a set of given premises, a thinking machine like the Kalin-Burkhart Logical Truth Calculator could work it out for you very much faster and with much less risk of error than any human being. Before long we may have mechanical devices capable of doing almost anything from solving equations to factory planning. Never-theless, no machine can create more thought than is put into it in the form of the initial instructions. In this respect it is very definitely limited by a sort of conservation law, the law of conservation of thought or instruction. For none of these machines is capable of thinking anything new.

A 'thinking machine' merely *works out* what has already been thought of beforehand by the designer and supplied to it in the form of instruc-tions. In fact, it obeys these instructions as literally as the unfortunate Casabianca boy, who remained on the burning deck because his father had told him to do so. For instance, if in the course of a computation the machine requires the quotient of two numbers of which the divisor hap-pens to be zero, it will go on, Sisyphus-wise, trying to divide by zero for ever unless expressly forbidden by prior instruction. A human computer would certainly not go on dividing by zero, whatever else he might do. The limitation imposed by the aforementioned conservation law has made it necessary to bear in mind what Hartree has called the 'machine-eye view' in designing such machines. In other words, it is necessary to think out in advance every possible contingency that might arise in the course of the work and give the machine appropriate instructions for each case, because the machine will not deviate one whit from what the 'Moving Finger' of prior instructions has already decreed. Although the limitation imposed by this conservation law on the power of machines to produce original thinking is probably destined to remain for ever, writers have never ceased to speculate on the danger to man from robot machines of his own creation. This, for example, is the moral of stories as old as those of Famulus and Frankenstein, and as recent as those of Karel Čapek's play, *R.U.R.*, Olaf Stapledon's *First and Last Men*.

It is true that as yet there is no possibility whatsoever of constructing Frankenstein monsters, Rossum robots or Great Brains—that is, artificial beings possessed of a 'free will' of their own. This, however, does not mean that the new developments in this field are without danger to mankind. The danger from the robot machines is not technical but social. It is not that they will disobey man but that if introduced on a large enough scale, they are liable to lead to widespread unemployment.

3

THE CALCULUS

THE knowledge that things, in spite of their apparent permanence, are really in a state of perpetual flux and change, is probably as old as human civilisation. This knowledge formed the basis of philosophical speculations about flux and change which preceded the mathematical formulations by whole millennia. Long before Greek civilisation, the mystical view of change, that nothing really exists but only flux or flow—a view revived more recently in Bergson's *Creative Evolution*—had evolved from commonplace observations. But the mathematics (as opposed to the mysticism of flux) originated during the second half of the seventeenth century, when mathematicians began to study the problem of change.

Speaking of the mathematics of flux, Leibnitz, who shares with Newton the honour of inventing it, said, 'My new calculus . . . offers truth by a kind of analysis and without any effort of the imagination—which often succeeds only by accident; and it gives us all the advantages over Archimedes that Vieta and Descartes have given us over Apollonius.' This claim was no exaggeration, for the calculus proved to be the master key to the entire technological progress of the following three centuries. What then is this calculus of Newton and Leibnitz which has had such momentous consequences?

Mathematically, the calculus is designed to deal with the fundamental problem of change, *viz.* the rate at which anything changes or grows. You cannot even begin to answer questions like this unless you know what it is that changes and how it changes. What you need, in fact, is a growth function—that is, a correlation of its growth against the flow of time. But growth and flow are rather vague terms and must first be given precise mathematical expression. Take first the flow of time. Our daily perception of events around us, not to speak of the physiology of our own bodies, makes us aware of what we call the flow of time. This means that every one of us can arrange the events that we perceive in an orderly sequence. In other words, we can tell which of any two events perceived occurred earlier and which later. By means of physical appliances, such as a watch, we can even say how much earlier or later. This enables us to particularise or 'date' the events by a number—the number of suns, moons or seconds that have elapsed since a certain beginning of time. This is the commonsense way of reckoning time, which both Newton and Leibnitz took for granted.

Now about growth. Here again, to fix ideas we may think of something concrete that grows, *e.g.* the weight of a newly born baby. We could say something about its growth if we weighed it on a number of different occasions. If we 'dated' the occasions we should get a succession of number pairs, one number denoting the date of weighing and the other the corresponding weight on that date. This table of number pairs, *i.e.* the weight and its corresponding date, is, in fact, the growth function—the mathematical representation of what we have called the correlation of growth against the flow of time.

We must first somehow infer the growth function of changing phenomena before the calculus can be applied, and our guide in this matter is mostly experience. Historically, the first phenomenon to be studied by the calculus was the motion of material bodies, such as that of a stone rolling down a hillside. Galileo inferred the growth function of rolling stones from studying its laboratory replica. He allowed balls to roll down inclined planes and observed the distances travelled by them at various times or 'dates' such as at 1, 2, 3, ... minutes after the commencement of the roll. If we repeat his experiment we may, for example, find the following table of corresponding number pairs:

Time (t) in seconds: 0, 1, 2, 3, 4, 5, ...
Distance rolled (y) in feet: 0, 1, 4, 9, 16, 25, ...

It is clear that we may replace this table by the formula $y = t^2$, which is called the growth function of the distance. The problem that the calculus is designed to handle is as follows. Knowing the growth function of the distance the ball rolls, can we discover how fast it moves? In other words, what is its speed?

It is obvious that the ball does not travel with the same speed during the whole course of its motion. The speed itself continually grows. The calculus seeks to derive the growth function of the speed, knowing that of the distance, and thus to give us a formula which enables us to obtain the speed at any given time. Suppose we wish to find its speed at the time denoted by $t = 2$. Take any other time close to $t = 2$ but slightly later, say $t = 2 \cdot 1$. The distances travelled in these two times are $(2)^2$ and $(2 \cdot 1)^2$ units of distance (say feet), respectively, for the distance rule, $y = t^2$, applies equally when t is a fraction. The distance travelled during the time interval from $t = 2$ to $t = 2 \cdot 1$ is therefore $(2 \cdot 1)^2 - (2)^2$. Hence, the *average* speed during the interval of time from 2 to $2 \cdot 1$, or $0 \cdot 1$ units of time (say seconds) is $\dfrac{(2 \cdot 1)^2 - (2)^2}{0 \cdot 1} = 4 \cdot 1$ feet per second. We could regard it as the *actual* speed of the ball at the time $t = 2$ for all ordinary purposes; but if someone insisted that it was only an average during the interval taken and not the actual speed at $t = 2$, we should have to concede the point. To satisfy our opponent, we might calculate the average speed during still shorter intervals of time, say, that during the intervals (2, 2·01), (2, 2·001), (2, 2·0001), . . ., ... In exactly the same way as before, we find that the respective averages during these shorter intervals are 4·01, 4·001, 4·0001, . . ., . . . feet per second. This shows that the average continually approaches 4 as we successively shorten the interval used for averaging. More precisely, we can make it come as close to 4 as we like provided only we take a sufficiently small interval of time for calculating the average. This value 4 is then the *limit* of the average—it is a sort of terminal point or bound below which the average can never fall. We are, therefore, justified in taking this limit as the actual speed at $t = 2$. If we try to work out the speed at other times such as $t = 1, 3, 4, 5, . . .$ in the same way, we find it to be 2, 6, 8, 10, . . . respectively. Thus the table of corresponding values of speed at various times is:

> Time (t) in seconds: 1, 2, 3, 4, 5, . . .
> Speed (v) in feet per second: 2, 4, 6, 8, 10, . . .

This suggests the formula $v = 2t$ for the growth function of speed.

We could treat the speed function in exactly the same way as we have treated the distance function. In other words, we could now enquire how fast does the speed change or what is the acceleration of the ball. We proceed in the same manner as before by taking an instant of time, say 2·1, very close to the instant 2. The speed changes from 2(2) to 2(2·1) so that the average increase in speed—or the acceleration—is

$$\frac{2(2·1) - 2(2)}{(2·1 - 2)} = 2.$$

We get the same result no matter how close to 2 we choose the second instant of time for averaging the increase in speed. We may, therefore, regard this average acceleration as the actual acceleration at $t = 2$. If we try to work out in the same manner the acceleration at other instants of time, such as $t = 1, 3, 4, . . .$, we find that it remains 2 all the time. Consequently, the table of corresponding values of acceleration at various times is:

> Time (t) in seconds: 0, 1, 2, 3, 4, . . .
> Acceleration (a):   2, 2, 2, 2, 2, . . .

This suggests the formula $a = 2$, which means that acceleration remains 2 at all instants of time. In other words, the ball rolls with uniform acceleration.

Although the calculus was originally devised to calculate time rates of change like speeds and accelerations of moving bodies, its technique is equally applicable to all sorts of rates of change. In everyday life we usually come across time rates, such as interest, speed, acceleration, growth, *etc.*, but whenever we have one variable quantity, y, depending on another variable, x, we may enquire about the rate of change of y per unit change of x. Thus the irrigation engineer is interested in the hydraulic pressure, y, that the dam surface has to endure at any given depth, x, below the water surface. Here, although the water pressure is assumed to remain static everywhere, we may, nevertheless, legitimately enquire how fast the water pressure rises with increasing depth. In other words, we may wish to determine the rate of change of pressure per unit change of depth.

Whenever we have a pair of magnitudes, y and x (such as hydraulic pressure and depth, freight or fare payable and distance of journey, income tax and income, *etc.*), so related that the measure of one depends on that of the other, the former is said to be a function of the latter. Symbolically, we denote this dependence by the expression $y = f(x)$, where f is only a shorthand for 'depends on'.

Now since dependence like so many other relations, such as friendship, is a reciprocal relation, we could equally regard the measure of x as depending on that of y. Thus, if y is a function of x, then equally x is some function F of y. The function $F(y)$ is called the inverse of $f(x)$. For instance, if $y = 3x - 6$, then $x = y/3 + 2$. Hence the inverse of y, $y = f(x)$ $= 3x - 6$ is the same as $x = F(y) = y/3 + 2$. Although, strictly speaking, either variable may be expressed as a function of the other, in most situations it is more natural to regard the variation of one as independent of and, in a way, 'controlling' that of the other. For instance, in the above-mentioned illustrations it is more natural to consider income tax as a function of income, railway freight as a function of distance, or hydraulic pressure as a function of depth, rather than the other way about. Income tax, railway freight and hydraulic pressure are therefore dependent variables, as they do, in a real way, depend on income, distance and depth, respectively. It is true that there are some cases in which it may not be quite obvious which of the two variables should be treated as independent, but in such cases we may make any choice to suit our convenience.

In most cases the dependence between the two variables y, x is far too complicated to be reduced to a formula. For example, the bitterness of Swift's satire or the pungency of Carlyle's invective may, perhaps, if they could be measured, be functions of the amount of bile secreted by their livers at the times of writing. But if so, no formula can be devised to express this dependence. In science, however, we mostly deal with functional dependences which can be reduced to a formula. For a closer peep into the working of nature, where we also come across dependences far too complex to be trapped in the neat expression of a single formula, mathematicians during the last century or so have been obliged to consider functional dependences, which may be expressed by a series of formulae instead of by a single formula. As an instance of such a function we may cite the relation between pressure, y, and volume, x, of a gas at constant temperature. For ordinary pressures such as could be applied in the time of Boyle, the formula, $y = constant/x$, usually known as Boyle's law, expresses this dependence. For other ranges of pressure such as are now possible, different formulae have to be used. The functional relation between pressure and volume of a gas is thus a multi-formula function. The same is true of most laws of nature.

Now, whatever y and x may be, and whatever the formula or formulae expressing their dependence, we may want to know the rate of change of one per unit increase of the other. Suppose $y = f(x)$ is a functional relation between any dependent variable y and an independent variable x. Let us calculate the rate of change of $f(x)$ per unit change of x. In general, this rate will itself vary and depend on the particular value of x chosen. Suppose we want it for $x = 2$, then, as in the case of time when we were discussing speed above, we choose a short range of the independent variable lying between 2 and $2 + h$, h being a small number. If the independent variable varies from 2 to $2 + h$, the dependent variable y will change from $f(2)$ to $f(2 + h)$. The net change of its value over the range $(2, 2 + h)$ is $f(2 + h) - f(2)$. The average rate of change over this range is, therefore,

$$\frac{f(2 + h) - f(2)}{h}.$$

If h is reasonably small, we may take this *average* rate as the actual rate of change of $f(x)$ at $x = 2$. But if someone insisted that this is only an average and not the actual rate, we should have to accept the validity of his objection. To avoid such an objection, we say that the actual rate of change of y at $x = 2$ is the *limit* of the average rate of change of y per unit change of x, when the interval of x, over which it is averaged, is decreased indefinitely. The idea is exactly analogous to that of speed, with x substituted for t. This limit is known as the differential co-efficient or derivative of y with respect to x at $x = 2$, and is usually written as $\left(\dfrac{dy}{dx}\right)_{x=2}$, or $f'(2)$.

We could do to $f'(x)$ what we did to $f(x)$, that is, we could enquire what is the rate of change of $f'(x)$ with respect to x at the value $x = 2$—or at any other value, for that matter. It is similarly the limit of the average rate of change:

$$\frac{f'(2 + h) - f'(2)}{h} \qquad \cdot \quad \cdot \quad \cdot \quad \cdot \quad (1)$$

as h tends to zero. This limit is the differential co-efficient or derivative of $f'(x)$ with respect to x at $x = 2$. Since $f'(x)$ itself was derived from $f(x)$ by the same process, we may also call the limiting value of (1) the *second* differential coefficient or derivative of $f(x)$ with respect to x at $x = 2$. We denote it by $\left(\dfrac{d^2y}{dx^2}\right)_{x=2}$, or $f''(2)$. There is no reason why we should stop here. We could repeat the process to derive successively the third, fourth ... differential coefficients of $f(x)$ for any specific value of x, such as 2. This successive generation of differential coefficients is no mere free

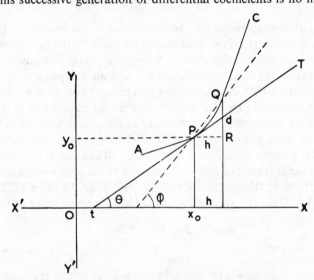

Fig. 2—As Q travels along the graph line to coincide with P, PQ takes the position of the tangent PT. The angle ϕ becomes the angle θ. In the limit therefore, $\dfrac{dy}{dx} = \tan \theta$.

creation of the curious mind. The railroad engineer has to employ second derivatives to calculate the curvature of the line he constructs. He needs a precise measure of the curvature to find the exact degree of banking required to prevent trains from overturning. The automobile designer utilises the third derivative in order to test the riding quality of the car he designs, and the structural engineer has even to go to the fourth derivative in order to measure the elasticity of beams and the strength of columns.

It is not difficult to see how the first and higher order derivatives arise naturally in problems like these. Suppose the curve $APQC$ represents the railway line. (See Fig. 2.) Before the calculus can be applied we must somehow represent it algebraically, that is, by some functional formula. This is done by taking any two perpendicular reference lines XOX' and

YOY' through a point of origin O and indicating the position of any point P by its distances (x, y) from these reference lines or 'axes'. The distances, x, y, are called co-ordinates. Now if we consider any point P on the line and measure its co-ordinates (x, y), we find that y is usually some function of x, such as $y = f(x)$.

This equation gives the variation of y due to a change of x. At any point $P(x_0, y_0)$ of the line the value of y_0 is $f(x_0)$. At another point $Q(x_0 + h, y_0 + d)$ very close to P and also on the railway line, the value of y is $f(x_0 + h)$. Hence the change in y is $f(x_0 + h) - f(x_0)$, or the length $QR = d$. The average rate of change of y per unit change of x is

$$\frac{f(x_0 + h) - f(x_0)}{h} = \frac{d}{h} = \frac{QR}{PR}.$$

If the angle QPR is denoted by φ, the ratio QR/PR is known as the *slope* or gradient of the line PQ, and is written as $\tan \varphi$, which is short for tangent of the angle φ. It follows, therefore, that the average rate of change of y at P with respect to x is the gradient or slope of the chord joining P to another point Q of the line very close to P. If we diminish h, the average rate of change of y during the interval $(x_0, x_0 + h)$ approximates more and more closely to the actual value of the instantaneous rate at P. The latter has already been defined as dy/dx. On the other hand, as h diminishes and Q approaches P along the graph line, the chord line PQ becomes a tangent to it at P, that is, the straight line PT which just grazes it. Hence the value of dy/dx at P measures the gradient or slope of the tangent to the curve at P.

In symbols, $\dfrac{dy}{dx} = \tan \theta$.

Now, at any point P the direction of the line is along the tangent PT. At another point Q on the line it is along the tangent QT' at Q. (See Fig. 3.) The measure of its bend as we travel from P to Q along the line is therefore the angle between the two tangents at P and Q. The rate of its bend or curvature is the ratio:

$$\frac{\text{total bend}}{\text{length of the curved line } PQ}.$$

This is, of course, the average curvature along the whole length of the arc PQ. If we want its precise curvature *at* the point P, we must, as before, find the limit of this ratio as the length of the arc PQ is indefinitely decreased by bringing Q infinitely close to P. In other words, we have to differentiate the angle θ of the tangential direction at P with respect to the length PQ of the arc. But, as we have seen already, $\tan \theta$ is dy/dx. It is,

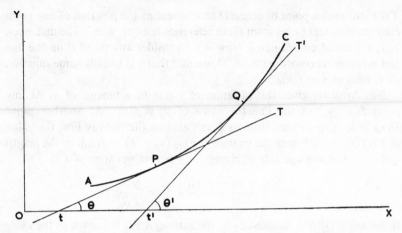

FIG. 3—The total bend of the line as we move from P to Q is the angle between the two tangents at P and Q, that is, $\theta' - \theta$.

therefore, inevitable that in the process of differentiation of the angle θ we should encounter second derivatives.

*   *   *   *

As we saw, the problem of calculating instantaneous speed and acceleration of moving bodies *at* any given instant of time from the mathematical formula connecting the distance (y) travelled and the time (t) of their fall or flight, gave rise to the differential calculus. The inverse problem of calculating the distance travelled, given a mathematical formula connecting speed or acceleration with time, led to the development of the *Integral Calculus*.

Suppose a particle moves with speed v which is given by the relation $v = t$. What is the distance, s, travelled during the time interval from say $t = 1$ to $t = 6$? If the speed had been uniform we could have got the desired distance by multiplying the uniform speed by the duration of the time interval during which it was maintained. But in the problem before us the speed does not remain the same even for two consecutive instants. How are we to apply the rule which is valid for a static speed to calculating the distance travelled when it is no longer so?

Essentially this is a problem of reconciling irreconcilables, of finding a method of resolving the inherent conflict between change and permanence. This conflict is universal—between the ever-changing material world and the static, permanent or quasi-permanent forms and categories that we invent and impose upon the world to understand it. Newton and Leibnitz were the first to devise a practical way of resolving this conflict in the case of moving bodies. The method proposed was simple in principle—once it was discovered. They divided the duration of motion into a large number of sub-intervals during each of which its velocity was assumed *not* to change. The rule for static speed could now be applied for each sub-interval *separately* and the distance derived by summing all the distances travelled in each sub-interval.

Suppose, for instance, we divide the interval of time from $t = 1$ to $t = 6$ into any number, say 10, of equal sub-intervals, each of duration $\dfrac{6-1}{10} = \tfrac{1}{2}$ second, by taking nine intermediate point-instants $1 + \tfrac{1}{2}$, $1 + \tfrac{2}{2}$, $1 + \tfrac{3}{2}$, $1 + \tfrac{4}{2}$, ..., $1 + \tfrac{9}{2}$ between the times $t = 1$ and $t = 6$. Consider now the first sub-interval from the initial instant $t = 1$ to $t = 1 + \tfrac{1}{2}$. Since the speed v is given by the formula $v = t$ the speeds at the beginning and end of the first sub-interval were 1 and $(1 + \tfrac{1}{2})$ respectively. The distance travelled during the first sub-interval is, therefore, greater than $1(\tfrac{1}{2})$ but less than $(1 + \tfrac{1}{2})\tfrac{1}{2}$. Similarly the limits between which the distances travelled during the second, third, fourth, ... and tenth sub-intervals lie, can be calculated. We tabulate these limits below:

Sub-interval of time	Lower limit of the distance travelled	Upper limit of the distance travelled
First	$1(\tfrac{1}{2})$	$(1 + \tfrac{1}{2})\tfrac{1}{2}$
Second	$(1 + \tfrac{1}{2})\tfrac{1}{2}$	$(1 + \tfrac{2}{2})(\tfrac{1}{2})$
Third	$(1 + \tfrac{2}{2})(\tfrac{1}{2})$	$(1 + \tfrac{3}{2})(\tfrac{1}{2})$
Fourth	$(1 + \tfrac{3}{2})(\tfrac{1}{2})$	$(1 + \tfrac{4}{2})(\tfrac{1}{2})$
Fifth	$(1 + \tfrac{4}{2})(\tfrac{1}{2})$	$(1 + \tfrac{5}{2})(\tfrac{1}{2})$
Sixth	$(1 + \tfrac{5}{2})(\tfrac{1}{2})$	$(1 + \tfrac{6}{2})(\tfrac{1}{2})$
Seventh	$(1 + \tfrac{6}{2})(\tfrac{1}{2})$	$(1 + \tfrac{7}{2})(\tfrac{1}{2})$
Eighth	$(1 + \tfrac{7}{2})(\tfrac{1}{2})$	$(1 + \tfrac{8}{2})(\tfrac{1}{2})$
Ninth	$(1 + \tfrac{8}{2})(\tfrac{1}{2})$	$(1 + \tfrac{9}{2})(\tfrac{1}{2})$
Tenth	$(1 + \tfrac{9}{2})(\tfrac{1}{2})$	$(1 + \tfrac{10}{2})(\tfrac{1}{2})$
Total =	s_1	S_1

Let s_1 and S_1 be the sums of the lower and upper limits of distances travelled during the ten sub-intervals. If we add up the ten terms shown in the above table of lower and upper limits we get,

$$s_1 = \tfrac{1}{2}\{10 + (1 + 2 + 3 + \ldots 9)\tfrac{1}{2}\}$$
$$= \tfrac{1}{2}(10 + \tfrac{45}{2}) = 16\cdot25;$$
$$S_1 = \tfrac{1}{2}\{10 + (1 + 2 + 3 + \ldots 10)\tfrac{1}{2}\}$$
$$= \tfrac{1}{2}(10 + \tfrac{55}{2}) = 18\cdot75.$$

The actual distance travelled, therefore, lies between 16·25 and 18·75 feet. It is not an exact answer but good enough for rough purposes. To improve the precision of our answer, we have to shorten further the sub-intervals by dividing the original interval $t = 1$ to $t = 6$ into a larger number of sub-intervals, say, 100 instead of 10 so that the duration of each sub-interval now is $\dfrac{6-1}{100} = \dfrac{1}{20}$. The only snag is that the calculation is longer but not different in principle. If we denote by s_2 and S_2 the corresponding sums of lower and upper limits for the distance travelled during these shorter sub-intervals, we find that

$$s_2 = \tfrac{1}{20}\{100 + (1 + 2 + 3 + \ldots \ 99)\tfrac{1}{20}\} = 17\cdot37;$$
and, $$S_2 = \tfrac{1}{20}\{100 + (1 + 2 + 3 + \ldots \ 100)\tfrac{1}{20}\} = 17\cdot63.$$

This gives us still closer limits within which the actual distance travelled must lie, *viz.* 17·37 and 17·63. If we repeat the calculation by dividing the interval into 1000 sub-intervals, the corresponding sums s_3 and S_3 will be found to be

$$s_3 = \tfrac{1}{200}\{1000 + (1 + 2 + 3 + \ldots 999)\tfrac{1}{200}\} = 17\cdot48;$$
and, $$S_3 = \tfrac{1}{200}\{1000 + (1 + 2 + 3 + \ldots 1000)\tfrac{1}{200}\} = 17\cdot51.$$

These are even closer limits for the actual distance travelled during the interval (1, 6). We may tabulate the successive values of the upper and lower limits:

$$
\begin{array}{ll}
s_1 = 16\cdot250 & S_1 = 18\cdot750 \\
s_2 = 17\cdot375 & S_2 = 17\cdot625 \\
s_3 = 17\cdot4875 & S_3 = 17\cdot5125 \\
s_4 = 17\cdot49875 & S_4 = 17\cdot50125.
\end{array}
$$

From the above table, we may infer that the sums of lower limits or, for short, lower sums $s_1, s_2, s_3, \ldots$ continually increase while the upper sums $S_1, S_2, S_3, \ldots$ continually decrease. Moreover, the difference between the two corresponding lower and upper sums continually decreases so that both sums approach the same limit as the number of sub-intervals is indefinitely increased. This common limit of the two sums which, as we may guess, is 17·5, is the actual distance travelled. It is also known as the integral of speed with respect to time over the interval (1, 6). We denote it by the symbol

$$\int_1^6 v\,dt, \quad \text{or} \quad \int_1^6 t\,dt, \ \text{as } v = t.$$

In this, the integral sign $\int$ is only a distorted form of S, short for sum, to remind us that it is really a limiting sum.

If we look at the speed-time graph (Fig. 4), the line ODC will represent the relation $s = t$. We can use this graph to simplify the calculation of the speed integral. Suppose we divide the time interval (1, 6) represented by the segment *AB* of the time-axis *Ot* into any number of equal sub-intervals. Consider now any such sub-interval, say, *LM*. At the instant of time represented by the point *L*, the speed is given by *LP* and at *M* by *MQ*. The distance actually travelled during the sub-interval *LM*, therefore, lies between the lower limit, *LM.LP*, and the upper limit, *LM.MQ*, since distance

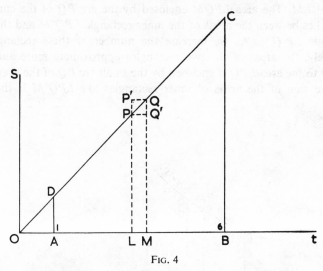

FIG. 4

= time × speed. These limits are obviously the areas of the two rectangles, *LPQ'M* and *LP'QM*. As we increase the number of sub-intervals, the sub-interval *LM* diminishes indefinitely. The areas of the two rectangles between which the actual distance travelled lies approximate more and more closely to the area of the trapezium *LPQM*. But as the total distance travelled during the interval (1, 6) is the sum of all such areas, it is naturally represented by the area of the trapezium *ABCD*. Now the area of the trapezium *ABCD*—

$$= \triangle OBC - \triangle OAD$$
$$= \tfrac{1}{2}(OB)(BC) - \tfrac{1}{2}(OA)(AD)$$
$$= \tfrac{1}{2}(6)(6) - \tfrac{1}{2}(1)(1)$$
$$= \tfrac{1}{2}(6^2 - 1^2) = 17\cdot5.$$

We may therefore write $\int_{1}^{6} t\,dt = \tfrac{1}{2}(6^2 - 1^2)$.

From this we may readily infer that the distance travelled during *any* interval (t_0, t_1) is the *definite integral* $\int_{t0}^{t1} t\,dt = \tfrac{1}{2}(t_1^2 - t_0^2)$.

We can further generalise this result. Instead of working with the speed-time graph $s = t$, we may start with the graph of any function $y = f(x)$, and enquire what is the area enclosed by the graph line BC, the two ordinate lines AB, DC and the segment AD on the x-axis, where OA is any length x_0 and OD any other length x_1. (See Fig. 5.) As before, we divide the segment AD into any number of equal sub-intervals. Take any sub-interval LM. The area $LPQM$ enclosed by the arc PQ of the curve evidently lies between the areas of the inner rectangle $LPQ'M$ and the outer rectangle $LP'QM$. As we increase the number of these rectangles indefinitely, the areas of the two rectangles approximate more and more closely to the area $LPQM$ enclosed by the small arc PQ of the graph line. But the sum of the areas of inner rectangles like $LPQ'M$ is the exact

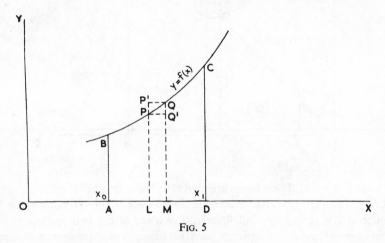

FIG. 5

analogue of the lower sum, and that of the outer rectangles like $LP'QM$ that of the upper sum we constructed earlier while integrating the speed graph $v = t$. These two upper and lower sums tend to a common limit as the number of sub-intervals is indefinitely increased. This common limit is the integral $\int_{x_0}^{x_1} f(x)dx$. Obviously, this integral is also the measure of the area $ABCD$ enclosed by the graph line BC as each of the inner and outer rectangles approximates, more and more closely, to the actual area $LPQM$ under an elementary arc PQ of the graph line.

Now although we have defined the definite integral $\int_{x_0}^{x_1} f(x)dx$ as the area of the curve $ABCD$, it is not the only possible interpretation that can be given to it. The essential idea behind it is that it is a *limiting sum* of an infinite series of terms. It thus comes to pass that whenever we have to add up an infinite number of values of a function corresponding to an infinite number of values of its independent variable, the definite integral plays an indispensable role. For instance, when the hydraulic engineer constructing a dam wants to ascertain the total water pressure likely to be exerted on the whole face of the dam, he divides the entire dam surface into an infinite number of point-bits. As the pressure at any arbitrary point-bit of the dam surface can be calculated, the calculation of the whole pressure on the dam surface is a summation problem and therefore amenable to integration.

Here pressure is a function of the depth of the point-bit below the water surface. It is the same with other problems facing the bridge engineer, the architect and the electrician in the calculations, respectively, of moments of inertia, centres of gravity of solids and surfaces, magnitudes of electromagnetic fields, *etc.* In all these problems the value of a quantity such as mass, moment, hydraulic pressure, electric or magnetic forces, is given at each of an infinite number of points in a region or space and it is desired to calculate their sum.

It is often inconvenient to calculate such limiting sums directly. The calculation is greatly simplified by the use of a theorem which links integration with differentiation. As we have seen, the integral $\int_{x_0}^{x_1} f(x)dx$ is the area enclosed by $ADCB$ (Fig. 5). If we treat x_1 in this integral as a variable, the integral itself becomes a function of x_1. Let us call it $F(x_1)$. Then

$$F(x_1) = \int_{x_0}^{x_1} f(x)dx.$$

What is the differential coefficient of $F(x_1)$? According to the rule we have established, it is the limit of

$$\frac{F(x_1 + h) - F(x_1)}{h}$$

as h tends to zero. Now $F(x_1)$ is the area $AQPB$ (Fig. 6) and $F(x_1 + h)$ is the area $AQ'P'B$ (Fig. 6). The difference is the area $QQ'P'P$ or approximately $QQ'.PQ = h.f(x_1)$. The smaller h is, the more nearly is the area $PQQ'P'$ equal to $h.f(x_1)$ and hence as h tends to zero, the limiting value of

$$\frac{F(x_1 + h) - F(x_1)}{h} = \frac{h.f(x_1)}{h} = f(x_1), \text{ is } f(x_1).$$

In other words, $\dfrac{dF}{dx_1} = \dfrac{d}{dx_1}\displaystyle\int_{x_0}^{x_1} f(x)dx = f(x_1).$

That is, the differential coefficient of the integral of a function is the function itself. Thus differentiation and integration are inverse operations like multiplication and division. Just as multiplying a number by another and

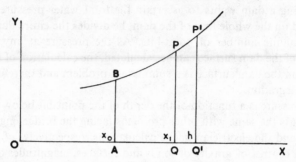

FIG. 6—$F(x_1)$ is the area $AQPB$, and $F(x_1 + h)$ the area $AQ'P'B$. The difference, therefore, is the area $QQ'P'P$ which is approximately $QQ'.PQ$, or $f(x_1).h$. The rate of change of area under the curve at P is the value of $f(x)$ at P.

then dividing the product by the same factor leaves the number unchanged, so integrating a function $f(x)$ and then differentiating the integral leaves the function as it was before.

The great power of the calculus depends on this fundamental theorem, for, in our study of nature we often assume a formula for the rate of growth

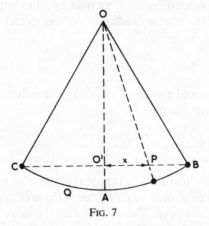

FIG. 7

of one variable per unit change of the independent variable, and we then want to know if the formula really works. For instance, in his study of motion Newton assumed that the temporal rate of change of velocity of a moving body, that is, its acceleration, is equal to F/m, where F is the force applied to it and m is its mass. He applied this formula to a wide class of phenomena, from the motion of the pendulum to that of the moon.

This formula always led to an equation in which occurred the rates of change or the differential coefficients. Thus, in the case of the motion of the pendulum bob, it led to the equation $dv/dt = F/m$.

Now, if we assume that the circular arc that the bob describes is small, we may consider it as oscillating along the straight line BC instead of the arc BAC (see Fig. 7). Let its distance in any position P be x from the central position O'. The force F acting on it in this position is that part or component* of the tension of the string holding the bob, which acts along the direction of its motion, that is, the direction $O'B$. It can be shown that this component is $-mg.x/l$, where l is the length of the string and mg the weight† of the bob. Newton's formula, therefore, leads to the equation

$$dv/dt = -gx/l.$$

But as we know, v itself is dx/dt, so that the equation of the motion of the pendulum bob is

$$\frac{dv}{dt} = \frac{d}{dt}\left(\frac{dx}{dt}\right) = \frac{d^2x}{dt^2} = -\frac{gx}{l}$$

or,
$$d^2x/dt^2 = -gx/l \qquad . \qquad . \qquad . \qquad . \quad (1)$$

Such an equation in which differential coefficients of the dependent variable with respect to the independent variable occur is known as a differential equation. All that Newton's famous laws of motion and gravitation do is to set up a system of one or more differential equations essentially of the type written above. The motion of the body or system of bodies is then known, if we can 'solve' the differential equations, that is, find x as a function of t so that it satisfies them.

If we consider the solution of any differential equation such as (1) we shall find that it is indeterminate in one respect. You will recall that in formulating it, we merely used Newton's law and took no account of the initial position from where the bob is let go. This means that the differential equation is indifferent to its initial position.‡ No matter where the bob is, one and the same differential equation (1) results, and yet what motion it actually executes must also depend on its initial position. Thus its motion when it is released at B is not the same as when it is released at P. In the one case it oscillates between B and C and in the other between P and Q (see Fig. 7). It therefore follows that the differential equation defines no

---

* See Chapter 4 for a fuller explanation of the component of a force.

† Weight, by the way, is the product of mass m and the acceleration g of gravity, the acceleration with which everything when released falls towards the surface of the earth.

‡ The assumption is still made that the arc BC is small and can be identified with the chord BC.

particular motion of the bob but the whole class of its possible motions corresponding to all its initial positions.

The precise solution of the equation therefore involves the choice of that particular solution out of this class which fits the prescribed initial condition. Here the initial condition is specified by a single magnitude denoting the position of the bob just at the commencement of its motion. But in more complicated cases there are usually many initial conditions which have to be specified by several magnitudes. Take, for instance, Lord Kelvin's theory of the transmission of signals in a submarine cable, whereby he established the theoretical feasibility of an Atlantic cable long before trans-oceanic cablegram became an accomplished fact.

In this case, not only is the differential equation more complicated but the initial conditions too are more numerous. The reason is that we need to know the state of the cable before one of its ends is suddenly connected to a battery terminal, and not merely at one particular point but all along its length. In other words, the initial conditions also include boundary conditions—that is, the initial state of affairs prevailing all over the boundary or surface of the cable.

From what has just been said it follows that even if we are somehow able to solve a differential equation, we must adjust it to suit the given initial conditions. A vast amount of new mathematics created during the past 200 years is merely the outcome of this search in physics and astronomy for solutions of differential equations satisfying prescribed initial conditions. After a prodigious number of special solution functions had been invented, it was found that in many cases a differential equation along with its accompanying set of initial conditions is equivalent to an *integral* equation, in which the unknown variable x appears under an integral sign instead of a sign of derivation. It is thus possible to reformulate certain physical problems in terms of integral rather than differential equations. The great advantage of such reformulation is that the passage from simpler to more difficult problems is not attended with any serious increase in complication as is the case with differential equations.

There are, however, phenomena in which the initial state of a system does not suffice to determine its subsequent evolution. Thus, the evolution of an elastic system is not always determined by its initial state alone but by all its previous states. The past is not completely obliterated and an ancestral influence of a sort controls the shape of things to come. For instance, the type of motion a pendulum bob executes depends solely on the initial position from where it is released no matter what oscillations it might have executed earlier, but if you try to twist an elastic wire, the result will largely depend on how and how many times it had been twisted before. It is because of this hereditary or ancestral influence that materials sometimes fatigue and break down under comparatively minor strains, whereas earlier they could have stood up to much heavier strains.

Such hereditary phenomena require a new mathematical instrument called the integro-differential equation, in which the unknown function x appears under a sign of both integration and derivation. It can be shown that such an equation is equivalent to an infinite number of ordinary differential equations, and this is the reason why they apply to hereditary phenomena. For the initial state which was determined earlier by a few magnitudes has now to be broadened to include the whole infinity of past states. It has, therefore, to be specified by an appropriate choice of an infinite number of magnitudes.

Now, if we view the elastic system atomistically, and thus consider an extremely large number of ordinary differential equations regulating the motions of an extremely large number of particles, each with its own set of initial conditions, we may disregard the hereditary feature which experience seems to require. But as an infinity of differential equations is mathematically intractable, it has to be replaced by a single integro-differential equation which is equivalent to this infinite set.

It is needless to add that it is easier to write equations, whether differential, integral or integro-differential, than to solve them. Only a small class of such equations has been solved explicitly. In some cases, when, due to its importance in physics or elsewhere, we cannot do without an equation of the insoluble variety, we use the equation itself to define the function, just as Prince Charming used the glass slipper to define Cinderella as the girl who could wear it. Very often the artifice works; it suffices to isolate the function from other undesirables in much the same way as the slipper sufficed to distinguish Cinderella from her ugly sisters.

*   *   *   *

(The following section may be omitted on first reading)

So far we have considered only functions which depend on a single independent variable x. This assumes that variation in a magnitude (y) can be explained by the variation in a single independent variable x. But in most situations, where we have to explain the variation of a magnitude (y), this assumption is simply not true. As a rule a dependent variable y is influenced by several independent variables simultaneously. Even in the simple case cited earlier, of the pressure of a gas enclosed in a cylinder, it is rather an over-simplification to say that pressure depends on volume, because in actual fact it also depends very materially on temperature too. In this case this extra dependence on temperature does not cause much inconvenience, for we can study the functional relationship between pressure and volume by keeping the temperature constant. But in many cases this device merely produces an abstract schema too far removed from reality.

For example, in econometrics the price of a commodity is considered to be a function of a whole host of variables, such as the incomes of various social groups, prices of other competing goods, production costs, the nature of the commodity itself (*e.g.* a necessity or a luxury), and the seasonal effects in the case of what are called 'anchovy goods', which are goods with very heavy supply-fluctuations, such as the catch of fish, which in one year may be a hundredfold that of another year. An econometric law which correlates one magnitude with another single variable under the *ceteris paribus* condition (other factors remaining the same) can never be actually tested, as the 'other factors' do not, in fact, remain the same. Nor would it be of much use even if it were 'true' in some imaginary market where other factors were assumed to remain the same. But long before econometricians felt the need for a calculus of multi-variate functions, that is functions depending on more than one independent variable, mathematicians had begun to create one in an attempt to study fluid motions.

Just as Newton was led to the notion of differential coefficients in his studies of the motion of projectiles and heavenly bodies, so also Euler was led to the notion of *partial* differential coefficients in applying Newton's equations of ordinary dynamics to fluid motion. In ordinary dynamics the projectile (or heavenly body) is considered to be a particle whose velocity is a function of the single independent variable time. But a fluid cannot be treated as a mere particle. It has bulk which cannot be disregarded even in an abstract schematic treatment. Moreover, a fluid in general not only varies in velocity from point to point in space at any one instant, but also from moment to moment of time at any one point.

A fluid motion, say on the Earth, has therefore two aspects: a geographic aspect and an historic aspect. When we consider the former, we fix our attention on a particular instant t of time and wish to study the velocity of a fluid particle as a function of its geographic position. In the latter we rivet our attention on a specified particle of the fluid and study its velocity as a function of time. A combination of both the aspects, that is, consideration of the velocity of a fluid particle as a function of both space and time simultaneously, will give an actual picture of the fluid flow as a whole. In this way we are led to the notion of a multivariate function; that is, a function depending on more than one variable. Such, for instance, is the case with the velocity of a fluid particle, as it is a function not only of time but also of the position co-ordinates of the particle in question.

To illustrate the idea of the partial differential coefficient to which Euler was led by a study of fluid motions, let us consider a function y which depends on two independent variables x_1, x_2, simultaneously:

$$y = f(x_1, x_2).$$

Consider first x_1. Since y is a function of x_1, x_2 only, that is, it depends on only x_1, x_2, any change (small or otherwise) in its value can arise only on account of a change in x_1 or x_2 or both. In science, whenever we have to investigate a phenomenon which depends on more than one cause (*e.g.* agricultural yield, which among other things is a function of the qualities of both manure and seed), we disentangle the complex field of influence of the two factors by studying the influence of each in isolation from the other. Once we have studied the effect of the causes singly, we can give due weight to each when the isolates are put back into their natural inter-relations.

This method is appropriate in the case under consideration. To begin with, we make a small change dx_1 in the value of x_1 only, making no change in that of x_2. What change would it cause in the value of $f(x_1)$? As we have seen, the derivative df/dx_1, is the rate of change of $f(x_1)$ or y per unit change of x_1. That is, for every unit change of x_1, the corresponding change in f or y is df/dx_1. Hence for a small change dx_1 in the value of x_1 the corresponding change in f or y would be the product:

$$\left(\frac{df}{dx_1}\right)dx_1.$$

Note carefully that f is a function of x_1 as well as x_2; but as we have assumed no change in the value of x_2 we must express this fact somehow in our notation. This is done by writing $\partial f/\partial x_1$ instead of df/dx_1 so as to remind us that the other variable, x_2, is not to be changed and is therefore to be treated *as if* it were a constant. The quantity $\partial f/\partial x_1$ is known as the partial derivative of f with respect to x_1. Hence the change in f or y corresponding to a small change in x_1, x_2 remaining unchanged, is $\left(\dfrac{\partial f}{\partial x_1}\right) dx_1$.

We now study the change in f (or y) corresponding to a small change in the value of x_2 while no change is made in the value of x_1. A similar argument shows that it is:

$$\frac{\partial f}{\partial x_2}dx_2.$$

What then is the change in f (or y) if both x_1, x_2 are changed simultaneously? We may assume that it is the sum of the changes induced by the small changes in x_1 and x_2 acting singly. It may not be an accurate assumption to make when the changes in x_1 and x_2 are combined to produce a new complex, but it is quite good, for all practical purposes, *if* the changes in x_1, x_2 are infinitesimally small. We therefore conclude that a small change in the value of y, induced by the joint operation of small changes x_1 and x_2 in the values of x_1 and x_2, is the sum of the changes induced by them singly. In other words,

$$dy = \frac{\partial f}{\partial x_1}dx_1 + \frac{\partial f}{\partial x_2}dx_2.$$

In obtaining his equations of motion of fluid flow Euler merely made use of an extension of this theorem, *viz.* a small change in any function $f(x_1, x_2, x_3, t)$ of four independent variables x_1, x_2, x_3, and time t is given by the sum:

$$df = \frac{\partial f}{\partial x_1}dx_1 + \frac{\partial f}{\partial x_2}dx_2 + \frac{\partial f}{\partial x_3}dx_3 + \frac{\partial f}{\partial t}dt.$$

With the help of this theorem and the application to a small volume of fluid element the second law of Newton—*viz.*, that the product of the mass and rate of change of velocity in any direction is equal to the resultant of forces (including fluid pressures) acting thereon in that direction, Euler obtained equations of motion connecting the partial differential coefficients (or rates of change) of velocities with respect to the three spatial co-ordinates and the time. In addition, he obtained another partial differential equation—the equation of continuity—from the consideration that the mass of any moving fluid element under consideration remains constant.

Now, since velocity at any point has three components, say (u, v, w), along the three co-ordinate axes, it is obvious that the above-mentioned equations would involve partial differential coefficients of three separate functions (u, v, w) with respect to x, y, z and t. It is possible in many cases to merge the search for the three separate functions, u, v, w, satisfying these equations into the search for a single function called the velocity-potential function, Φ, of four independent variables x, y, z, t satisfying the same equations. But the equations by themselves do not specify the velocity-potential function uniquely. They are satisfied by a far more general class of functions of which the potential function Φ is just one. To narrow down further the margin of indeterminateness we have to make use of boundary conditions which the velocity potential must satisfy. What this means may be explained by an example.

Suppose we consider the flow of water in a canal between two parallel banks. In this case the velocity potential Φ must not only satisfy the Eulerian equations of motion and continuity referred to above but also the further condition that at the boundary of the fluid (that is, at the bank) the velocity component perpendicular to the bank is zero. In other words, Φ must satisfy the boundary condition that the fluid flow at the bank is parallel to it and that no particle flows past it, which could happen only if the banks were breached. Thus, if the Eulerian equations of motion and continuity are the slippers that isolate our Φ-Cinderella from the undesirable members of its family, the boundary condition is the censor that forbids the banns should a relation attempt to masquerade as Φ-Cinderella by wearing the slippers. Unfortunately it has not been possible to discover a Φ-Cinderella who could wear the slippers and satisfy the censor in the general case of fluid motions. And yet Euler had made the slippers rather loose, since he disregarded an important property of real fluids, *viz.* their viscosity.* Seventy years later Navier and Stokes made them much tighter by adding what may be called the *viscous terms* to the three Eulerian equations of fluid motion. That made matters more difficult.

Largely owing to the great inherent difficulties of the subject, hydrodynamical theory was obliged to make a number of simplifying assumptions and thus became more a study of 'perfect' or 'ideal' fluids than that of actual fluids in the real world. This was natural. But unfortunately the next step—that of proceeding to a more realistic state of affairs by embroidering variation from the ideal on the theory—was long delayed. As a result the theoretical development of fluid mechanics did not lead to such perfect harmony between theory and observation as in other branches of mathematical physics such as optics, electricity, magnetism, thermodynamics, *etc.*

For instance, soon after the formulation of the Eulerian equations of fluid motion a paradox, known as D'Alembert's paradox, emerged. The paradox arose because hydrodynamical theory seemed to prove that any body completely immersed in a uniform, steady stream of fluid would experience no resistance whatever—a result quite contrary to experience. D'Alembert's paradox was not an isolated case where plausible hydrodynamical argument led to a conclusion contradicted by physical observation. Before long it appeared that classical hydrodynamical theory was replete with paradoxes which it was unable to rationalise. Such, for instance, were Kopal's paradoxes, paradoxes of airfoil theory, the reversibility paradox, the rising bubble paradox, the paradox of turbulence in pipes, Stokes' paradox, the Eiffel paradox, the Earnshaw paradox, the DuBuat paradox, *etc.*

The emergence of such a multitude of paradoxes clearly showed that hydrodynamical theory did not conform to experimental reality. One

* We commonly call liquids like tar or treacle viscous as they exhibit a tendency to resist change of shape.

consequence of this failure was particularly unfortunate. It made hydrodynamics increasingly abstract, academic and removed from actuality.
The engineers, who felt that the mathematicians had left them in the
lurch by producing hydrodynamical results largely at variance with reality,
began to create a new science of their own. This science of hydraulics was
designed to give approximate solutions to real problems. But, lacking a
sound theoretical foundation, hydraulics rapidly degenerated into a morass
of empirical and semi-empirical formulae. So, by the close of the nineteenth century, both the mathematicians engrossed in 'pure' theory devoid
of fresh physical inspiration, and the engineers absorbed in accumulating
experimental data without adequate rationalisation by deductive theory,
seemed each to have reached blind alleys of their own.

With further progress thus blocked, it was now time to think of a way
out. Naturally the very first question that arose was to consider whether
these paradoxes were due to the neglect of viscosity or to some more
serious flaw deep in the fundamental assumptions of traditional hydrodynamics. We shall examine this basic question of the foundations of
hydrodynamics a little more fully. To begin with, let us take viscosity.

It will be recalled that hydrodynamical theory considered *ideal* fluids of
zero viscosity rather than *real* fluids which show more or less tendency to
resist change of shape. But this is only part of the story. The complete
story is that in deriving the equations of fluid motion we must make one of
two approximations. Either we neglect viscosity, which leads to Euler's
equations for a so-called non-viscous fluid, or we neglect compressibility
and assume that fluid density remains constant to obtain Navier-Stokes'
equations for an incompressible fluid. Can we attribute the paradoxes of
hydrodynamics *solely* to one of these two approximations? The answer is
no, because it can be shown that many of the paradoxes are not due to one
or other of these two approximations. But even if it were shown that they
were, there seems to be no way of avoiding them, for even with these
simplifying approximations the mathematical problem in most cases is
quite involved if not intractable. Besides, these approximations have to be
made because we do not actually know how viscosity acts in a fluid under
rapid compression. Without them hydrodynamical theory, therefore,
would have to stop almost at the very threshold. How are we then to
rationalise hydrodynamical paradoxes?

Birkhoff has recently shown that there can be no simple answer to this
difficult question. What is required is a profound analysis of the entire
body of mathematical, logical and physical assumptions (tacit or otherwise) of hydrodynamical theory *in the light of experimental data*, for, as
Birkhoff has demonstrated, hydrodynamical paradoxes are *not solely* due
to the single 'unjustified' neglect of viscosity. They are equally due to

faulty physical and logical assumptions underlying hydrodynamical reasoning. Take, for instance, the rising bubble paradox. If we consider a small air bubble rising in a large mass of water under its own buoyancy, conditions of symmetry* require that it should rise vertically. And yet in most cases it ascends in a vertical spiral instead. This paradox arises because of the assumption that 'symmetric causes produce symmetric effects'. But as Birkhoff rightly suggests, the symmetric effects that symmetric causes produce need not necessarily be stable. If they happen to be unstable, a slight deviation would tend to multiply and the symmetry in effects would be too short-lived to be noticeable: the observed effects therefore would be non-symmetric. In other words, while exact symmetric causes would, no doubt, produce exact symmetric effects, nearly symmetric causes need not produce nearly symmetric effects *if they happen to be unstable*. Hence before we can legitimately use the arguments of symmetry, we must first show that symmetric effects deduced are stable. But a demonstration of the stability of a mathematical problem is far more intricate than the deduction itself. It therefore happens that the deduction is often made before its stability can be proved. This leaves only one alternative, *viz.*, to make use of the symmetric argument but to test the stability of deduction in practice—that is, by experiment. This is one illustration of the way in which deductive theory must be interwoven with experimental practice if hydrodynamics is to be freed from paradox.

Another cause of paradox is the assumption that small causes produce small effects. Since every physical experiment is actually affected by innumerable minute causes, we should be quite unable to predict the result of a single such experiment if we did not continually make this assumption; and yet it is not universally true. An obscure fanatic's bullet—as at Sarajevo—may precipitate a global war, a slight fault in the earth's crust may cause a devastating earthquake or a deep-sea explosion, and a single mutation in a gene may alter the entire genetic mould of an individual or even a race. In the limited field of fluid mechanics there are cases where arbitrarily small causes do produce *significant* effects which cannot be neglected. For instance, a small change in viscosity (though not a small change in compressibility) may drastically affect fluid flow. This is because in Navier-Stokes' equations of fluid motion viscosity is the coefficient of the highest order derivative appearing in the equations. Now in systems of differential equations, the presence of arbitrarily small terms of higher order can entirely change the behaviour of the solutions. It is *not* always the case that as the coefficient of a term in an equation tends to zero, its solution tends to the solution obtained by deleting that coefficient. It may also

---

* Under surface tension the air bubble will be spherical and there is no cause which does not operate symmetrically through the centre of the bubble.

happen that as this coefficient tends to zero the solution suffers an abrupt change of nature at some stage. Thus, for example, in the equations of motion of a sphere through a fluid, a minute change in the value of the viscosity coefficient from a small value of 10^{-5} to 0.5×10^{-5} causes a sudden and radical change in the nature of the whole solution, resulting in Eiffel's paradox.

We must therefore learn to discriminate between 'right' and 'wrong' approximations. The only way we can do so is boldly to use them in our deductive theory but to *test* the conclusions so derived by experiment. If any of its conclusions is contradicted by physical observation, we should have to examine all the approximations made to discover those at fault. This is the only way we can proceed and even then we may not always be able to overcome the trouble. For it is quite possible for a paradox to arise because *too many* approximations have been assumed* and it is almost impossible to say which of them was 'wrong'.

These considerations show that hydrodynamical paradoxes are due to over-free use of approximations, non-rigorous symmetry considerations and physical over-simplifications. But this does not mean that we can rescue hydrodynamical theory by supplanting mathematical deduction by the more 'physical' reasoning so popular with practical engineers—quite the contrary. For recent developments in fluid mechanics have shown that mathematics is not just a useful device for presenting results whose broad outlines were suggested by physical intuition, as Archimedes and Newton used geometry to present results derived by 'analytical' or 'fluxion' methods. In many cases mathematical deduction gives correct results verified by later experiments which physical intuition is not only unable to derive but would, in fact, straightway reject as grossly absurd. The moral of all this is that what we need to build (that is, a paradox-free fluid mechanics) is a happy blend of the practice of the hydraulic engineer with the deductive theory of the mathematician—a complete interpenetration of deductive theory and practical experience at all levels. It is fortunate that during the past fifty years, mainly as a result of the impetus provided by the needs of aviation, this gap between hydrodynamical theory and experiment has been progressively bridged by just such a happy blend.

*　　　*　　　*　　　*

Lewis Mumford has divided the history of the Western machine civilisation during the past millennium into three successive but over-lapping and interpenetrating phases. During the first phase—the eotechnic phase—trade, which at the beginning was no more than an irregular trickle, grew to such an extent that it transformed the whole life of Western Europe. It

* This is the case with the Earnshaw paradox.

is true that the development of trade led to a steady growth of manufacture as well, but throughout this period (which lasted till about the middle of the eighteenth century) trade on the whole dominated manufacture. Thus it was that the minds of men were occupied more with problems connected with trade, such as the evolution of safe and reliable methods of navigation, than with those of manufacture. Consequently, while the two ancient sources of power, wind and water, were developed at a steadily accelerating pace to increase manufacture, the attention of most leading scientists, particularly during the last three centuries of this phase, was directed towards the solution of navigational problems. The chief and most difficult of these was that of finding the longitude of a ship at sea. It was imperative that a solution should be found as the inability to determine longitudes led to very heavy shipping losses. Newton had tackled it, although without providing a satisfactorily practical answer. In fact, as Hessen has shown, Newton's masterpiece, the *Principia*, was in part an endeavour to deal with the problems of gravity, planetary motions and the shape and size of the earth, in order to meet the demands for better navigation. It was shown that the most promising method of determining longitude from observation of heavenly bodies was provided by the moon. The theory of lunar motion, therefore, began to absorb the attention of an increasing number of distinguished mathematicians of England, France, Germany and America.

Although more arithmetic and algebra were devoted to Lunar Theory than to any other question of astronomy or mathematical physics, a solution was not found till the middle of the eighteenth century, when successful chronometers, that could keep time on a ship in spite of pitching and rolling in rough weather, were constructed. Once the problem of longitude was solved it led to a further growth of trade, which in turn induced a corresponding increase in manufacture. A stage was now reached when the old sources of power, namely wind and water, proved too 'weak, fickle, and irregular' to meet the needs of a trade that had burst all previous bounds. Men began to look for new sources of power rather than new trade routes.

This change marks the beginning of Mumford's second phase, the palaeotechnic phase, which ushered in the era of the 'dark Satanic mills'. As manufacture began to dominate trade, the problem of discovering new prime movers became the dominant social problem of the time. It was eventually solved by the invention of the steam engine. The discovery of the power of steam—the chief palaeotechnic source of power—was not the work of 'pure' scientists; it was made possible by the combined efforts of a long succession of technicians, craftsmen and engineers from Porta, Rivault, Caus, Branca, Savery and Newcomen to Watt and Boulton.

Although the power of steam to do useful work had been known since the time of Hero of Alexandria (A.D. 50), the social impetus to make it the chief prime mover was lacking before the eighteenth century. Further, a successful steam engine could not have been invented even then had it not been for the introduction by craftsmen of more precise methods of measurement in engineering design. Thus, the success of the first two engines that Watt erected at Bloomfield colliery in Staffordshire, and at John Wilkinson's new foundry at Broseley, depended in a great measure on the accurate cylinders made by Wilkinson's new machine tool with a limit of error not exceeding 'the thickness of a thin six pence' in a diameter of seventy-two inches. The importance of the introduction of new precision tools, producing parts with increasingly narrower 'tolerances', in revolutionising production has never been fully recognised. The transformation of the steam engine from the wasteful burner of fuel that it was at the time of Newcomen into the economical source of power that it became in the hands of Watt and his successors seventy years later, was achieved as much by the introduction of precision methods in technology as by Black's discovery of the latent heat of steam.

A natural consequence of the introduction of higher standards of refinement in industry and technology was that mathematical language itself became increasingly exact, subtle, fine, intricate and complex. The greatest change in this direction occurred in the language of the infinitesimal calculus, as the reasoning on which its technique had been based was shockingly illogical.

The methods of calculus were accepted, not because their reasoning was logically impeccable, but because they 'worked'—that is, led to useful results. For instance, in calculating the 'fluxion', or—as we now say—the differential coefficient of x^2, the founders of the calculus would substitute $x + 0$ for x in a term like x^2, expand the resultant expression $(x + 0)^2$ as if the zero within the bracket was a non-vanishing quantity, and then let 0 disappear in the final step. In other words, they believed that there existed quantities known as the 'infinitesimals', or 'fluxions', which could be treated as zero or non-zero according to the convenience of the mathematician.

This glaring illogicality of the calculus did not escape unnoticed. It was exposed with masterly skill by a non-mathematician, Bishop Berkeley, the famous idealist philosopher. While the mathematicians, with unerring instinct, ignored the attack and went on piling formulae upon formulae like Ossa upon Pelion, these methods brought the mathematicians into bad repute. Swift's caricature of the mathematicians of Laputa and his denunciation of them as 'very bad reasoners' was possibly inspired by Berkeley's withering critique of the 'fluxions'. In his *L'Homme Aux*

Quarante Ecus, Voltaire, too, had a dig at the analysts when he remarked that a 'geometer shows you that between a circle and a tangent you can pass an infinity of curved lines but only one straight line, while your eyes and reason tell you otherwise.'

It is true that some gifted mathematicians of the eighteenth century had realised that although the calculus towered over mathematics like a colossus, its feet were made of clay. Thus, D'Alembert neatly epitomised the whole situation when he remarked that 'the theory of limit is the true metaphysics of the differential calculus'. Lagrange, another eminent mathematician of the same century, tried, though unsuccessfully, to cut the Gordian knot by jettisoning the infinitesimals, 'fluxions' and 'limits' as so much useless lumber, and by representing all functions as sums of a power series—that is, a non-terminating series like $a_0 + a_1 x + a_2 x^2 + \ldots$. No doubt, in this way he escaped the mysterious infinitesimals which were both zero and not zero at the same time, but he thereby ran into another difficulty, no less serious—that of summing an infinite series.

The way out of the difficulty was first pointed out by a French mathematician, Cauchy, who, during the second decade of the nineteenth century, set out to purge the calculus of all its illogicalities, loose methods of reasoning, and premises of doubtful validity. He thus virtually introduced a 'New Look' that has come to stay in mathematical reasoning. New Look was suspicious of all arguments based on vague analogies and geometric intuition. So it began to examine more precisely all those vague notions and concepts which had hitherto been taken for granted. Take, for instance, the idea of mathematical limit which we explained earlier. The founders of the calculus thought they knew what they meant by a limit. In ordinary language, we use it to mean a sort of a terminal point or a bound that may not or cannot be passed. Thus, the Phoenician navigators considered the Pillars of Hercules as the *limit* of navigation, because in those days few ships that sailed beyond them ever returned. When mathematicians began to define speed *at* an instant as the limit of the average speed when the time interval over which it is averaged tends to zero, they pictured it in much the same manner as the Phoenicians thought of the Pillars—as a sort of barrier past which the average cannot go. But this explanation of the limit notion, though it may be a good enough start, is not sufficient. For, while the mathematical meaning of limit is vaguely suggested by its linguistic usage, it is not precisely defined thereby. Cauchy gave the first genuinely mathematical definition of limit, and it has never required modification. The need for it had been realised earlier by many, but it came into being nearly 150 years after mathematicians had been manipulating with limits. So if you do not get it right first reading, do not be disappointed. It took the mathematicians themselves the best part of a century to frame it.

To grasp the idea underlying Cauchy's definition of a limit, let us consider once again the distance function $y = t^2$ by which we introduced the limit notion, but to give greater generality to our results we shall make our symbols y and t meaning-free. That is, we shall no longer think of y as distance and t as time but let them denote any two variables whatever related by the same functional formula. It would help in this abstraction if we changed the notation a little and worked with the formula $y = x^2$ instead of $y = t^2$. Earlier we defined dy/dx at $x = 2$ as the *limit* of

$$\frac{(2 + h)^2 - (2)^2}{(2 + h) - 2} \qquad . \qquad . \qquad . \qquad . \quad (1)$$

when h tends to zero. Here the expression (1) is really a function of h as it depends on the value of h taken. For any given value of h (except $h = 0$*), we can calculate the corresponding value of (1) by substituting it for h in the expression (1). In fact, if we simplify this expression, we find that it is

$$\frac{4 + 4h + h^2 - 4}{h} = \frac{(4 + h)h}{h} = 4 + h$$

for *any* value of h other than zero.

This shows that if we assume h successively to be ·1, ·01, ·001, ·0001, . . ., the corresponding values of (1) are 4·1, 4·01, 4·001, 4·0001, . . . respectively.

We thus conclude that the limit of (1) is 4 as h is indefinitely decreased. This may seem to suggest that the limiting number 4 is a sort of barrier point beyond which the value of (1) cannot pass. Actually this is not always true. Thus if we assign h a succession of negative and positive values ·1, −·1, ·01, −·01, ·001, −·001, ·0001, −·0001, . . . we obtain for (1) the values 4·1, 3·9, 4·01, 3·99, 4·001, 3·999, 4·0001, 3·9999, ...

Here even though the value of (1) continually crosses the barrier number 4 backward and forward, nevertheless it keeps on coming closer and closer to 4. We are, therefore, still justified in calling 4 as the limit of (1). The essential idea of the limit is not that it is a sort of impassable terminal but that it is a point of continually closer and closer approach. For this purpose it is, of course, necessary that we should be able to make the value of (1) approach as near its limiting value 4 as we like by making h sufficiently small. But what is even more important is that once we have *so*

* For $h = 0$, the formula gives the meaningless expression 0/0. Division by zero is a taboo in mathematics as it leads to *no* result. For instance, if you had hundred rupees you could issue ten cheques of Rs. 10 each in all. But if you issued cheques of zero value, you could issue *any* number of them for all the difference it would make to your bank balance.

brought the value of (1) *within any arbitrarily chosen small range of closeness to its limiting value, it should continue to remain as close, if not closer, for all other numerically smaller values of h.* Thus, suppose we wish the value (1) to differ from its limiting value 4 by less than an arbitrarily small number, say, ε. In order that the difference

$$(4 + h) - 4$$

may remain numerically less than ε, all that we need do is to make h less than ε. So long as we keep h smaller than ε the difference between 4 and the actual value of the expression (1) will never exceed ε. In other words, the value of (1) always lies between $4 - \varepsilon$ and $4 + \varepsilon$ for *all* values of h less than ε.

In general, we say that a function $f(x)$ tends to a limit l as x tends to zero, if we can *keep f(x)* as close to l as we like by keeping x sufficiently close to zero. This means that given any arbitrarily small number ε another small number δ depending on ε alone can be found such that the difference $l - f(x)$ *always* remains numerically below ε as long as x stays below δ. In other words, if you wish to imprison *all* the values of $f(x)$ within the two corners of a number mesh of any given degree of narrowness— no matter how small—round the number l, you have to devise *another* number mesh round zero within which the value of x must remain confined. We usually denote the narrowness of the number mesh round l by ε, where ε is *any* arbitrarily small number we care to nominate.

In the same way we could examine if $f(x)$ tends to a definite limit, say, L when x tends to any other value, say, $x = a$. The examination may be done in steps as follows. In the first place, we nominate an arbitrarily small number to specify the narrowness of the number mesh round L within which we wish to trap all values of $f(x)$. Let it be ε. This means that we wish to confine all values of $f(x)$ within the number mesh $L - \varepsilon, L + \varepsilon$.

The second step is to discover what restriction should be imposed on the values of x in order to confine $f(x)$ within $L - \varepsilon, L + \varepsilon$. This means that we are to examine whether we can discover another small number δ depending on ε, such that if we confine x to the number mesh, $(a - \delta, a + \delta)$ round a, the function $f(x)$ remains within $(L - \varepsilon, L + \varepsilon)$. If this condition is satisfied, we say $f(x)$ tends to L as x tends to a.

This is Cauchy's famous 'epsilon-delta' definition of limit; he defined it in purely arithmetical terms and thus freed it from vague associations with its counterpart of everyday speech.

*   *   *   *

The motive force of modern mathematics is abstraction: in fact, abstraction *is* power and the reason is this. Whenever we treat some real problem mathematically, whether in physics, astronomy, biology or the social sciences, there is only one way in which we can proceed. We must first simplify it by having recourse to some sort of an abstract model or replica representing those features of reality considered most essential for the problem in question. Take, for instance, Eddington's famous elephant problem. Here an elephant weighing two tons slides down a grassy hillside of 60° slope. How long did he take to slide down? If we strip the problem of its 'poetry' (Eddington's word), that is, if we make an abstract mathematical model embodying its essential features, all that we have is a 'particle' sliding down an 'inclined plane'. Here 'particle' is a mathematical abstraction, which just retains that essential property of material bodies we call 'ponderosity' or 'inertia' common to them all. Similarly, 'inclined plane' is a geometrical abstraction of hillsides embodying its essential feature 'steepness'.

Abstract mathematical models of this kind not only simplify the real problem by retaining only the bare essentials without those encumberances, the irrelevant details which complicate matters and make it intractable, but also apply to a much wider class of problems than the original. This explains the paradoxical statement sometimes made: the more modern mathematics departs from reality (that is, grows abstract) the closer it comes to it. For no matter how abstract it may appear, it is in the ultimate analysis an embodiment of certain essential features abstracted from some sphere of reality.

Now an important group of problems in the most varied fields of science may be reduced to the consideration of systems changing with time. We have already given instances of it—the motion of material bodies such as stones and elephants rolling down grassy hill-sides, and the changing positions of planets in the sky. Other instances of this kind are: in biology, the development of a population or the growth of an organism; in astronomy, the changing luminosity of variable stars; in economics, the varying demands and prices of a market; in actuarial science, the changing number of claims for payment on an insurance company; in telephone theory, the changing incidence of telephone calls, etc. In any of these problems we may imagine an abstract mathematical model where 'something' changes with time. In some problems, such as the development of a biological population or the growth of insurance claims, this 'something' changes at discrete moments and in whole numbers. That is, it jumps from one whole number to another.* In others, such as the positions of planets in the sky or the growth of a biological organism, the essential feature of the change is that it takes place incessantly by small degrees.

* For instance, if the population of a species is, say, 10 at any given moment, it cannot possibly attain fractional values like 10·5 or 9·2 at any other time. The birth and death of its individuals can only alter it in whole numbers and at discrete moments.

There are thus two main kinds of mathematical models. In the one designed for handling continuous processes, this 'something', or the dependent variable (y), changes continuously with time. In the other, designed for treating discontinuous processes, it *jumps* discontinuously. We could, in fact, carry the process of abstraction still farther and consider a set of related variables—a dependent variable y depending on an independent variable x which need not necessarily represent time. The application of the fundamental laws of the process usually leads to one or more differential, integral or integro-differential equations. The question then arises whether such equations can be solved. In many cases it can be proved that there is a unique solution, provided it is assumed that y is a continuous function of x. It is, therefore, important to examine under what conditions a function is continuous.

Our naïve intuitive idea of a continuous function is this. It is continuous when its graph can be drawn without ever lifting the pencil from paper. If the graph jumps at any particular point, the function is discontinuous at that point. Consider, for instance, a rocket fired vertically upwards with an initial speed of, say 440 feet per second. Suppose further that the rocket carries an explosive charge, which on explosion during the course of its ascent is enough to impart to it instantaneously a further vertical speed of 88 feet per second. If the charge explodes, say five seconds after the initial start, the speed function (y) will be given by the formulae

$$y = 440 - 32x, \qquad . \qquad . \qquad . \qquad . \quad (1)$$

for all x up to and equal to 5,

and, $$y = 528 - 32x, \qquad . \qquad . \qquad . \qquad . \quad (2)$$

for all values exceeding 5 till the vertical ascent of the rocket ceases.

If we draw its graph, it will be represented by the line AB for the values of x lying between 0 and 5. (See Fig. 8.) At $x = 5$ the rocket charge explodes and increases its speed instantaneously by 88 feet per second. The speed graph is now represented by the parallel line CD for values of x exceeding 5. The speed function as a whole is, therefore, represented by the two lines AB and CD. At the point B we have to lift the pencil to draw it. This shows that the speed function is discontinuous at B.

We can make this notion of continuity more rigorous, if abstract, by disregarding the graph of the function and considering only the functional relation between y and x. If we are considering a range of values of x from 0 to 5, formula (1) applies. Hence if x tends to 5 from below, that is, by always remaining less than 5, y tends to $440 - 32(5) = 280$. On the other hand, if x tends to 5 from above, that is, by remaining always more than 5, formula (2) applies and the value of y tends to $528 - 32(5) = 368$.

There is thus a hiatus at $x = 5$. If we approach from below y tends to 280, while if we approach it from above it tends to 368. This is merely a mathematical translation of the physical fact that at the instant $x = 5$ the speed of the rocket jumped by 88 feet per second due to the instantaneous explosion of the charge during its ascent.

Whenever we have a function whose value tends to two different limits as x tends to any given value a from above and below, the function is said

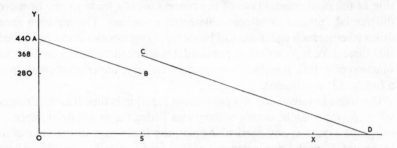

FIG. 8—The speed function of the rocket jumps at B to C and is therefore discontinuous at $x = 5$.

to be discontinuous at $x = a$. We thus have three values of a function at any point $x = a$: its actual value at $x = a$ and the two limits as x approaches a from two sides—above and below. The function is continuous at $x = a$ if and only *if* all three are equal. If *any* two happen to differ, the function is discontinuous at $x = a$.

All ordinary functions defined by one or more formulae are generally continuous except, possibly, for some isolated values. Thus, for instance, the speed function cited above was continuous everywhere except at $x = 5$. Is it possible to have functions defined by regular mathematical formulae, which are discontinuous everywhere throughout the range of the independent variable? The earlier mathematicians would have unhesitatingly replied no. We now know better. Suppose we have a function $y = f(x)$ defined by the following two formulae for all values of x in the range $(0, 1)$:

$$y = f(x) = 0, \text{ whenever } x \text{ is a rational fraction;}$$

and

$$y = f(x) = 1, \text{ whenever } x \text{ is an irrational fraction.}$$

Thus for $x = \frac{1}{2}$, y is zero; but for $x = \sqrt{\frac{1}{2}}$ y is unity.

Such a function is everywhere *discontinuous*. For, take *any* value of x such as $x = \frac{1}{2}$. Within any range round $\frac{1}{2}$ no matter how small, there are any number of rational and irrational values. Thus in the range $(\frac{1}{2} - \frac{1}{3000}, \frac{1}{2} + \frac{1}{3000})$, the value $\frac{1}{2} + \frac{1}{6000}$ is rational and $\frac{1}{2} + \frac{1}{3000\sqrt{2}}$ is irrational.

For the rational values of x in the range, $f(x)$ will be zero while for its irrational values $f(x) = 1$.

The function $f(x)$ therefore continually oscillates between 0 and 1 however narrow the range of values of x round $x = \frac{1}{2}$ we may choose. It cannot, therefore, tend to any limit as x tends to $\frac{1}{2}$. But there is nothing in the above argument special about $x = \frac{1}{2}$. What has been shown about the limiting behaviour of $f(x)$ when x approaches $\frac{1}{2}$ holds equally for *any* other value of x in the interval (0, 1) for which we have defined the function $f(x)$. The function $f(x)$ is therefore discontinuous *everywhere*. It gives us an absolute discontinuity.

If we have functions which are discontinuous somewhere as well as everywhere, what about their differential coefficients at these points of dis-

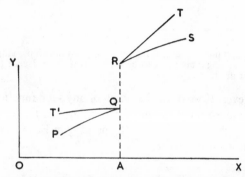

Fig. 9—Here $f(x)$ is discontinuous. As x approaches A the graph line suddenly jumps from Q to R. At this point no single line which can be regarded as a tangent to the graph line as a whole can be drawn. The derivative of $f(x)$, therefore, does not exist at A.

continuity? Clearly a discontinuous function can have no differential co-efficient at a point of discontinuity. For, if we draw the graph of the function (see Fig. 9), there will be a sudden jump at a point of discontinuity, so that it is impossible to draw a tangent to it at that point. But since the slope or gradient of the tangent is the measure of the differential coefficient, the latter cannot exist if the tangent cannot be drawn. If the tangent cannot be drawn at a point of discontinuity, can it be drawn at all points of continuity? Not necessarily. Examine, for instance, a continuous curve like the one drawn in Fig. 10. It is continuous everywhere even at the kinky point Q. But no tangent can be drawn to the graph as a whole at Q. For the portion of the graph to the right of Q, the tangent at Q is QT, for that on the left, QT'. Since there is no unique tangent to the graph at Q, the differential coefficient cannot exist although the graph is continuous.

It might appear that Q is an exception and that in general a continuous function is bound to have a tangent everywhere except possibly at a few isolated points. This, however, is one instance where our geometric intuition leads us astray. For we can construct functions which, though continuous for *every* value of the independent variable, have no derivative for

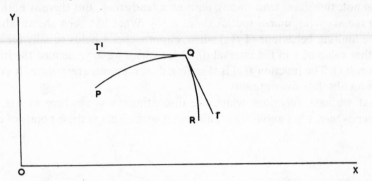

FIG. 10—The graph line suddenly turns turtle at Q Although the graph line is clearly continuous at Q, no single tangent can be drawn to it there. Hence the derivative of $f(x)$ does not exist at Q.

any value whatever. If we draw the graph of any such function, it will be a continuous line at *no* point of which we can draw a tangent. Every point on the graph will thus be a kink like the point Q in Fig. 9. That is why it is impossible to draw it on paper.

*　　　　*　　　　*　　　　*

A similar attempt was made to rigourise and refine the theory of integration. The refinement, no doubt, made it more abstract, complex and subtle, but it was worthwhile. It paved the way for a much wider generalisation of the theory of integration, which has been extensively applied. We shall deal with this generalisation and its applications in Chapter 6.

To understand the refinement, let us recall the essential steps in the calculation of the speed integral $y = t$ over the time interval (1, 6). First, we divided the interval (1, 6) into a large number of equal sub-intervals. Second, we multiplied the duration or length of each sub-interval by the lower and upper limits of the speed function *in* that sub-interval. We thus obtained the lower and upper limits within which lay the actual distance travelled during that sub-interval. Third, we added the products corresponding to all these sub-intervals and formed two sums, the lower sum s and the upper sum S, between which lay the total actual distance travelled. Fourth, we computed the limits of s and S as the number of sub-intervals was indefinitely increased. This common limit—the speed integral over time—was then the actual distance travelled.

Consider now any function $y = f(x)$ of any independent variable x, which may or may not represent time. Suppose we wish to find the integral of y over any given range of x, say from $x = 0$ to $x = 5$, and suppose further that $f(x)$ is bounded in the interval (0, 5). This merely means that all values of $f(x)$ remain within two definite bounds, say m and M as x varies within the interval (0, 5). Thus, for instance, the speed function $y = 440 - 32x$ of the rocket cited earlier is bounded within the interval (0, 5). For all values of $f(x)$ lie within the bounds $m = 280$ and $M = 440$ as x varies from 0 to 5. On the other hand, the function $y = x/(5 - x)$ for $0 < x < 5$ is infinite for $x = 5$. Here, as x varies in the range 0 to 5, y bursts all bounds and assumes bigger and bigger values. The table below shows this clearly if you have any doubts:

x	y
4	4
4·5	9
4·75	19
4·9	49
4·99	499
4·999	4999
4·9999	49999
etc.	etc.

The function $y = x/(5 - x)$, therefore, is unbounded. In other words, there is no definite number or upper bound M below which all the values of y remain as x varies in the range (0, 5).

It is in this sense that the function $f(x)$ is assumed to be bounded in the interval (0, 5). Now suppose we divide the interval (0, 5) into three parts, thus:

You may picture this by breaking a rule (AB) five inches long into three pieces at the two points marked 2 and 3 inches. We may, for convenience, call these three pieces i_1, i_2, i_3. If $f(x)$ is bounded in the original interval (0, 5), it is *a fortiori* bounded in all the three sub-intervals i_1, i_2, i_3 into which it has been divided. More, the upper and lower bounds for the sub-intervals are usually much closer than those for the original interval. Thus, for instance, in the case of the function, $f(x) = 440 - 32x$, it is easy to see that the lower and upper bounds are 344 and 376 respectively for the second sub-interval i_2 as against 280 and 440 for the whole interval. We tabulate the upper and lower bounds of the function for each of these three sub-intervals:

Sub-interval	Lower bound of $f(x)$	Upper bound of $f(x)$
i_1	376	440
i_2	344	376
i_3	280	344
Main interval	280	440

With these preliminaries we may now revert to the integration of $f(x)$, and proceed as before. First, we divide the interval (0, 5) of the independent variable x into a number m of sub-intervals which need *not* be all equal. As before, you may picture this by breaking a straight stick equal in length to the line AB into a number of pieces of unequal lengths and then piecing together the stick by laying the pieces end to end. Let i_1, i_2, i_3, ... be the lengths of the first, second, third, ... pieces so juxtaposed. Then the stick is the physical analogue of the total interval AB and the pieces of the sub-intervals i_1, i_2, i_3, ... Since $f(x)$ is bounded in the whole interval, it is naturally bounded in every sub-interval also though these bounds are different for each sub-interval. Consider the first sub-interval i_1. Let m_1, M_1 be the upper and lower bounds of $f(x)$ in i_1. Likewise, let (m_2, M_2), (m_3, M_3), ... be the respective bounds of $f(x)$ in the sub-intervals i_2, i_3, ... We are now ready for the second step. We multiply the length of each sub-interval by the upper and lower bounds of $f(x)$ in that sub-interval. Thus, for the first sub-interval i_1, we form the products $m_1 i_1$ and $M_1 i_1$, for the second sub-interval the products $m_2 i_2$ and $M_2 i_2$, for the third sub-interval the products $m_3 i_3$ and $M_3 i_3$ and so on for the nth and last sub-interval. We now add these products and form two sums

the upper sum, $S = M_1 i_1 + M_2 i_2 + M_3 i_3 + \ldots M_n i_n$,

and the lower sum, $s = m_1 i_1 + m_2 i_2 + m_3 i_3 + \ldots m_n i_n$.

This is the third step. It is obvious that to every method of splitting the interval (0, 5) there corresponds an upper sum S and a lower sum s. These two sums, S and s, are known as the upper and lower Darboux sums, after the French mathematician, J. G. Darboux, who first constructed them.

Finally, we take the fourth step and see what happens when the number of sub-intervals is increased indefinitely in such a way that the length of the *longest* sub-interval tends to zero. This is equivalent to breaking our stick into smithereens in such a way that the biggest smither is vanishingly small and no bigger than a pin point. It can be proved that when the number of sub-intervals is increased in this manner the sums S and s tend to two definite limits I and I' respectively. (The proof is omitted here.) If these two limits are equal, then the common limit I is known as the integral of $f(x)$. It is written as $I = \int_0^5 f(x)dx$.

The necessary and sufficient condition that the function $f(x)$ be integrable is that the upper and lower Darboux sums S and s tend to the same limit.

In mathematics we have often to distinguish between the necessary and sufficient conditions for securing an object. This distinction sometimes appears mysterious to the layman, although in the days of rationing it should have been obvious that a condition necessary for a certain purpose need not be sufficient for it. Thus in order to buy your ration it was necessary but not sufficient to have the money required to pay for it. The ration card and the buying power were the necessary and sufficient conditions for securing it, though either in isolation was not sufficient.

It can be shown that if $f(x)$ is a continuous function, the two limits of Darboux sums I, I' are equal. A continuous function is, therefore, always integrable, though, as we have seen, not necessarily differentiable. Before the introduction of the New Look style of reasoning in the calculus, it was generally believed that the two properties of a function, differentiability and integrability, went hand in hand. About 100 years ago it was shown that there exist functions which are non-differentiable though continuous and, therefore, integrable. Thirty years later it was proved that the discord between the two properties—differentiability and integrability—is reciprocal when bounded differentiable functions whose derivatives are not integrable were constructed.

4

VECTORS

F OR about 150 years after Newton the study of the motion of material bodies, whether cannon balls and bullets, to meet the needs of war, or the moon and planets to meet those of navigation, closely followed the Newtonian tradition. Then as it was just about beginning to end up in high-brow pedantry it was rescued by the emergence of a new science—electricity—in much the same way as cybernetics was to rescue mathematical logic a century later.

Though known from earliest times electricity became a quantitative science in 1819, when Oersted accidentally observed that the flow of an electric current in a wire deflected a compass needle in its neighbourhood. This was the first explicit revelation of the profound connection between electricity and magnetism, already suspected on account of a number of analogies between the two. A little later Faraday showed that this connection was no mere one-sided affair. If electricity in motion produced magnetism, then equally magnets in motion produced electricity. In other words, electricity in motion produced the same effects as stationary magnets and magnets in motion the same effects as electricity.

This reciprocal relation between electricity and magnetism led straightway to a whole host of new inventions from the electric telegraph and telephone to the electric motor and dynamo. In fact, it is the seed from which has sprouted the whole of heavy electrical industry which was destined to transform the paleotechnic phase of Western machine civilisation with its ugly, dark and satanic mills, into the neotechnic phase based on electric power. But before this industry could arise results of two generations of experiments and prevailing ideas in different fields of physics—electricity, magnetism and light—had to be rationalised and synthesised in a mathematically coherent theory capable of experimental verification. Now the results of the mathematical theory depended for their verification on the establishment of accurate units for electricity—a task necessary before it could be commercialised for household use. The theory, thus verified, in turn formed the basis of electrical engineering—itself the result of a complete interpenetration of deductive reasoning and experimental practice. It reached the apogee of its success when Hertz experimentally demonstrated the existence of electromagnetic waves, which Maxwell had postulated on purely theoretical grounds, and from which wireless telegraphy and all that it implies was to arise later.

Maxwell's theory was actually a mathematisation of the earlier physical intuitions of Faraday. In this he used all the mathematical apparatus of mechanics and calculus of the Newtonian period. But in some important and puzzling respects the new laws of electromagnetism differed from those of Newton. In the first place, all the forces between bodies that he considered as, for example, the force of earth's gravity on falling bodies, acted along the line joining their centres. But in the case of a magnetic pole it was urged to move at *right angles* to the line joining it to the current-carrying wire. Secondly, electromagnetic theory was differentiated from the earlier gravitation theory of Newton in its insistence that electric and magnetic energy actually resided in the surrounding empty space. According to this view the forces acting on electrified and magnetised bodies did not form the whole system of forces in action but only served to reveal the presence of a vastly more intricate system of forces acting everywhere in free space.

The theoretical consequence of this innovation was that while the gravitational effect of a system of material bodies could be fully described by assigning only a numerical value to each point of free space, that of a system of electrified and magnetised bodies required *in addition* the specification of the *direction* associated with the numerical value at each point. Consequently, mathematical theory had to take account not only of pure magnitudes but also their associated directions.

This was the first clear intimation of the inadequacy of the older arithmetic in which number alone counted and reigned supreme. It forced the recognition that the real number system, which man had evolved out of counting discrete collections and measuring fields, was incomplete in that its number vocabulary was not rich enough to express fully certain types of magnitudes which the new science of electromagnetism was creating. In fact, the inadequacy of the real number was already being felt even in the older mechanics which had to handle such magnitudes as velocity, acceleration, force, *etc.* For the peculiar thing about these quantities was that they too had a *direction* as well as a magnitude. In arithmetic we treated two lengths as equivalent in all respects if they were measured by the same number. Not so in the new science of electromagnetism, or even in the older Newtonian mechanics. A force, for example, that acted in one direction did not have the same effect in the material world as a force of equal magnitude that acted in a different direction.

To specify completely a force, a velocity, an acceleration, an electric current, a magnetic pull, or more generally any vector, we need two things. First a way of quantising its magnitude or intensity, and second, a way of indicating the direction in which it acts. We have already seen how the real number system suffices to measure any magnitude. So there is no difficulty in fulfilling the first requirement. As for the second, we fix two perpendicular lines (see Fig. 11) *SON* and *EOW* as our lines of reference. Any vector, like *OP*, may now be denoted by the length *OP* and the angle *PON* it makes with one of the fixed lines of reference. It may be expressed geometrically

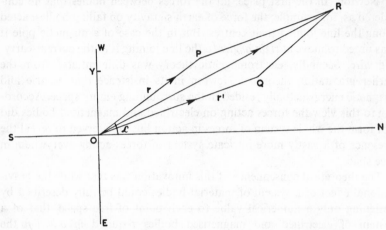

Fig. 11—The vector sum of the vector *OP* and the vector *OQ* is the vector *OR*, the diagonal of the parallelogram *OPQR*.

by the arrowed straight line *OP*, or analytically by a magnitude *r* measuring its intensity, and angle *α* that its line of action *OP* makes with one of the reference lines such as *SON*.

Now suppose we have two such vectors *OP* and *OQ* (Fig. 11); how shall we add them? If they act along the *same* line, there is no difficulty. The resultant of the two vectors will continue to act in the same line and its intensity will be the algebraic sum of the intensities of the two given vectors. Thus, if the intensities of *OP* and *OQ* are *r*, *r'*, that of the sum of *OP* and *OQ* will be *r* + *r'*. But how shall we add two vectors whose directions are different—like the two vectors *OP* and *OQ* in Fig. 11? In real life we have often to add two such vectors which act in different directions. For instance, a boat which a rower is attempting to row across a river has two velocities. One is the velocity with which the river is carrying the boat along with everything else down the stream, and the other is the still-water velocity (the velocity if there were no current) given to it by the rower at right angles to the direction of the river's flow. The velocity with which the boat actually moves under the combined effect of these two velocities may naturally be taken as the *sum* of the two velocities. (See Fig. 12.)

To discover a rule for adding any two vectors like velocities let us first consider the simplest case of a vector, *viz.* the case of a pure displacement of a rigid body from one position to another without change of orientation. Suppose, for example, we have a rectangular lamina *ABCD*. (See Fig. 13.) We may denote its position with respect to two reference lines *NOS* and *EOW* by specifying the position of any point fixed in the lamina, such as the corner *A*, and any line fixed therein such as the edge *AB*. That is, if we knew the position of the corner *A* and the direction of the edge *AB*, we could fix precisely the location and orientation of the lamina. Let the corner *A* be at any point A_1 and let the edge *AB* make any angle α with the reference line *NOS*. The lamina will then be in the position I shown in Fig. 13. If we shift the lamina to another position in such a way that the corner *A* now occupies the position A_2 while the edge *AB* continues to make the same angle with *NOS* as before, we have a case of pure displacement* of the lamina without any change of orientation. We may denote this displacement by the directed line A_1A_2 as the line A_1A_2 indicates the direction of the shift of the corner *A*

FIG. 12—The top two illustrations show the still-water velocity of the rower and that of the stream, respectively. The third shows the resultant (or vector sum) of these two velocities.

and the length A_1A_2 the magnitude thereof. But there is no special virtue in the line A_1A_2. As far as the displacement of the lamina is concerned, we may as well denote it by any other line whatsoever so long as its direction is parallel to A_1A_2 and its magnitude is equal to the length A_1A_2. One such line is the line *OP* where *OP* is parallel to A_1A_2 and $OP = A_1A_2$

---

* Such a displacement without change of orientation is called a *translation*.

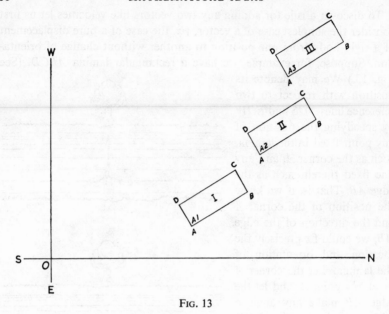

FIG. 13

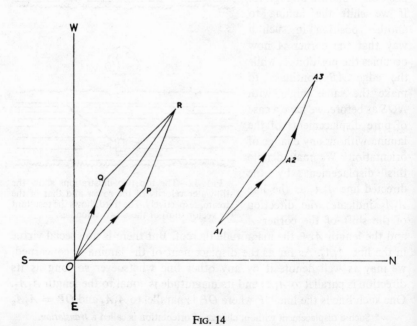

FIG. 14

(Fig. 14). We may therefore equally denote the displacement of the lamina from A_1 to A_2 by the vector OP.

Suppose now we shift the lamina corner A from A_2 to A_3 (Fig. 13) again without any change of orientation. The shift from A_2 to A_3 may be denoted by A_2A_3 or by the vector OQ (Fig. 14) where OQ is parallel to A_2A_3 and the length OQ is equal to A_2A_3. Obviously the two combined shifts of the corner from the position A_1 to A_2 and then from the position A_2 to A_3 are equivalent to the single shift of the corner A from the initial position A_1 to the final position A_3. This last shift is the vector OR where OR is parallel to A_1A_3 and the length $OR = A_1A_3$. Now it can be proved by a little high-school geometry that OR is the diagonal of the parallelogram with OP and OQ as its adjacent sides. It therefore follows that if OP represents one displacement and OQ another, their resultant is represented by the diagonal OR of the parallelogram constructed by taking OP and OQ as adjacent sides. The displacement OR is thus the *vector sum* of the displacements OP and OQ.*

Since velocity is only a rate of change of displacement, it follows that the same rule would apply for adding velocities in two different directions. But rate of change of velocity is acceleration and acceleration according to Newton's law is proportional to the force producing acceleration. It is therefore obvious that the parallelogram law of vector addition applies not only to displacements but also to all vectors such as velocities, accelerations, forces, *etc.*

Now what about the multiplication of two vectors like OP and OQ? In the case of integers, it is merely the repeated addition of one integer to itself a certain number of times. Thus, multiplication of four by three is the addition of four to itself thrice. In the case of fractions and irrationals, too, we found it useful to introduce multiplication in order to be able to handle continuous magnitudes. But coming to vectors, it is not clear what meaning should be attached to multiplication. If multiplication is to be defined as a sort of repeated addition of vectors, it can only mean addition of a vector to itself a certain number of times, say, five. But this would give us the product of a vector by a pure member such as five and not by a vector, which is what we wanted to define. Before we set out to search for a suitable multiplication rule for vectors, it may perhaps be asked why we should bother about multiplication of vectors when *prima facie* no physical meaning can be attached to it. The reason is as follows:

If we dispense with the multiplication of vectors, we cannot build up a generalised number system, of which the real number is only a special case. *Some* analogue of multiplication as used in the real number system

* Conversely OP and OQ are the components of OR.

is necessary. For otherwise the vector would not be a genuine extension of the real number system. As we shall see, there are several ways of defining a multiplication rule for vectors. One such way was suggested by the activities of craftsmen, engineers and technicians. During the second half of the eighteenth century, Watt, who drew patent royalties in the form of a percentage of the saving of fuel through the efficiency of his improved steam engine, was led to careful measurements of work done by steam engines. This scientific rating of their performance gave rise to the useful notion of physical work, which could in some way be regarded as the product of two vectors. Following Watt, work was defined as the product* of force and the displacement produced by it.

For example, work done by a torrent of water flowing down a steep hillside or a precipice is the product of the force of gravity moving it and the displacement of water which it causes. It is this work which is the source of hydro-electric power. But from the theoretical point of view this multi-

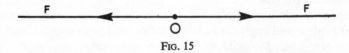

Fig. 15

plication rule has its limitation. It cannot serve as a basis for a generalised vector calculus, because the multiplication rule which defines work as the product of two vectors, force and displacement, yields a *pure* number, as there is no particular direction which can be associated with it. What the generalised vector calculus needs is a rule which preserves the vectorial quality of the vectors it multiplies. The rule must be such that when two vectors are multiplied the result is a vector and not something less general, such as a pure number.

To discover such a multiplication rule we must examine how vectors behave in real life. If two equal but opposite forces act at any point O of a rigid body, the two forces balance so that the body remains at rest. A tug of war, when the rival teams pull the rope with equal might in opposite directions, is a case in point (Fig. 15).

If, however, the two forces do not act at the same point but at two different points of the rigid body, they will not completely balance but tend to rotate it. You may easily verify this by a simple experiment. Place a pencil on a table and push it simultaneously at its two ends with equal force but in opposite directions. The effect of these two equal but opposite pushes applied at its two extremities will be to rotate it in the plane of the table (Fig. 16). Such a pair of forces is called a 'couple'.

In practice it is difficult to apply equal pushes at the two ends. But if you fix one of them by nailing it to the table loosely, the nail will auto-

* Such a product is known as the *scalar* product.

matically produce a reaction equal and opposite to the push you apply at the other. Suppose now you apply a gentle push F at the end B at right angles to the pencil. This will call forth an equal and opposite reaction at the other end A. The effect of the two forces will be to rotate the pencil in the plane of the table. This rotatory effect is a vector, and it is expressed

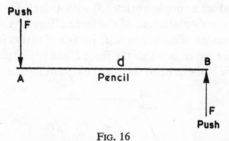

FIG. 16

as a measure along the axis of rotation of the pencil, that is, a line at right angles to the table at A. The direction of rotation is indicated by the 'screw law',* and the magnitude of the moment by the length of the line. This quantity depends on two things—the intensity of the push and the length of the pencil. The greater the push and the longer the pencil, the greater is this rotatory effect. It may, therefore, be measured by the product Fd, of the push F and the length d of the pencil.

In making an abstract mathematical model out of this situation, the pencil, nail and table fade out of the picture and all that we need retain is a couple of equal and opposite forces separated by a certain length or

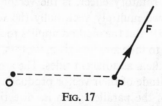

FIG. 17

distance. The rotatory effect of this couple is a vector because it has a magnitude and direction. Its line of action is perpendicular to the plane of the parallel forces, and its magnitude is the product of the forces and their distance apart. It is because of this rotatory effect of equal and opposite forces that it is possible to transform the linear motion of the piston rods of a steam or internal combustion engine into a rotatory motion of the wheels.

With these preliminaries we can now devise a suitable multiplication rule for vectors. Suppose a force F acts at any point P of a rigid body (Fig. 17). Can we displace it to some other arbitrary point O of the body?

* If rotation is to the left the vector stands up above the table; if it is to the right the vector points downwards beneath the table.

In other words, can we combine a vector-force F at P with a displacement-vector OP in any manner that is physically significant? Since two equal and opposite forces acting at the same point balance, we may imagine that two equal and opposite forces F act at O as shown in Fig. 18.

Now if the force F at P and the equal but opposite force F at O are combined to produce a couple, we are left with a single force F acting at O. In other words, the displacement of the force F from P to O introduces a couple whose rotatory effect is a vector. Its line of action is at right angles to the plane of the paper in which the force F and the displacement OP lie. Its magnitude Fd is the area of the parallelogram $OPQA$, whose adjacent

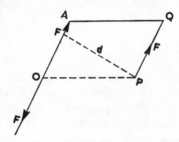

FIG. 18

sides are the vector-force F and the vector-displacement OP. You may recall that the area of a parallelogram is the product of the base PQ and the perpendicular distance d from the base to the opposite parallel OA.

This operation of displacing a force from P to O, which introduces a vector couple with a rotatory effect, is the vector analogue of ordinary multiplication. Here we multiply vectorially the force-vector F and the displacement-vector OP. But the method applies to all vectors whatsoever. All that we need do is to take any two given vectors, p and q, and construct a parallelogram with them as adjacent sides. The area of the parallelogram will give us the magnitude of their vector product and the line OP at right angles to the plane of the parallelogram its direction (see Fig. 19). This method gives us a multiplication rule, which does not destroy the vectorial quality of the vectors it multiplies.

The aforementioned rule of vector multiplication is no mere 'free creation' of the curious mind. It is a precise expression of an aspect of vector behaviour in real life. We have seen how in mechanics the operation of displacing a force a certain distance is a generalisation of ordinary multiplication in which pure numbers are replaced by vectors like force and displacement. It is also the way in which electromagnetic vectors such as electric current and magnetic force actually combine as was first observed by Ampère. In fact, Ampère's rule for deriving the mechanical force exerted by a magnetic pole on a small element of current carrying wire is the same as the rule for multiplying in the manner described above the two vectors concerned, *viz.* the electric current and the magnetic force exerted at that point. (See Fig. 20.)

However, this multiplication rule is rather complicated. For one thing if

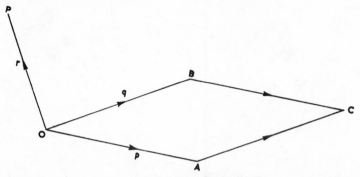

FIG. 19—The product vector *OP* is at right angles to the plane of the parallelogram *OACB* and is equal in magnitude to its area.

we have two vectors in one plane, say, the plane of the paper, the multiplication rule gives their product vector along a line at right angles to the plane of the paper, that is, out in three-dimensional space. We shall

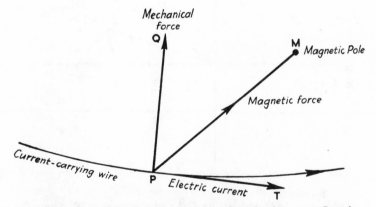

FIG. 20—At any point *P* of the current-carrying wire the current flows in the direction of the tangent *PT*. The magnetic force exerted by a magnetic pole *M* acts along *PM*. The mechanical force exerted on the wire in the neighbourhood of *P* acts along *PQ* at right angles to both *PT* and *PM*.

examine later the consequences of adopting it, as it leads to a further fascinating generalisation of vectors. Meanwhile let us confine ourselves to vectors lying in one plane, for example, the plane of the paper and observe if we can define a simpler multiplication rule for combining two vectors.

Let OP and OQ be any two such vectors that we wish to multiply (see Fig. 21). Let OA be the unit vector, that is, the unit length OA along the reference line SON. Join AQ and construct on OP a triangle OPR similar to OAQ. This can easily be done by making $\angle POR = \angle AOQ$

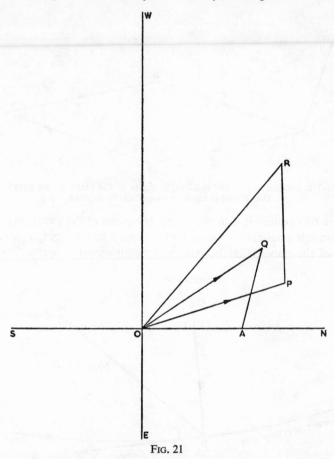

Fig. 21

and $\angle OPR = \angle OAQ$ and letting OR and PR meet in R. Since the triangles OPR and OAQ are similar, the sides opposite equal angles are proportional, and hence

$$\frac{OR}{OP} = \frac{OQ}{OA} = OQ, \text{ (as } OA = 1)$$

Hence $OR = OP \cdot OQ$.

We may, therefore, take OR as the product of the two vectors OP and OQ. Note carefully that we obtain the same vector OR if we reverse the roles of *OP* and *OQ*. You may verify it by describing a triangle on *OQ* similar to the triangle *OAP*. An exactly equivalent way of obtaining the product vector *OR* is to rotate the line of action of one of them through an angle made by the line of the other with the reference line *SON*. (See Fig. 22.) Cut off on the direction so obtained a length *OR* equal to the product of the lengths *r*, *r'* of *OP* and *OQ* respectively. Thus, we may rotate *OP* through an angle equal to *NOQ* and bring it to a new position

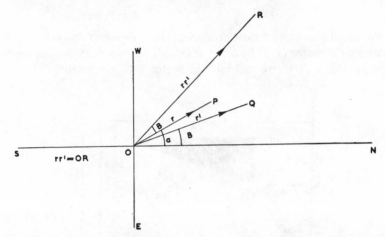

Fig. 22—Vector product. Rotate one of the vector lines, say *OP*, through the angle made by the second, *OQ*, with *SON*, to obtain *OR*. The product vector is then *OR* where the length of the line *OR* is *rr'*.

OR. This gives the direction of the product vector. We would have obtained the same line of action had we rotated instead the line of action of *OQ* through an angle made by *OP* with the reference line *SON*.

With the aforementioned rules for the addition and multiplication of vectors, we can construct a new and more general arithmetic for handling them. First, take the vector *OP*. We expressed it analytically by the length *r* of the segment *OP* and the angle α it made with the reference line *SON* (see Fig. 23). Draw perpendiculars *PM* and *PL* from *P* on the two reference lines. Now, by the parallelogram law of addition the sum of the vector *OM* along *NOS* and the vector *OL* along *EOW* is the vector *OP*.* We could, therefore, equally denote the vector *OP* by two vectors *OM* and *OL* along the two reference lines. If the magnitude of *OM* is *x* and of *OL y*, then the vector sum of *x* and *y* is equal to the vector *OP*. We may thus replace the vector *OP* by its components *x* and *y*. In other words, a vector may also be represented by a number couple (*x*, *y*).

How shall we add and multiply these number couples? We can deduce

---

* A rectangle is only a special case of a parallelogram.

the rules for adding and multiplying such number couples from the corresponding rules already given for two vectors. Thus, suppose we have another vector OQ represented by the number couple (x', y'). Then, as we have seen, the vector sum of OP and OQ is the vector OR, the diagonal of the parallelogram with OP and OQ as adjacent sides. Now a glance at Fig. 23 will show that the vector OR has components $x + x'$, $y + y'$. Hence the addition rule:

$$(x, y) + (x', y') = (x + x', y + y').$$

We could also express their product in terms of vector components. To multiply the vector (x, y) by a vector (x', y') we may carry out the operation in stages. First, we multiply the vector x by the vector x'. This is

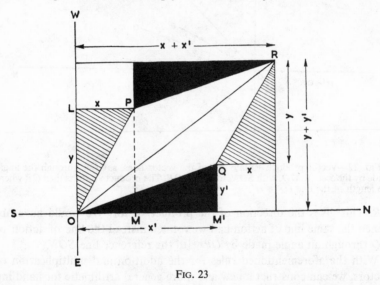

FIG. 23

simple, for the line of action of both is the reference line SON, and, therefore, their product is xx' along SON. We now multiply the vector x along ON by the vector y' along OW. The second vector makes an angle of $90°$ with the reference line SON. If, therefore, we apply the product rule for multiplying two vectors whose directions differ, we shall have to rotate x through $90°$ so that its line of action becomes EOW. We now multiply the magnitudes x and y' and the result is the vector xy' along EOW.

Next in order is the product of y and x'. Here, again, the two vectors are at right angles to one another and, following the same rule, their product is yx' along EOW. Finally, we have to multiply y and y'. Applying the same rule, we have to rotate y through an angle that the other, $viz.$ y', makes with SON, $i.e.$ through $90°$. If we rotate the vector y through $90°$, its line of action becomes OS. Hence the product of the vectors y, y' is the vector yy' along OS. We thus have four vectors, xx' along ON and yy' along OS, on the one hand, and xy' and yx' along OW, on the other. Hence the components of the product vector are $(xx' - yy')$ along ON,* and $(xy' + yx')$ along OW. This leads to the multiplication rule for the number couples (x, y) and (x', y')

$$(x, y)(x', y') = (xx' - yy', xy' + yx').$$

We have thus a new arithmetic of generalised numbers, vectors, which are denoted by number couples such as (x, y). The laws of addition and multiplication of two such generalised numbers like (x, y) and (x', y') are:

Addition law: $(x, y) + (x', y') = (x + x', y + y')$;

Multiplication law: $(x, y)(x', y') = (xx' - yy', xy' + x'y)$.

In learned treatises on the subject, they sometimes start with number couples like (x, y) and then attempt to build up an algebra by assuming various types of addition and multiplication laws. If the object is to construct the theory of rational fractions, the laws assumed are

(I)
Addition law: $(x, y) + (x', y') = (xy' + x'y, yy')$
Multiplication law: $(x, y)(x', y') = (xx', yy')$.

If the object is to develop the theory of complex numbers, the laws assumed are

(II)
Addition law: $(x, y) + (x', y') = (x + x', y + y')$.
Multiplication law: $(x, y)(x', y') = (xx' - yy', xy' + x'y)$.

This way of starting the subject is apt to give rise to a sense of mystery. One wonders why one set of laws is adopted for the one and quite another for the other. If mathematics were merely a game or a pastime for the ingenious, any such explanation would also be quite out of place. But as mathematics, like any other branch of science, arises out of man's attempt to build a basis for civilised life, such questions are of great social importance. We saw† how the former set of laws (I) arose quite naturally out of the need for combining fractional residues. We now see how the latter set (II) arises equally naturally when we have to combine directed magnitudes like forces and velocities. This also gives a natural explanation of the so-called "imaginary" numbers, which puzzled even learned mathematicians right up to the first half of the nineteenth century and still mystify the student and the layman. The real devil of the piece is the 'square root of minus one'. What is that number which when multiplied by itself gives

---

* The minus sign has been used since yy' is along a direction opposite to that of xx'.
† See page 12.

the result −1? It cannot be 1 for 1 × 1 is 1. Nor is it −1, for −1 × −1 is plus 1 and *not* −1. The answer is: *there is no such number.*

However, if we extend our number concept to include vectors—generalised numbers—which are given by number couples in the manner described above, we can find a number which answers to the specification. This is the *unit vector*, OY, directed along the line OW (See Fig. 11). Since its component along ON is zero and along OW, 1, it is given by the number couple (0, 1). If we multiply it by itself we get by the multiplication rule (II) quoted above

$$(0, 1) \times (0, 1) = (-1, 0).$$

We get the same result if we apply the product rule directly as is, of course, natural. Let us try to multiply the vector OY by itself. We rotate OY through an angle equal to the angle it makes with ON, that is, through 90°. The direction of the vector is thus along OS. Its magnitude is $1 \times 1 = 1$. Hence the square of the unit vector along OW gives a unit vector along OS, or −1 along ON. It is this unit vector along OW, the number couple (0, 1), whose square is the unit vector along OS, which may be interpreted as the 'square root of minus one'. In fact, it has a second 'square root' also. It is the unit vector along OE, or, the number couple (0, −1), which when multiplied by itself gives rise to the unit vector along OS, or the number couple (−1, 0). It is usual to denote the number couple (0, 1) by i, and the number couple (0, −1) by $-i$. It is only when we bear in mind that these letters i and $-i$ are *in fact* abbreviations of their corresponding number couples that we may write the equality $i^2 = -1$, and call i the 'square root of minus one'. But so long as we remain in the domain of the real number field, it is better to discard unhesitatingly as non-existent the imaginary i, as our Hindu ancestor, Mahavira, did eleven centuries ago.

The number couple (x, y) used to denote a vector is also known as a complex number and is often written as a single letter z. When we do so, we write $z = x + iy$, which means that z is a plane vector, whose components along the two perpendicular reference lines are x and y. We express the vector component along ON in ordinary units, but that along OW in 'i' or 'imaginary' units to indicate the fact that any vector along this line when multiplied by itself undergoes a rotation through one right angle. The theory of complex numbers is extremely complicated and is still fast developing. It has reaped a rich harvest largely because of extensive applications to a wide diversity of phenomena, from the theory of electric currents to map making and fluid flows.

For example, the theory of conformal transformation, depending as it does on a basic theorem of complex variables, is absolutely fundamental in the practice of making geographic maps. In mapping a surface such as the surface of the earth, all that we do is to devise a one-to-one matching process whereby to each point of the earth there corresponds one and only one point of the map and vice versa. There is an infinite number of ways of doing this but if the map is to be of any use to a navigator or an explorer it must ensure conservation of directions. In other words, if a direction PT at any point P on the earth makes an angle α with some other direction PT' at the same point, then the matching process which leads to

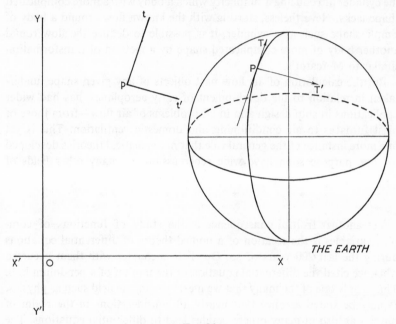

FIG. 24

corresponding directions pt and pt' on the map must be such as to leave the angle tpt' unaltered. (See Fig. 24.) Without this conservation of angles between corresponding directions the map is of no use. Such a map is known as conformal.

Again, there is an infinite number of types of conformal maps. One such map is the well-known Mercator's projection used in school atlases. A Mercator's projection transforms the curved surface of the earth into a flat sheet of paper in which the meridians and the parallels of latitude on the sphere correspond to straight lines parallel to the axes of co-ordinates on the flat map. In fact, the whole spherical surface of the earth is represented by an infinitely long strip of finite breadth. The mathematical theory of complex variables which permits this feat—the transformation of a spherical surface into a plane—equally permits others still more remarkable.

For example, in the design of aircraft it is necessary to know the lift exerted on a shaped body, such as an aerofoil,* by an air stream flowing past it. A well-known theorem of aerodynamics enables this calculation to be made if the strength of the circulation of the air stream were known. But this latter quantity can be calculated only in the simple case when the body in question happens to be a long circular cylinder, the symmetry of the cylinder introducing a simplicity which a body with a more complicated shape lacks. Nevertheless, starting with the known flow around a body of simple shape such as a cylinder, it is possible to deduce the flow round another body of more complicated shape by a method of transformation similar to Mercator's.

But the calculation of air flow past objects of any given shape, undertaken in relation to the development of early aeroplanes, has had wider applications in ship design and in all problems of air flow—from those of blast furnaces to air conditioning and domestic ventilation. This is yet one more instance of the general rule that mathematical theories developed for one purpose soon have wide repercussions in many other fields of daily life.

*   *   *   *

Yet another fruitful consequence of the study of functions of complex variables is the creation of a unified theory of differential equations during the last 200 years or so. Our first encounter with them occurred when we cited the differential equation of the motion of a pendulum bob. This is only one of the many that we meet even in one field such as physics. It may be stated at once that nearly all considerations in the realm of physics as also in many others besides lead to differential equations. The reason is this. In physics we wish to ascertain changes in the values of certain magnitudes corresponding to changes in those of others. The data of the problem are usually a system of physical laws, which express rates of change of the former per unit change of some other as a function of these variables. This gives rise to one or more equations involving all or some of the variables as well as their rates of change, which are nothing but differential coefficients of some variables with respect to others. That is why most riddles in physics when interpreted in mathematical symbolism lead to one or more differential equations. To divine the riddle is to 'solve' the differential equations to which it leads.

Sometimes a whole class of riddles leads to the same differential equation. For instance, the differential equation of wave motion arises as

* The shape of the wing of an aeroplane is called an aerofoil.

naturally in the electro-magnetic theory of light as in the theory of sound, elastic vibrations or radio waves. Such equations have, therefore, been intensively studied during the past two centuries by some of the greatest mathematicians of all times, whose abilities were taxed to the utmost in creating complicated mathematical functions satisfying them. Their creation would have been all but impossible without recourse to the theory of complex variables. The reason is the limitation of the real number field. We saw earlier (page 14) how the limitation of the integral domain prevented us from carrying out universally the operations of subtraction and division. To make all these operations universally possible we had to extend the domain of integers to include fractional and negative numbers. This made the field of rational numbers closed under the four fundamental operations of elementary arithmetic.

But the field of rational numbers, though closed, was still not large enough to prevent anomalies from arising. One such anomaly was that it did not permit inverses of certain operations like squaring, cubing, *etc.* Thus, while every rational number had a rational square, many rational numbers like 2 and 3 had no rational square roots. The field of rational numbers had therefore to be further extended. But the extended real number field is still *not* large enough to prevent anomalies in the behaviour of functions of real variable. For instance, even such a simple equation as $x^2 + 1 = 0$ has no solution so long as we operate only with real numbers.

Again, the solution of certain types of differential equations of mathematical physics requires the integration of complicated functions. Such integrations in turn lead to their own crop of anomalies. At bottom they all arise for one and the same reason—the limitation of the real number field. What we need to remove them is not merely a number field closed under the four arithmetical operations but a number field *which is the largest so closed*. It happens that the class of complex numbers is such a number field. That is why mathematicians have been obliged to have recourse to functions of complex variables to get rid of these anomalies.

The theory of complex functions shows that many of the differential equations of physics and astronomy may be derived from a single generic form: if the latter could be solved, so could its derivatives, but to solve it we have to employ new functions previously unknown. One of the methods adopted in inventing these new functions is an extended use of the process of inversion which, as we have seen, is such a prolific source of new mathematical entities. Thus if y is a function of x we may derive its inverse by expressing x as a function of y. Suppose, for instance, y is defined by the equation

$$y = \int_0^x x\,dx.$$

Here we may readily compute the integral on the right-hand side and replace the equation by

$$y = x^2/2.$$

This leads to the inverse $x = \sqrt{2y}$, but if we had a more complicated expression under the integral sign, we might not be able to complete the integration. Nevertheless, we may still consider the equation in its unsimplified form *without* performing the integration and use it to define x as an inverse of y exactly as we did before. The only difficulty is that we cannot now write x as a function of y in the ordinary way, but it is not insurmountable, and in fact, it is overcome in a ridiculously simple manner. We merely invent a new function x, the inverse of y, and give it a new name. Surprising as it may seem, this artifice really works, and it enables us to fix the behaviour pattern of many a newly invented function. The study of their behaviour shows that some of them singly or in suitable combinations satisfy differential equations of physics and astronomy. In other words, they provide solutions of these differential equations.

Although the initial impulse to their creation came from the needs of the physicist and astronomer, many of them were later refined, generalised, or even invented outright with no particular applications in view. This transition from the immediately applicable to the abstract with no application in sight is an ever-recurring theme in the progress of modern mathematics. Like Shelley's 'Skylark'—

> Higher still and higher
> From the earth thou springest
> Like a cloud of fire;
> The blue deep thou wingest.

But sooner or later it must descend to earth again to derive fresh inspiration for a new flight. This is exactly what actually happened also in the theory of differential equations.

Useful as these solution functions were, many of the newer ones were too abstract and 'flighty' to be of much use to the physicist. To get on with the job he has in view, the physicist needs not some abstract way of defining the solution by means of complicated functions, but some practicable procedure of calculating their numerical values, involving nothing more recondite than ordinary addition, subtraction, multiplication and division. For the sake of such simple procedures he is prepared to pay a price. He is willing to be much less exacting and is quite prepared to relax the rigorous standards to which the mathematician wants to cling. The greatest exponent of this relaxed down-to-earth point of view nowadays is R. V. Southwell, whose introduction of Relaxation Methods in Engineering and Physics has inaugurated a truly amazing advance in both.

The idea underlying Southwell's Relaxation Methods is simple. It rests on an important distinction in the determination of a function in mathematics and physics. In the former a variable y is defined as a function of a variable x by means of one or more formulae. For any given value of x within its range of variation, say 0 and $10a$, we can determine the corresponding value of y by means of the given formula or formulae. All that the mathematician requires is that this determination be theoretically possible. He is not deterred by the practical difficulties encountered in exploiting this theoretical possibility. He is quite satisfied if he can *assume* merely that the value of the function y is known for each one of the infinite values of x within the given range. But in physics the data must be obtained by physical measurement, and, therefore, can be neither exact nor complete. It cannot be exact because all measurement is subject to error, and it cannot be complete because the infinity of observations required for the purpose cannot be obtained in finite time.

In physics, therefore, a function y is said to be known if its approximate values are known, corresponding to a finite and discrete set of values of the independent variable (x). For instance, when we study y, the magnitude of current passing through a resistance coil, as a function of x, the voltage difference across the resistance, we deem y to be known if we can measure its approximate values corresponding to a number of discrete values of x such as

$$x = 0, a, 2a, 3a, 4a, 5a, 6a, 7a, 8a, 9a, 10a.$$

Consequently the determination of a wanted function has for the physicist a meaning quite different from what it carries to the mathematician. The former will accept a function, if he can ascertain its approximate numerical values at a discrete and finite set of values of the independent variable x. The latter, on the other hand, must have its exact value for *every* value of x. But the relaxation introduced by the physicist has its merit in that it enables solution of many problems quite intractable by the exact and rigorous methods of orthodox mathematics.

Instead of trying to trap all the infinity of its values at one blow in one or more functional forms, which is quite often impossible, relaxation methods of computation endeavour to evaluate *approximate* values of the required function for a discrete and finite set of values of x such as

$$x = 0, a, 2a, 3a, 4a, 5a, 6a, 7a, 8a, 9a, 10a.$$

$$0 \quad a \quad 2a \quad 3a \quad 4a \quad 5a \quad 6a \quad 7a \quad 8a \quad 9a \quad 10a$$

Such a set defines what Southwell calls a relaxation mesh or net. The values of $x = 0, a, 2a, \ldots, 10a$, for which the values of the wanted function are to be computed, are known as *nodal* points and the mesh length *a*, that is, the length of the interval separating the nodal points, is called the mesh-side. It is possible to obtain increasing accuracy, at the cost of proportionately increased labour, by utilising the results obtained on one size of mesh as a starting assumption in relation to a smaller mesh. In other words, having calculated the values of the function for the nodal points $x = 0, a, 2a, \ldots, 10a$, we can proceed to calculate its values for any number of other intermediate points on the basis of computations already made. For instance, we may subdivide each of the original ten meshes defined by $x = 0, a, 2a, \ldots, 10a$ into three sub-meshes each defined by

$$x = 0, \tfrac{1}{3}a, \tfrac{2}{3}a; a, 1\tfrac{1}{3}a, 1\tfrac{2}{3}a; 2a, 2\tfrac{1}{3}a, 2\tfrac{2}{3}a; 3a, \ldots, 10a.$$

We thus secure thirty meshes from the original ten. This device is known as *advance to a finer mesh*.

The next relaxation adopted by Southwell is the replacement of differential co-efficients occurring in the differential equation under assault by their *finite-difference approximations*. To explain the idea of *finite-difference approximations* of differential coefficients, consider a function $\varphi(x)$, which we shall consider *sufficiently* determined if we know its values at only three nodal points, *viz.* $x = 0, a$ and $2a$. Let these three values be $\varphi(0)$, $\varphi(a)$ and $\varphi(2a)$. Its differential co-efficient $\dfrac{d\varphi}{dx}$, too, will have different

values at these three nodal points. Let the three values of $\dfrac{d\varphi}{dx}$ at these three nodal points $x = 0, a, 2a$ be $\left(\dfrac{d\varphi}{dx}\right)_{x=0}, \left(\dfrac{d\varphi}{dx}\right)_{x=a}$ and $\left(\dfrac{d\varphi}{dx}\right)_{x=2a}$. It can be shown that these three values are given approximately by the equations

$$\left(\frac{d\varphi}{dx}\right)_{x=0} = \frac{1}{2a}\left\{-3\varphi(0) + 4\varphi(a) - \varphi(2a)\right\} \qquad . \qquad . \quad (1)$$

$$\left(\frac{d\varphi}{dx}\right)_{x=a} = \frac{1}{2a}\left\{-\varphi(0) + \varphi(2a)\right\} \qquad . \qquad . \qquad . \qquad . \quad (2)$$

$$\text{and} \qquad \left(\frac{d\varphi}{dx}\right)_{x=2a} = \frac{1}{2a}\left\{\varphi(0) - 4\varphi(a) + 3\varphi(2a)\right\} \qquad . \qquad . \quad (3)$$

Now the differential equation is true for all values of x within the range of its variation and, in particular, for the three nodal values of x under consideration, *viz.* $x = 0, a$ and $2a$. If we substitute for x its nodal value $x = 0$, the independent variable x disappears from the equation altogether and the equation contains only $\varphi(0)$ and $\left(\dfrac{d\varphi}{dx}\right)_{x=0}$. But the latter term can be replaced by the right-hand side of equation (1). Consequently the differential equation is transformed into an algebraic equation involving only the three 'wanted' values $\varphi(0)$, $\varphi(a)$ and $\varphi(2a)$ and nothing more. We could repeat the same process with the two remaining nodal values $x = a$ and $2a$. This would give two more similar algebraic equations containing only the three unknowns $\varphi(0)$, $\varphi(a)$, $\varphi(2a)$.

We therefore obtain three algebraic simultaneous equations in three unknowns, viz., $\varphi(0)$, $\varphi(a)$, $\varphi(2a)$. The single differential equation is thus reduced to a *set* of three simultaneous algebraic equations. In general, it is much easier, both in theory and practice, to solve a series of simultaneous equations than a differential equation. We shall deal with them later in Chapter 7.

* \qquad * \qquad * \qquad *

Our survey of arithmetic has led us from positive integers to negative integers, rational fractions, irrationals and finally to complex numbers. Is that the end? Alas! there is no end to number making. About 100 years ago, the celebrated Irish mathematician, Sir William Hamilton, began to consider space vectors in the manner in which we have considered plane vectors in our foregoing account. We assumed that all the directions of the vectors with which we had to deal lay in one plane, the plane of the paper, on which we drew our diagrams. But in the real world, the forces, velocities, accelerations and other directed magnitudes need not be and, in fact, often are not, in one and the same plane. How shall we deal with such vectors? In the case of two-dimensional vectors lying in the plane of our paper we denoted them by number couples like (x, y) giving their components along two mutually perpendicular reference lines. The numbers x, y were the co-ordinates of P with respect to the two reference lines. The extension of this theory to space vectors is now obvious.

We take three mutually perpendicular reference lines OX, OY, OZ and denote the space vector OP by a number triple (x, y, z), giving its three components along the three reference lines, the numbers x, y, z being the co-ordinates of P with respect to the three mutually perpendicular reference lines (see Fig. 25). If you have difficulty in imagining such reference lines, have a look at the inside of your study. Take any corner of the floor as your origin and call it O. Two edges of the floor meeting in O give you two straight lines at right angles to one another. Call one of them OX and the other OY. The vertical line through O to the ceiling, which you may call OZ, is obviously a line at right angles to OX as well as OY. These three lines meet in O and are mutually at right angles to one another. You cannot draw all three of them on paper as they do not lie in one plane; but you may create an *illusion* thereof by means of a drawing in perspective like Fig. 26. Now imagine any point P in the room, such as a bulb hanging from the ceiling. What is its height from the floor? To find it you may drop (in imagination) a plumb-line from P till the lead at its lower end touches the floor at M. The height of P above the floor is clearly the length, say, z, of the string PM when taut with the lead just touching the floor at M. Consider now M, the point where the plumb-line touches the floor. Let its

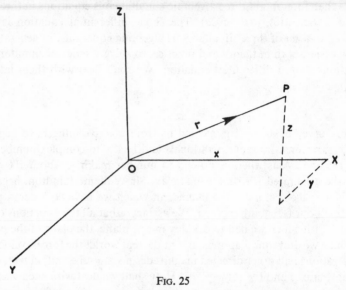

Fig. 25

distance from the reference line OX be y and that from OY, x. Then the numbers x, y, z define the location of the point P with reference to the three mutually perpendicular reference lines OX, OY, OZ. They are known as the co-ordinates of the point P but they may also be used to define the space vector denoted by the line OP. The space vector OP is simply defined by the number triple (x, y, z) where x, y, z are the co-ordinates of P as well as its three components along the three reference lines OX, OY, and OZ. We may also write it as $\mathbf{I}x + \mathbf{J}y + \mathbf{K}z$, where the symbols $\mathbf{I}$, $\mathbf{J}$, $\mathbf{K}$ printed in bold type do not represent numbers but operations or certain acts that you are required to perform. Thus, $\mathbf{I}x$, or the operation $\mathbf{I}$ on number x means that you are to move a distance x in the direction of OX. Likewise, the other two operations $\mathbf{J}$ and $\mathbf{K}$ mean:

Operation $\mathbf{J}$ on number y or $\mathbf{J}y$ = Move a distance y in the direction of OY;

Operation $\mathbf{K}$ on number z or $\mathbf{K}z$ = Move a distance z in the direction of OZ.

You may perform these three operations in succession. Thus you may move first a distance x along OX—operation $\mathbf{I}x$. This brings you to A. (See Fig. 26.) Having arrived there, you may perform the next operation, $\mathbf{J}y$ by moving a distance y from A in the direction of OY. You are now at M. From M the third operation $\mathbf{K}z$ means moving a distance z in the direction of OZ or vertically upwards. This third operation sends you to P. Now

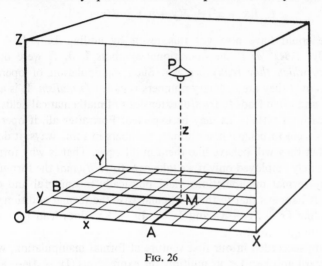

Fig. 26

does it matter in what order these three operations are performed? We cannot answer the question *a priori*. There are many operations where the order in which they are performed is *not* immaterial to the end-product. Suppose $\mathbf{I}$ represents an operation of shuffling a pack of cards in a particular way, while $\mathbf{J}$ is another such way. Obviously the final shuffle will ordinarily depend on whether the operation $\mathbf{I}$ or $\mathbf{J}$ is performed first. We have to discover in each case by actual trial whether the order in which operations are performed matters or not. In this particular case, you may easily verify that in whatever order the three operations $\mathbf{I}$, $\mathbf{J}$, and $\mathbf{K}$ are performed, you finally reach the same point P. Since the line OP defines the vector OP, we may represent it by the operational 'sum'

$$\mathbf{I}x + \mathbf{J}y + \mathbf{K}z.$$

This means that you can obtain the terminus P of the vector OP by successively performing the three operations $\mathbf{I}x$, $\mathbf{J}y$ and $\mathbf{K}z$. Now consider another vector OQ (Q being, say, another lamp). In a like manner, we may also represent it by the operational sum

$$\mathbf{I}x' + \mathbf{J}y' + \mathbf{K}z'.$$

What is the vector sum of the two vectors OP and OQ? As we have already seen, it is given by the parallelogram law. That is, we construct a parallelogram with OP and OQ as adjacent sides and take its diagonal OR as the vector sum of OP and OQ. Can we represent the vector OR also as an operational sum? Yes, we can. It can be shown that the vector OR is the sum

$$\mathbf{I}(x + x') + \mathbf{J}(y + y') + \mathbf{K}(z + z').$$

But we could have also got this result by adding $\mathbf{I}x + \mathbf{J}y + \mathbf{K}z$ to $\mathbf{I}x' + \mathbf{J}y' + \mathbf{K}z'$ *as if* the operational symbols $\mathbf{I}, \mathbf{J}, \mathbf{K}$ were ordinary numbers *which they really are not*. Such manipulation of operational symbols *as if* they are ordinary numbers is called *formalism*. It is a useful process and often leads to fruitful extensions of mathematical fields. But it is not without pitfalls, as may be expected. For, after all, if operational symbols seem to behave like ordinary numbers in some ways, it does not mean that they will behave like them in all ways. That is why formalism can be safely employed only if we take care to *verify that* the formulae derived by formal manipulations yield physically meaningful and correct results. It happens that in this particular case the formal addition of two vectors like $\mathbf{I}x + \mathbf{J}y + \mathbf{K}z$ and $\mathbf{I}x' + \mathbf{J}y' + \mathbf{K}z'$ does lead to a correct and meaningful result.

Having succeeded in our first venture at formal manipulation, we may now hazard another. Let us multiply the expressions $(\mathbf{I}x + \mathbf{J}y + \mathbf{K}z)$ and $(\mathbf{I}x' + \mathbf{J}y' + \mathbf{K}z')$ *as if* $\mathbf{I}, \mathbf{J}$ and $\mathbf{K}$ were ordinary numbers I, J and K. We obtain

$$\mathbf{I}^2 xx' + \mathbf{J}^2 yy' + \mathbf{K}^2 zz' + \mathbf{IJ}xy' + \mathbf{JI}yx' + \mathbf{IK}xz' + \mathbf{KI}zx' + \mathbf{JK}yz'$$
$$+ \mathbf{KJ}zy'.$$

Here we encounter symbols I^2, J^2, K^2, IJ, JI, etc., to which we have given no meaning. How are we to interpret the result of this mathematical abracadabra? We overcome the difficulty by defining—

$$\mathbf{I}^2 = -1, \quad \mathbf{J}^2 = -1, \quad \mathbf{K}^2 = -1;$$
$$\mathbf{IJ} = \mathbf{K}, \quad \mathbf{JK} = \mathbf{I}, \quad \mathbf{KI} = \mathbf{J};$$
$$\mathbf{JI} = -\mathbf{K}, \quad \mathbf{KJ} = -\mathbf{I}, \quad \mathbf{IK} = -\mathbf{J}.$$

Our product then becomes:

$$-(xx' + yy' + zz') + \mathbf{I}(yz' - y'z) + \mathbf{J}(zx' - z'x) + \mathbf{K}(xy' - x'y).$$

We can now interpret this result. It can be shown that the first term, *viz.* $-(xx' + yy' + zz')$, is the negative of the scalar product and the remaining three terms, viz.:

$$\mathbf{I}(yz' - y'z) + \mathbf{J}(zx' - z'x) + \mathbf{K}(xy' - x'y),$$

the vector product of the two given vectors. In other words, our formal multiplication yields at one blow both the scalar and vector products.

Sir William Hamilton, who first suggested this procedure, called the aforementioned product of the two vectors a *quaternion*. So a quaternion is a sort of generalised vector just as a vector is a generalised number. Hamilton, however, had great difficulty in securing official recognition for his quaternions. The stumbling block was the hypothesis he made concerning the products like IJ and JI. It was bad enough equating I^2, J^2, K^2 to -1, but in his day that could be tolerated, as the introduction of complex or 'imaginary' quantities had inured mathematicians to such sights. But to set IJ equal to K and JI to $-K$, that surely was a 'howler', which no respectable mathematician could accept even on the authority of Sir William! Of course, if I and J were pure numbers like 2 and 3, it would be foolish to claim that 2×3 is 6 while 3×2 is -6. But Hamilton's I, J, K are *not* ordinary numbers. They are symbols for certain operations and, as we have already seen, the end-product in many cases does depend on the order in which these operations are performed. There was, therefore, no inherent absurdity in Hamilton's equations, in which IJ and JI were unequal.

Hamilton was so fascinated with his own discovery of the quaternions that he devoted the remainder of his working life solely to their study. He thought that he had found in the quaternions the master key to geometry, mechanics and mathematical physics, just as Pythagoras, 2400 years before, had thought the whole number to be the essence and principle of all things. Both were disappointed and essentially for the same reason. We create a number language to describe the mysteries of nature, but we find them too deep for its vocabulary even though enriched by the acquisition of fractions, irrationals, imaginaries and quaternions. The difficulty will certainly remain even if the number language is extended still further.

While Hamilton was creating quaternions in the hope that he had at last touched the *ultima Thule* of number extension, one of his contemporaries, Grassmann, was developing a method of generating still more generalised numbers in which Hamiltonian quaternions figured as a minor detail. Long before Einstein popularised the fourth dimension, mathematicians were working with purely fictitious spaces of four, five, and even n dimensions. Grassmann took a vector in such an n-dimensional space. As we saw, to specify a two-dimensional plane vector we needed a number couple: for a space vector we required a number triple. For an n-dimensional vector we should naturally want to have an n-ple number such as $(x_1, x_2, \ldots x_n)$. Just as Hamilton denoted a space vector by the expression $Ix + Jy + Kz$, so also Grassmann denoted an n-dimensional vector by the expression

$$E_1 x_1 + E_2 x_2 + \ldots + E_n x_n.$$

Like Hamilton's **I, J, K,** Grassmann's E_1, E_2, ..., E_n are not ordinary numbers but operational symbols put in just to remind us that the magnitudes of the vector components x_1, x_2, ..., x_n in this expression are not to be added as we might be tempted to add, say, $2 + 3$. Having created n-dimensional vectors, or *hypercomplex* numbers as they are generally called, Grassmann had now to set up rules for combining such numbers. The addition of two hypercomplex numbers presented no difficulty. It was a simple extension of the addition law for space vectors. But he had some difficulty in defining a multiplication law to complete the calculus of hypercomplex numbers. Here he found that he was literally choked with an *embarras de choix*. He gave several such laws, one of which was, in principle, similar to Hamilton's rule described above.

The theory of hypercomplex numbers of Grassmann includes a host of other theories such as the theory of quaternions, determinants, matrices and tensors, which were beginning to be developed about the same time and the last two of which were to be applied extensively in quantum and relativity mechanics about seventy years later.

While the extension of the real number system to complex numbers led to an extensive development of a new theory, the theory of functions of complex variables, the extension of complex numbers to Hamilton's quaternions or Grassmann's hypercomplex numbers has not led to any corresponding development of a theory of functions of hypercomplex variables. The reason is that while the theory of complex variables, in spite of its 'purity', has had extensive applications from the theory of electric currents and map making to that of fluid flows, the 'pure' theory of Grassmann numbers has hitherto withered for want of a similar application. In fact, we remember him now chiefly because his theory has at last been applied to relativity and quantum physics.

*   *   *   *

The new and latest generalisation of number to vector is not a 'free creation' of the spirit; mathematicians dealt with various kinds of directed magnitudes for over two centuries before they even thought of devising a simple symbolism to denote them. When, at last, some of them did create a calculus adequate for the manipulation of the new symbols, the majority of mathematicians ignored it. It seems that every major extension of number system receives official recognition with the greatest difficulty. When Pythagoras came across incommensurable magnitudes, he decided to suppress his great discovery rather than give up his pre-conceived idea that integer is the essence of all things. Later, when the discovery became generally known, the numbers expressing incommensurable magnitudes were called 'irrationals', devoid of reason! It took centuries before the stigma attaching to the *irrationals* was washed away, and they were admitted among the comity of numbers as respectable numbers. The next great extension of the number system occurred when complex numbers were recognised as essentially a simple case of multiple algebra, in fact, a double algebra. Here again, the recognition came slowly and tardily, long after they had begun to occur in the solution of various problems such as those of trigonometry—so vital for navigation. In the beginning the mathematicians tried to suppress their existence too, by rejecting them out of hand as 'imaginary' whenever they appeared. It took a long time to understand that the complex numbers also reflect certain aspects of the external world, which are as 'real' as those represented by the integers and irrationals. In our own day we find the vector struggling for official recognition.

During the nineteenth century, while Grassmann's hypercomplex numbers were hardly noticed, Hamilton's quaternion calculus fell flat on the mathematical world. Except for Tait and Gibbs, the majority of the scientists preferred to work with the old-fashioned Cartesian methods. Even as recently as about thirty years ago the vector could hardly be said to have come into its own. In the preface to his *Treatise on Vectorial Mechanics* published in 1948, Milne records that he did not at first believe his teacher Chapman, when he told him (1924) that 'vectors were not merely a pretty toy, suitable only for elegant proof of general theorems, but were a powerful weapon of workaday mathematical investigation, both in research and in solving problems of the types set in English examinations.' Since then mathematicians have ceased to look upon vectors as a 'mere shorthand for sets of Cartesian expressions' and have begun to realise that there is all the difference in the world between the old-fashioned methods of working with Cartesian co-ordinates and the new vectorial methods. The former is like picturing a building by looking at its plan and elevation, while the latter is like seeing it stereoscopically, that is, in its three-dimensional solidity. For, as Milne has remarked, the old-fashioned Cartesian method diverts attention from lines and surfaces, which are of primary interest, to their projections on the three axes, whereas vector analysis provides a kinematic picture of the motion in question that gives far more insight into the phenomenon than the corresponding Cartesian analysis. Vector analysis views the phenomenon as a whole, and to that extent therefore it is more in tune with gestalt methodology. Because of its great power it is now being extensively applied to a whole gamut of diverse fields from econometrics to quantum mechanics.

ZENO AND INFINITY

THE web that men weave with words sometimes ensnares their own minds. Thus for a long time intelligent men were astonished that Zeno's verbal dialectics should seem to 'prove' that Achilles pursuing a tortoise could never overtake it. Some of them might perhaps have even believed that motion was an illusion. If, nevertheless, things did seem to move, so much the worse for the gross senses that lead men astray! And yet when Zeno first astounded Athens with his paradoxes, it is all but certain that he meant them to be taken as mere parables with a moral. Probably he wanted to harass either the Pythagoreans or the Atomists. Whatever the motive, the moral he wanted to deduce was that apparently water-tight reasoning could lead to manifest absurdity. But the moral of a paradox as of a parable may be misunderstood. Thus, when La Fontaine related the fable of the gay grasshopper, he (himself a carefree vagabond) never intended to extol the avarice of the hard-fisted ant and disparage the song and dance of the grasshopper. Quite the contrary; and it was the same with Zeno.

Gradually, as the power of the words waned and the astonishment wore off, men not only disregarded the moral but began even to consider the paradoxes as sophistries—mere word play. One might think that nowadays, over 2300 years after Zeno, we should be rid of such sophistries and cease to be concerned about them. Yet, as Bertrand Russell has remarked, 'the arguments of Zeno have, in one form or another, afforded grounds for almost all the theories of space, time and infinity, which have been constructed from his day down to our own.' Indeed, if the ancient world had one Zeno to contend with, we have several, each one of whom has given his own specific paradox. They are the Italian Burali-Forti, the English Bertrand Russell, the German König, and the French Richard. We shall not state their respective antinomies here as some of them can be expressed only in highly technical language; but we may add that they are in one way or another all concerned with the nature of infinity and infinite processes.

But what is infinity? Among the several meanings listed in the Concise Oxford Dictionary, 'very many' is shown as one of its synonyms, and historically this is precisely the sense in which the word infinite was originally used. The technique of counting had not yet been perfected, men could count numbers up to only a limited number, and what could not be counted was 'infinite', 'very many', or 'numerous', be they the stars in the sky or the grains of sand on a beach. Infinite, then, was not the 'uncountable but the yet uncounted'. Later, the technique of counting advanced and human ingenuity invented numbers for counting bigger and bigger collections. Our Hindu ancestors actually reached the colossal figure of 10^{13} which they called *pradha*, and which they considered as the ultra-ultimate number beyond which human mind could not advance. Several centuries later, Archimedes invented even a bigger number, of the order of 10^{52}, to represent the number of grains of sand in a globe of the size of the celestial sphere! Yet bigger and bigger numbers were devised when finally man realised that there can be no limit to human thought. However large a collection you may have, you can at least always *imagine* a bigger one by adding one more item to it. If, therefore, we require our number system to be adequate for counting any collection that we may think of, we cannot close the system of integers with a last integer, however large. We must keep the domain of integers open in order that we may always find a number to represent the plurality of collections of any size whatever. In this act of keeping the number domain open, of not closing it with a last integer, lies the genesis of the infinite. But alas! the creation of the infinite, the never-ending repetition of an act or an operation that is once possible, has turned out to be a snare from which the mathematicians have been trying to extricate themselves for the past 2500 years—from the time of Pythagoras and Zeno to Hilbert and Brouwer in our own day. Perhaps that is why, as Eddington once remarked, mathematicians represent infinity by the sign of a tangled love knot, ∞.

The root cause of the trouble lies in the fact that the laws of ordinary logic, such as we derive from an intuitive appreciation of our experience, inevitably confined to only finite classes, do not apply to infinite collections. For instance, we know that a whole is necessarily bigger than any genuine part of itself. Thus, the class of all Asiatics is necessarily smaller than the class of all Homo Sapiens, for the former is only a sub-class or a part of the latter. But when we apply this law to an infinite class we fall into error; for, as we shall presently see, the infinite class of *all* integers is *exactly* equal to any infinite part of it, as, for example, the infinite class of only *even* integers.

The failure to understand that an infinite class can be equal to a proper part of itself led Zeno to his paradox of Achilles and the tortoise. (See Fig. 27.) Suppose Achilles to be at a point A and the tortoise at a point B of the course at the beginning of the race. If, subsequently, Achilles overtakes the tortoise at a point C, then the infinite set of points on the line AC would have to be exactly equal to that on the line BC. For, to every position, say *P*, of Achilles between *A* and *C*, there corresponds one and only one, *viz. Q*, of the tortoise between *B* and *C*, and vice versa. The series of point-positions occupied by Achilles has, therefore, the same number of terms as the series of the point-positions occupied by the tortoise. Accordingly the aggregate of point-positions occupied by Achilles, *viz.* the line *AC*, is exactly equal to the aggregate of the corresponding or simultaneous point-positions occupied by the tortoise, that is, the line *BC*.

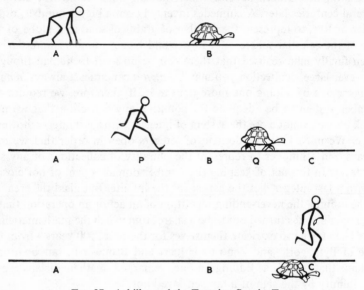

FIG. 27—Achilles and the Tortoise. See the Text.

But this seems to lead to the paradoxical conclusion that the line *AC* is equal to the line *BC*!

Zeno's paradox remained unresolved for about 2000 years till Galileo observed that, unlike the finite classes, an infinite class must necessarily have as many things in some part of it as there are in the whole of it. Just as primitive men thought all pluralities which they could not count as equivalent as, for instance, the primitive Tasmanians for whom all numbers bigger than two were 'plenty' in the sense that they transcended the limit of their ability to count, so Galileo thought that all infinite classes were equal. Thus, in his book entitled *Dialogues Concerning the New Sciences*, he argued that the number of points on one line is the same as on any other because both are infinite, and the conception of one being bigger than the other was not applicable to infinite collections but only to finite. All infinities, therefore, were equivalent.

But even primitive man must have noticed that there are 'manys' and 'manys'—the difference, for instance, between the 'many' soldiers in an invading host, the 'many' stars in the nocturnal sky, or the 'many' grains of sand on the beaches. So, too, the scientists began to feel vaguely that there are infinities and infinities. How did they discriminate between the various infinities if none of them could be counted? It was indeed by an extension of an old idea which had preceded even the practice of counting, the idea underlying the matching process that we explained in Chapter 2. The idea is so important that it may be explained again by a simple illustration.

If you had a large audience in a hall and a number of chairs, how would you decide which of the two was more numerous without actually counting them? You could ask your audience to take their seats, one person to a chair, and watch the result. If, after everybody was seated, there were still some chairs left, obviously the number of chairs was larger than the number of persons. If, on the other hand, after all the chairs were occupied you still had some persons left, the number in the audience was the bigger. And if, finally, neither any chair was left unoccupied nor any person left standing, the two collections were exactly numerically equal. Now this process of 'marrying' the items of one group to those of another is called *matching*, and if the matching between the items of two groups is essentially 'monogamous' so that one item in any set has a unique partner in the other and vice versa, the two groups are obviously equal. If the same idea is applied to two infinite groups, obviously you cannot complete the matching process because the number of items in either group is inexhaustible. But you may be able to set up a general formula whereby, given any item of one infinite group, you could discover its unique mate or counterpart in the other. For instance, the collection of integers

$$1, 2, 3, 4, 5, \ldots$$

and that of even integers

$$2, 4, 6, 8, 10, \ldots$$

are both infinite. But we could formulate that to every number, say 5, in the first group, corresponds its double, *viz.* 10, considered as a number of the second group, and to every number in the second, say 10, corresponds its half, *viz.* 5, in the first. We have thus specified a general rule whereby the members in the two groups are uniquely 'married' or correlated and we are thus justified in calling the infinite set of integers and that of the even integers as equal though infinite.* Similarly, we could devise a formula for 'marrying' monogamously the points on a straight line to those of any

* In passing, we may note that any infinite set or collection, whose items can be thus monogamously married off to those of the infinite set of integers, 1, 2, 3, 4, ..., is known as an *enumerable* or a *countable* set.

other line. Let AB and CD be any two such lines (Fig. 28). Let AC, BD meet in O. If P is any point on CD we can obtain its corresponding mate Q on AB by joining OP and letting it intersect AB in Q. Conversely, given Q, we obtain its corresponding mate P in CD by joining OQ and letting it intersect CD at P. The rule is general and the correspondence it establishes between the points of CD and AB is 'monogamous'. The infinite number of points on the line AB is, therefore, the same as that of points on the line CD.

Since AB, CD are any pair of lines arbitrarily selected, we find that the number of points on *any* straight line is exactly 'equal' to that on any other, the word 'equal' here meaning merely that the points of one can be

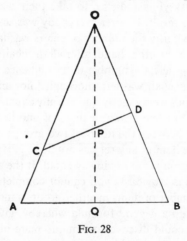

FIG. 28

married off monogamously to those of the other. In particular, the infinite number of points on any line is the same as that on a line of unit length. Now how does the infinity of points on a unit segment AB compare with the infinity of the unending set of positive integers

$$1, 2, 3, 4, 5, \ldots?$$

Since both the collections—the collection of points on the unit segment AB and the collection of positive integers—are infinite, we can compare them only by seeing whether the elements of one can be married off monogamously to those of the other. If so, we could reasonably call the two infinite collections equal in the sense defined above. If not, one of the two would be bigger. Now if the monogamous marriage of the points on AB and the unending series of integers 1, 2, 3, ... were possible, this would simply mean that we could have a roll-call of the points of the line in much the same way as a foreman might muster his gang. The essence of the roll-call process is that each one of the gangmen, Bill, Fred, Harry, Casey, ...

has one and only one roll number assigned to him, so that when any number, say 3, is spoken, only one person in the gang, say Harry, answers the call. Now before we can roll-call the points on the line AB we have to devise a way of naming them. Since, unlike the number of men in the gang, the number of points is inexhaustible, we cannot hope to name each one of them individually. All that we can do is to devise a rule or a formula whereby, given any point, we could manufacture its name. One such way would be to call them arbitrarily P_1, P_2, P_3, ..., P_n, ... But the trouble with such an indefinite naming scheme is that it gives us no grip on the points named. We have no way of telling which particular point has been named P_3 for example. A better procedure, which gives a firmer grip on the points named, would be to name the point by means of the number denoting its distance from one end. In other words, if we choose a point, say P_3, whose distance from A is $\frac{1}{2}$ we can call it the point '$\frac{1}{2}$' alias P_3. Similarly if the distance of another point, say P_2, from A is $\frac{1}{3}$, we could name it the point '$\frac{1}{3}$'. We shall find it more convenient to express the distances in the decimal notation, that is, by the non-terminating decimals $0.50000...$, and $0.333...$ instead of the fractions '$\frac{1}{2}$' and '$\frac{1}{3}$' respectively. If we adopt this system of naming points on the line, every 'name' such as the number '$\frac{1}{7}$' or its decimal equivalent, $0.142857142857...$, for example, leads to a definite point P_1 whose distance from A is (in this case) the length $\frac{1}{7}$.

Since, by hypothesis, the entire length of the line AB is unity, we shall never have occasion to use a number exceeding 1 in order to name the points. Having devised a scheme of naming the points, we are now ready for the roll-call. If the infinite set of points on the line AB is 'equal' to the unending series of integers

$$1, 2, 3, 4, ...,$$

it would be possible to assign to every name one and only one number of the series 1, 2, 3, 4, ... In other words, we could express the distance of every point on the line AB in a roll-call order of, say, the following type:

Roll Number	Indefinite name of the point	Distance name of the point from A
1	P_1	$\cdot$ 1 4 2 8 5 7 1 4 2 8 5 7...
2	P_2	$\cdot$ 3 3 3 3 3 3 3 3 3 3 3 3...
3	P_3	$\cdot$ 5 0 0 0 0 0 0 0 0 0 0 0...
4	P_4	$\cdot$ 6 6 6 6 6 6 6 6 6 6 6 6...
5	P_5	$\cdot$ 7 2 1 8 9 6 5 2 4 7 1 4...
6	P_6	$\cdot$ 8 1 2 3 5 2 9 1 4 1 2 3...
7	P_7	....
8	P_8	....
etc.	etc.	etc............................

Now consider the number in the decimal notation formed by taking only the digits in bold type in the above scheme. This number is

$$\cdot130692\ldots$$

From it we could manufacture a series of other numbers. This way of generating a number is known as Cantor's diagonal process. The reason is that we pick up the digits in the diagonal of the array and produce another by changing each. For instance, we could substitute for '1' in the first digit the next number 2, for the number '3' in the second digit the number 4 and for '0' in the third place 1, for '6' in the fourth place 7, for '9' in the fifth place 0 (not ten), for '2' in the sixth place 3 and so on. We thus obtain from the number $\cdot130692\ldots$ another number $\cdot241703\ldots$ It can be shown that this last number is not included in the aforementioned roll-call scheme. For it is not the first number, because by the very process of its manufacture it was made to differ from the first number P_1 in the first digit. This difference in the first digits of the two numbers suffices to make them different, even if every one of their succeeding digits tallied. Likewise, it is not the second number, for its second digit was made to differ from the second digit of the second number. It is obvious that the argument can be repeated indefinitely so that the number $\cdot241703\ldots$ that we have manufactured differs in at least one place from *every* one of the numbers listed in the roll-call scheme. The point on the line whose 'name' or distance from A is the number $\cdot241703\ldots$ is thus left out of the muster that we have attempted to design.

The assumption that we can specify a formula whereby the points on the line AB can be married off monogamously to the members of the infinite set of integers 1, 2, 3, . . . leads to a contradiction. Whatever we do there is always at least one point on the line AB left without any partner in the integral number fold. We thus regard the infinity of the aggregate of points on a straight line like AB as bigger than that of the aggregate of integers. If we denote the 'infinity' or 'power' of the latter by the symbol a and the bigger infinity or power of the former by c, we have what are called *transfinite numbers*.* We might conceivably have a transfinite of a higher power than a but lower than c. No one has yet produced an intermediate transfinite of this kind. It is believed that there are none such, although no one has yet succeeded in clinching the issue by a strict mathematical demonstration of the non-existence of such intermediate transfinites. On the other hand, infinite aggregates with a power greater than c can be constructed. In fact, Cantor proved that if an infinite aggregate M with any

---

* An infinite aggregate like the unending set of positive integers is sometimes called a-infinite, enumerable, or countable, while the infinite aggregate of points on a line is called c-infinite.

power a exists, another infinite aggregate M' with power p exceeding a can always be constructed.

Suppose we have an infinite aggregate M of the lowest power a, viz. the a-infinite set of integers

$$1, 2, 3, 4, \ldots m \ldots$$

we may consider the elements m of the set M as arranged in their natural order as written above. Now this ordered set of elements can be made to generate a whole group of other sets. To do so we *may make use of Confucius's dualistic notion* of yin (female) and yang (male). Let us then denote a yin by the symbol ☉ and a yang by the symbol 玉. We start with the given set M of integers m, thus:

$$1, 2, 3, 4, \ldots m \ldots$$

Suppose we replace every element m of this set by either a yin ☉ or a yang 玉. We thus obtain an arrangement or permutation of yins and yangs. For instance, suppose we replace 1 in M by a yin, 2 by a yang, 3 by yin, 4 by yang, 5 by yin, 6 by yin, and so on. We have then an arrangement of yins and yangs such as:

☉ 玉 ☉ 玉 ☉ ☉ 玉 玉 玉 · · ·

Proceeding in this way we can generate a whole infinity of permutations of yins and yangs. Consider now the infinite aggregate M' of *all* such permutations or arrangements of yins and yangs obtained by replacing the elements m of the set M. We can show that the infinity of the aggregate M' is higher than that of M. In other words, if we tried to match the elements of M' over those of M, some elements of the former would be left over. Suppose, if possible, we matched the following permutations of yins and yangs—elements of M'—over the elements of M, that is, the set of positive integers 1, 2, 3, ...

Elements of M	Elements of M' or Arrangements of yins and yangs
1	⊙ 玉 ☉ 玉 ☉ ☉ 玉 玉 玉 · · ·
2	玉 ⊙ 玉 玉 ☉ 玉 ☉ ☉ 玉 · · ·
3	玉 ☉ ⊙ ☉ 玉 ☉ 玉 玉 ☉ · · ·
4	☉ ☉ 玉 𝐓 ☉ 玉 ☉ ☉ ☉ · · ·
5	. . . .
	. . . .

Now consider the permutation of yins and yangs obtained by picking out those in bold type in the above scheme. This is the arrangement:

⊙ ⊙ ⊙ 玉 · · ·

If, now, we change every yin into a yang and vice versa in this arrangement, we have the arrangement

$$\text{丑} \quad \text{丑} \quad \text{丑} \quad \odot \quad \cdots$$

This arrangement clearly differs from every one of the arrangements listed in the above matching scheme. For in the first place we have put a yang against the yin appearing in the first place of the first arrangement. It therefore differs from the first arrangement. In the second place we have substituted a yang as against the yin in the second place of the *second* arrangement. It therefore also differs from the second arrangement, and so on. We have thus manufactured a permutation of yins and yangs, that is, an element of M' which cannot be matched over the elements of M. Consequently M' is of higher power than M. In fact, it can be shown that the power of M' is the same as that of the unit line AB (page 98).

Consider now the set M' of the points of the unit line AB. We could do to the elements of this set (that is, the points of the line AB) just as we did to the elements m of the first set M, *viz.* the set of infinite integers. In other words, we could replace each element or point P of the line AB, by a yin or yang. We thus obtain again an infinite arrangement of yins and yangs. Notice carefully that the arrangement of yins and yangs thus obtained would now consist of an infinite number of yins and yangs as before but that the power of this infinity would be higher. The reason is that formerly we replaced the '*a*-infinite' set of integers by yins and yangs, whereas we are now replacing the '*c*-infinite' set of points of the line AB by yins and yangs.

If this way of distinguishing the two types of arrangements appears too abstract, you may try to picture them in another way. This alternative way, too, needs a good deal of imaginative effort, but it is well worth making as it will show you how vastly greater *c*-infinity is compared to *a*-infinity. As we have seen, an *a*-infinite arrangement of yins and yangs is simply an unending sequence such as:

$$\odot \quad \text{丑} \quad \text{丑} \quad \odot \quad \text{丑} \quad \text{丑} \quad \odot \quad \odot \quad \cdots \qquad \qquad (1)$$

Now we can also construct an arrangement of yins and yangs which is a *doubly* unending *mosaic* of the type

$$
\left.
\begin{array}{l}
\odot \;\; \text{丑} \;\; \text{丑} \;\; \odot \;\; \text{丑} \;\; \text{丑} \;\; \odot \;\; \odot \;\; \cdots \\
\odot \;\; \odot \;\; \text{丑} \;\; \odot \;\; \text{丑} \;\; \odot \;\; \text{丑} \;\; \odot \;\; \cdots \\
\text{丑} \;\; \text{丑} \;\; \odot \;\; \text{丑} \;\; \odot \;\; \odot \;\; \odot \;\; \text{丑} \;\; \cdots \\
\odot \;\; \text{丑} \;\; \odot \;\; \odot \;\; \text{丑} \;\; \text{丑} \;\; \text{丑} \;\; \odot \;\; \cdots \\
\cdot \;\; \cdot \;\; \cdot \;\; \cdot \;\; \cdot \;\; \cdot \;\; \cdot \;\; \cdot \;\; \cdots \\
\cdot \;\; \cdot \;\; \cdot \;\; \cdot \;\; \cdot \;\; \cdot \;\; \cdot \;\; \cdot \;\; \cdots \\
\cdot \;\; \cdot \;\; \cdot \;\; \cdot \;\; \cdot \;\; \cdot \;\; \cdot \;\; \cdot \;\; \cdots
\end{array}
\right\} \qquad (2)
$$

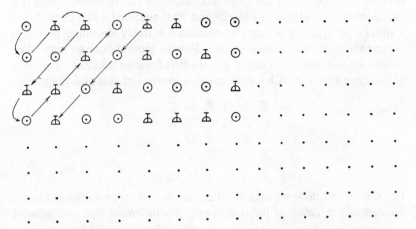

FIG. 29—If you start counting the yins and yangs in the figure, following the direction of the arrows, you will be able to match all of them over the infinite series of integers 1, 2, 3, 4, ... This doubly-unending mosaic of yins and yangs is therefore enumerable.

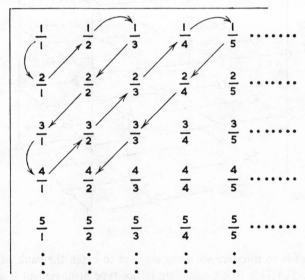

FIG. 30—Fig. 29 shows that a doubly unending mosaic of yins and yangs is enumerable. Now, we could arrange *all* fractions (or rational numbers) in a similar mosaic-type arrangement and match them with the series of integers in the same manner. Hence the infinite set of all fractions or rational numbers is also enumerable.

Contrary to what our intuition may tell us, this mosaic-type arrangement is really no bigger than the single-line sequence (1). In other words, it is still *a*-infinite. A glance at Figs. 29 and 30 should make this clear. From a mosaic of type (2) we can manufacture a trebly unending lattice—a never-ending flight of mosaic storeys. For the ground floor we may take a mosaic arrangement such as (2). For the first floor we take another mosaic of the same type but with a different arrangement of yins and yangs, *e.g.*

$$
\begin{array}{cccccc}
壬 & ☉ & ☉ & 壬 & ☉ & 壬 \cdots \\
壬 & 壬 & ☉ & ☉ & ☉ & 壬 \cdots \\
☉ & 壬 & ☉ & 壬 & ☉ & ☉ \cdots \\
\cdot & \cdot & \cdot & \cdot & \cdot & \cdot \cdots \\
\cdot & \cdot & \cdot & \cdot & \cdot & \cdot \cdots
\end{array}
$$

For the second floor we take another mosaic and so on endlessly like the builders of the Tower of Babel with only this difference that no confusion

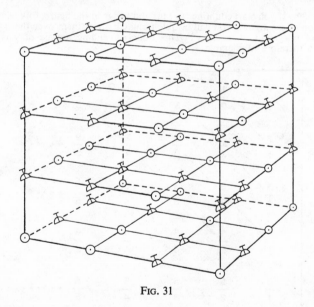

Fig. 31

of tongues is to interfere with our attempt to reach the vault of heaven. (See Fig. 31.) This trebly unending lattice-type arrangement is again no bigger than the simple sequence (1).

We are now at the end of our tether, for we can move only backwards or forwards, right or left, and up or down. We used the first type of motion to obtain an infinite sequence of the type (1). We used both, the first and second kinds of motions, to create an infinite mosaic such as (2) and all three to manufacture an infinite flight of mosaic floors or a trebly unending lattice-type arrangement. To proceed farther we need at least one more 'degree of freedom'—a new dimension. Although our physical space allows just three dimensions for movement and no more, the mathematician, unlike Alexander, never cries that he has no more worlds to conquer. From a three-dimensional lattice of yins and yangs he constructs (in imagination now) a quadruply unending 'super-lattice' arrangement of yins and yangs in exactly the same way as we built a mosaic such as (2) out of a sequence like (1) and a Babel-type lattice out of a mosaic like (2). But even so, we have only an arrangement as a-infinite as the simple sequence (1) ever was. So you see it is no easy matter to soar out of the prison walls of even a-infinity, though it is the lowest type of infinity among mathematical infinities.

There is, however, no limit to a mathematician's imagination. He soars higher and higher and builds a whole series of super-lattice arrangements of yins and yangs. He evolves a whole hierarchy of them, beginning with a singly unending sequence such as (1), then a doubly unending mosaic such as (2), then a trebly unending lattice, like a never-ending flight of mosaic floors, then a quadruply unending super-lattice, and so on *for ever and ever*. This last 'for-ever-and-ever' type of super-lattice arrangement of yins and yangs, which is so to speak *unendingly* unending, is at long last bigger than the infinite sequence (1) and has the power c. We have now flown out of the prison walls of a-infinity: but what a flight!

Consider now a c-infinite arrangement of yins and yangs, an unendingly unending 'super-lattice' of the type described above. As we saw, we derived it by replacing each point P of the line AB by either a yin or a yang. Obviously we could produce an infinite number of other arrangements of yins and yangs similar to this super-lattice arrangement. The aggregate of all such possible super-lattice types of arrangements of yins and yangs is itself an infinite set M''. It can be shown that the power γ of the set M'' is higher than the power c of the infinite set M'. In other words, if the elements of M'' were to be matched over those of M', a number of the elements of M'' would be left without any partners in M'. The power γ of the infinite aggregate M'' is therefore higher than c, the power of M. We might call M'' the super-lattice manifold but it is more usually called the functional manifold. As before, there is nothing to indicate in Cantor's theory that there are no transfinites lying between c and γ; but it is generally believed that there are none such. We could repeat the same process by starting with the elements of the aggregate M'' and generate another aggregate with still higher power and so on *ad infinitum*. As the process is interminable there can be no last transfinite. Musing over the glory of heavens, St. Paul once said, 'There is one glory of the sun and another glory of the moon and another glory of the stars; for one star differeth from another star in glory.'* Musing over infinity a mathematician might well exclaim 'there is one infinity of the integers, another of the real numbers, and another of the functional manifold; for one aggregate differeth from another in the power of its infinity!'

This act of transcending the infinite and weighing it, as it were, in the balance, was a veritable *tour de force* on the part of Cantor, the creator of the theory of transfinite numbers. But, alas! no sooner was this beautiful structure of thought reared than it began to appear that somewhere deep down in its foundations there was a serious flaw. For while, as we have seen, one train of Cantor's arguments led to the conclusion that there is no last transfinite, another apparently equally valid seemed to prove that this cannot be so! For the power of the aggregate of all possible aggregates must be a transfinite, which is the greatest conceivable or the *ultima Thule* of number evolution. It is, therefore, the last transfinite! This contradiction is allied to certain other antinomies of the infinite, which sprang up from the ambiguous manner in which Cantor and others had used the word 'all' in their reasoning. Such, for instance, was the paradox of the Italian Burali-Forti, who showed that 'the ordered series of *all* ordinal numbers defines a new ordinal number which is not one of the "all"'. This might appear a little too recondite but Burali-Forti's point would be understood by considering the paradox of the village barber who shaves *all* men who do not shave themselves. Either the barber shaves himself or he does not. If he does not shave himself, he is one of the non-selfshavers and is, therefore, shaved by the barber, that is, himself. If, on the other hand, the barber shaves himself, he is one of the men who shave themselves, and hence he is not shaved by the barber, *i.e.* he does not shave himself. In either case there is a contradiction, which arises on account of the illegitimate inclusion of the barber himself in the word 'all' of the original enunciation.

With the emergence of the paradoxes of the infinite discovered by Burali-Forti and others, the ghosts of Zeno, Eudoxus and Cavalieri, that had apparently been laid to rest by the work of Galileo and Bolzano, stirred to life again and began to mock the analysts' attempt to comprehend the nature of the infinite. It seemed as though mathematics could steer clear of the Scylla of a paradoxical infinite only by perishing in the Charybdis of a severely restricted arithmetic. Amazing as it may seem, the neo-Pythagorean L. Kronecker did openly suggest that mathematics practically scuttle itself in such a Charybdis of a narrow arithmetic. This mathematical Samson threatened to demolish the real number system, the calculus and every branch of mathematics which employed the infinite. He demanded that the infinite must be banished outright from mathematical thought and every

* 1 Corinthians xv, 41.

theorem in analysis stated as a relation between integral numbers only, thus eliminating entirely the terminology involved in the use of negative, fractional and irrational numbers. 'God made the integers, all else is the work of man,' he said, and insisted that man must operate with God-given integers and nothing else. If Kronecker could have had his way, very little of mathematics as we know it today would be left. It would, perhaps, be a delightful prospect for the high-school student, but you would also have to do without numbers to express many magnitudes of everyday use, such as the diagonal of a square in terms of its side or the area of a circle in terms of its radius. Obviously, therefore, Kronecker's *nihilistic* programme could not be practical mathematics.

Nevertheless, Kronecker's lead was followed by some of the greatest mathematicians of our time, at least in precept, even though their practice did not always conform to it. Poincaré, Brouwer, Weyl, one after the other, thundered against Cantor's theory of the infinite and condemned it as a 'disease' of which mathematics had to be cured. They held that for a concept to be admissible in mathematics, it is not enough that it be 'well defined' in words. It must also be 'constructible', that is, obtainable by a finite number of processes, or at least by such infinite processes as are reducible to finite by means of a finite number of rules. For instance, if we require the square root of 625, the ordinary process of root extraction is 'admissible' because, after a finite number of divisions, we obtain the square root 25. But if, on the other hand, we require the square root of 2, the same process had to be repeated *ad infinitum*. It is true that by stopping the process after a finite number of steps, we can secure as close an approximation to the square root of 2 as we like. But, however far we may go, a finite number of steps will *never* lead us to exactly the number whose square is 2. Hence, conclude the rigid finitists, that $\sqrt{2}$ is not an 'admissible' number. The finitist's argument is that since the infinite sometimes leads to contradiction, it should be completely banished from mathematics. One is reminded of the pious Sur Das who blinded himself because the sight of beautiful girls occasionally caused him concupiscence.

If, then, the infinite cannot be banished from mathematics without destroying nearly all of it, how are its contraditions to be resolved? There are some who, like Tobias Dantzig, are willing to accept the 'illusion' of the infinite as a 'mathematical necessity' on the ground that it preserves and furthers the intellectual life of the race. On that score, they agree to 'counterfeit' the universe by number, by the infinite and accept an otherwise 'false' judgment, justifying the acceptance by Nietzsche's aphorism, 'renunciation of all false judgment would mean a renunciation, a negation of life'! If it means anything, it is presumably this. Since we literally think with concepts which are expressed in language, the picture we make of the universe around us is refracted by our linguistic medium very much as the stars are displaced by the atmosphere through which we see them.

The progressive elimination of this falsifying of the universe by language, number and symbolism is one of the major tasks of the philosophy of science. We cannot shirk it on the plea that it is one of those inevitables that must be endured because it cannot be cured. For instance, one species of aberration is the creation of static and permanent forms and categories that we invent and impose on the ever-changing universe to understand and transform it. We correct this aberration later by breaking the old categories and synthesising new ones. In the case of infinity its introduction appears to some 'illusory' or 'false' because they tend to think in terms of too rigid categories which they are unable to synthesise. Accordingly, the concept of the infinite, in which the mind conceives of an infinite number of single objects and at the same time treats the whole as an individual object, appears self-contradictory and therefore 'false'. This is an old and persistent error of idealist philosophy. About 2300 years ago, noticing the existence of opposite qualities in matter (as, for example, the union of the quality of hardness with that of softness in a piece of wood), Plato was led to the idealist belief that matter is 'self-contradictory' and therefore 'unreal'. The only difference between him and his present-day successors is that the latter, with their sharper sense of fullness of life, agree to accept a 'false' judgment so as to further the 'intellectual life of the race'.

6

THE THEORY OF SETS

IN the study of mathematics we come across the term 'point' for the first time in geometry, where it appears as quite a humble sort of item. The text-book definition, that a point is that which has position but no magnitude—meaning thereby that it is a disembodied dot—sounds rather like a riddle. And yet within the last century and a half this scarcely visible speck of ink—the 'point'—has grown into a giant, a veritable Atlas, that now supports the entire mathematical world! How has this miracle come to pass?

Two independent lines of development have contributed to the exaltation of the point as the sovereign entity of mathematical thought: dynamics and the theory of heat. The need of traders for safe and reliable methods of navigation led to the study of dynamics during the sixteenth and seventeenth centuries. The need of manufacturers for a new source of power to meet the ever-growing demands of trade, and its partial satisfaction by the invention of the steam engine during the following century, gave rise to more precise studies of thermal phenomena and theories of heat. The beginnings of the first line of development, the evolution of dynamical theory, can be traced to Descartes, who turned to good account the practice of the medieval cartographers whereby the location of terrestrial places could be indicated on charts or maps. The idea underlying their practice is simple. If we draw two reference lines on the surface of the earth such as, for example, the equator and the Greenwich meridian, the position of every place on the earth's surface can be specified by giving two numbers indicating its longitude and latitude. Applying the same idea to a flat surface like the plane of paper, we take any two reference lines XOX', YOY' through a point of origin O and indicate the position of any point P by its distances from these reference lines. (See Fig. 32.)

We can also extend this procedure to indicate the location of points in air, such as positions of flying aeroplanes, by taking the vertical through the origin O as our third reference line in addition to two mutually perpendicular lines in a horizontal plane. Thus on a flat surface, such as a plane field, the position of every point is given by two numbers—its two distances from two given reference lines. In space its position would be given by three numbers, its three distances from three given reference planes (see Fig. 33).

The discovery that points on a flat plane can be represented by a number couple and those in space by a number triple was revolutionary enough, but still more revolutionary was the converse idea that *any* number

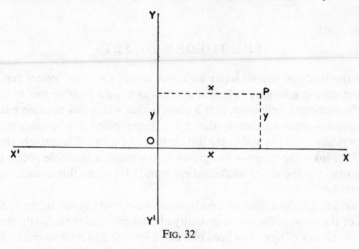

FIG. 32

couple could be represented by a point on a plane such as the surface of a graph paper. This idea is nowadays a mere commonplace and is used extensively by people other than mathematicians when they wish to keep an eye on some state of affairs. For instance, take a business man who is

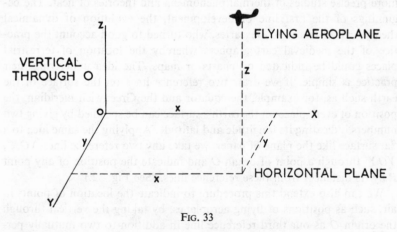

FIG. 33

usually concerned with the growth of his turnover. He has in mind a sequence of number pairs, the amount of his turnover and the date thereof. He represents all such number pairs by a series of points on graph paper. Drawing a continuous line through them he obtains a graph, which gives him a vivid idea of the growth of his turnover. Similarly, if he is interested in the correlation between his sales and prices, he draws the sales and price diagram, in which each point is represented by a number pair denoting the sale and the corresponding price. But if he wanted to study the inter-relation between a larger number of items such as his sales, prices, number of salesmen engaged, their wages bill, *etc.*, by graphical method, he would have to devise a way of representing all these sets of corresponding numbers by a single 'point', whose 'graph' would then be a picture of this inter-relationship.

Although business men hardly use it, a way exists whereby a set of three or more numbers, which, taken together, indicate some state of affairs, can be represented by a single point. It was used for the first time in dynamics. Take, for example, the simple case of the motion of a single particle in a straight line. To know the state of its motion completely we require two quantities, its distance from a fixed origin and its speed at any particular time. We can represent this number couple—its distance and speed—by a point in a plane. A series of such points corresponding to different positions and speeds of the particle at different times would give a graph of its motion. Now suppose we had a system of two particles moving in a straight line in such a way that they have the same speed at any time. The state of this dynamical system is given by three numbers, two for the positions of the two particles and one for their common speed. If we want to draw a graph line to represent the dynamical state of this system, we cannot proceed as in the case of a single particle. The reason is that while a point in the plane of a graph paper can deputise for a number couple, it cannot do so for a number triple. To be able to draw a graph in such a case we should have either to omit one number out of the three required for the description of the state of the system, or find a way of drawing a three-dimensional graph. Obviously such a graph cannot be drawn on an ordinary two-dimensional graph paper. But it can be drawn in the perceptible three-dimensional space around us. Taking three mutually perpendicular reference lines in space, say, two horizontal lines and the vertical through a point of origin O, we can represent any state of our dynamical system as given by its three specification numbers by a point in space. The line joining all such points would thus be the analogue of our graph line in a two-dimensional space like that of the graph paper.

Now dynamical systems are of various degrees of complexity and may need one, two, three, four or more specification numbers for a complete description of their state. In the case of systems consisting of a large number of particles, this number may indeed be very large, although, due to the existence of mutual interlocking and constraints subsisting between the particles of the system, it may not be as large as we may be led to imagine at first sight. For instance, consider the motion of a rigid door about its hinges. We might imagine that to specify completely the state of its motion we have to know the positions and velocities of the innumerable particles of which it is composed. Actually, however, its motion is completely described by two specification numbers just as a smart tailor is able to fit your clothes like a glove by taking a few measurements over selected portions of your body. Your suit has to fit your body over a large number of parts, over the sleeves, the chest, the back, the neck, etc. Thanks to the mutual interlocking between the sizes of the various parts of the human body, the suit would fit all over if it fitted the measured parts.

In the same way, we could deduce the position of any particle of the door whatever if we knew just one specification number, *viz.* the angle made by

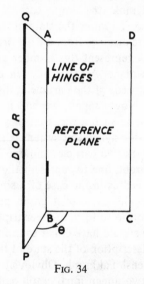

Fig. 34

the plane of the door with any selected reference plane through its hinges (see Fig. 34). Likewise, the speed of any particle of the door could be worked out if we knew the angular velocity of the door round the hinges. Just as ordinary velocity is the distance travelled per unit time, so angular velocity is the angle through which the door swings in a unit time. Thus two specification numbers suffice in this case, although the door consists of innumerable particles. On the other hand, a single particle moving freely in space would need as many as six specification numbers for a complete description of its motion—three to indicate its position in space and an equal number for the three components of its velocity along the three co-ordinate axes. A plane disc moving in any manner in space would require six specification numbers to fix only its position in space for the following reasons.

You would need three numbers just to locate its centre of gravity. Suppose its centre of gravity was at a point marked *G* in Fig. 35. You could specify it by three numbers, *viz.* its three distances from the three co-ordinate axes; but fixing *G* does not fix the disc. With its centre of gravity at *G* the disc could have any *orientation* whatever in space. Fig. 35 shows just two of them but you could have any number of them. To fix its orientation you would need to know the direction of a line at right angles to the plane of the disc. But to specify the direction of this line you have to measure

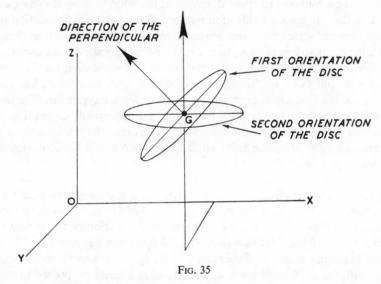

Fig. 35

the three angles it makes with the three reference lines *OX, OY, OZ*. This makes a total of six specification numbers—three for the centre of gravity and three for the perpendicular line—required to fix in space the position of the disc.

To fix its *velocity* another set of six specification numbers would be required, three for the three components of the velocity of its centre of gravity and three for the components of rotations of the disc around the three co-ordinate axes. The state of the motion of the disc is thus described completely by a set of six specification numbers for its position and another set of six for its velocity—twelve specification numbers in all. In general, any dynamical system could be described completely by a set of *n* specification numbers in so far as the position and configuration in space of its constituent particles are concerned, and another set of an equal number of specification numbers for its 'velocities', in all 2*n* specification numbers. In the case of the moving disc, as we saw, *n* is 6.

Now, if $n = 1$, as is the case with the motion of a single particle moving in a line or a door swinging on its hinges, we need only two specification numbers to describe the state of motion of the dynamical system. No difficulty arises in drawing a graph to represent this state, as the two numbers can be adequately represented by a point on a graph paper. The graph line joining the sequence of such points is a picture of the dynamical state of the system. But when the number n becomes 2, we need four specification numbers to describe the state of the system. As we saw before, we could draw a space-graph to represent the state of systems requiring *three* specification numbers for their description, but what shall we do when we require four or more specification numbers for a complete description of the dynamical state? Since our physical space ends with three dimensions, we can invent spaces of four, five, six, . . . or, in general, n dimensions to enable us to represent states with four or more specification numbers by means of 'points'. Such 'spaces' are, of course, pure phantasies, like the castles which Don Quixote imagined when he saw a wayside inn. The important difference, however, is that while the renowned Knight De la Mancha always came to grief because of his dreams, the mathematician's phantasy creates ordnance maps which help man to mould his environment to his heart's desire.

For instance, Hamilton derived his famous principle of least action by representing the configuration of any dynamical system at any time t as specified by its n specification numbers by means of a 'point' in an imaginary space of n dimensions. Since every dynamical situation of the system is given by specifying a set of n numbers, the latter can be represented by a point in a super-space of n dimensions. As the system changes its configuration with the passage of time, we can visualise a series of 'points' in our super-space to correspond to each new situation of the system. The history of the system is then epitomised by a set of 'points' in our imaginary super-space, generally known as 'configuration space'. This set of 'points' is virtually the 'path' followed by the dynamical system as a whole in our super-space. Hamilton showed that the system as a whole moves along a 'path' such that the time integral of 'action'* over the set of 'points' constituting this 'path' is a minimum.

While Hamilton, and following him Jacobi and others, were turning dynamics into an abstract theory of 'point' sets in imaginary spaces of several dimensions, a further impetus to the study of 'point' sets in such spaces was given by the successful operation of the steam engine. If the scientists can claim any credit at all for the creation of the steam engine, it must be admitted that they, like Seeley's Englishman, created it in a fit of

---

* Every dynamical system has two kinds of energies—one due to the motion of its particles and the other in virtue of their position. The difference of the two is known as *action*.

absent-mindedness. For a working model of the steam engine came into being long before there was any understanding of its basic theory—*i.e.* the nature of heat, its conduction and conversion into mechanical work, the behaviour of gases and vapours like steam, *etc.* These questions now began to receive attention. The theory of heat conduction, for example, was studied by Fourier, and this led him to the remarkable conclusion that any arbitrary function whatever could be represented by a sum of trigonometrical series since known as Fourier's series. This discovery was the veritable germ from which grew a general theory of 'point' sets. The study of the behaviour of gases and vapours like steam gave rise to the kinetic theory of gases. A gas enclosed in a cylinder or chamber was visualised as a swarm of a definite number N of molecules moving at random. As the dynamical state of each molecule can be specified by six numbers, three for its position co-ordinates and three for its velocity components, the dynamical state of the gas as a whole would be specified by a set of $6N$ numbers. To represent it by a single point we should require a super-space of $6N$ dimensions, technically known as 'phase space'. Working in such a space and using simple hypotheses such as that of molecular chaos, Liouville, Boltzmann, Gibbs and others proved about gases the fundamental theorems that are named after them, and that have now become classical.

Although the beginnings of the theory of point sets are to be found in the studies of mathematical physicists occupied with theories of heat, molecular motion, *etc.*—studies directly inspired by the successful working of the steam engine—the theory was taken up by pure mathematicians. Soon they presented it in a finished form of its own, obliterating all traces of any contact with questions of dynamics, thermal flow, molecular motion—questions which actually gave birth to it. By the end of the nineteenth century, the pure mathematicians came into their own and began even to criticise violently the reasoning of the mathematical physicists as shockingly imperfect and illogical. For example, the German mathematician, Zermelo, initiated himself into the methods of the mathematical physicist by translating into German Gibbs's book on *Statistical Mechanics*. He was shocked by the scandalous state of Gibbs's reasoning and raised a powerful objection against it—an objection which was not cleared till two distinguished physicists, Paul Ehrenfest and his wife Tatjana, took the trouble to learn enough mathematics to hoist the pure mathematician with his own petard.

The theory of point sets has manifold applications in modern mathematics. It owes its great power to the fact that members of the set—generalised 'points'—can be made to represent almost any measurable thing from the dynamical state of a system of moving particles to that of Ford's business administration. That is how the 'point', the dimensionless and scarcely visible dot, so insignificant by itself and yet so powerful in league, has come to establish its hegemony over almost the entire realm of modern mathematics.

A set in mathematics means exactly what it does in ordinary speech, *viz.* a collection or an aggregate of objects having certain specified properties. In real life we come across numerous kinds of sets, such as the set of stars of fifth magnitude, the set of horses participating in a race, the set of days in a week, *etc.* The objects of the set need not be concrete; they may be any entities of human thought whatever so long as they are well defined.

Now a set may be defined by cataloguing all its members as, for instance, the set of first five integers 1, 2, 3, 4, 5 or the set of telephone subscribers listed in a directory. Alternatively, it may also be defined by specifying some common property or properties of all its members as, for instance, the set of U.N. soldiers who fought in Korea, the set of babies born in India in 1950, the set of even integers, or the set of points on a given straight line. The most important sets from the point of view of the mathematician are sets of points on a straight line, plane, space or super-space of four or more dimensions. Before we describe the theory of such point sets, we may in passing note a few general characteristics of sets, whose elements may be objects of any kind whatever and not necessarily 'points'.

Suppose we have a set S of any kind, clearly we can form another sub-set, S', by taking only some of its elements. Thus if S is the set of babies born in India in the year 1950, we may construct a sub-set S' by taking only the female babies belonging to the set S. In general, S' is said to be a sub-set of a set S if *every* element of S' is an element of the set S. When this happens, we write

$$S' \subset S,$$

where $\subset$ means 'belongs to'. Naturally 'belongs to' is a two-way relation, for if S' belongs to S, then equally S 'owns' or 'contains' S'. Another way of paraphrasing the above statement would therefore be

$$S \supset S',$$

where $\supset$ means 'owns' or 'contains'.

Like most other possessive relations, the 'belonging to' and 'owning' relations, $\subset$ and $\supset$, can be passed on from one set to another. In other words, if a set S'' belongs to S' and S' to S, then S'' belongs to S. In symbols,

$$\text{if } S'' \subset S' \text{ and } S' \subset S, \text{ then } S'' \subset S.$$

An equivalent way of stating the same thing would be if S 'owns' or contains S' and S' contains S'', then S contains S''. In symbols, if $S \supset S'$, and $S' \supset S''$, then $S \supset S''$. For example, the set S'', consisting of all journalists living in New Delhi, belongs to the set S' of its literate adults, which itself belongs to the set S of all adults of the city.

Now suppose we have a set S' of literate adults living in New Delhi and a set s of the first five integers 1, 2, 3, 4, 5. If we combine the two sets S' and s, we shall get a hybrid set one part of which will consist of literate adults and another part of the first five integers. Obviously such sets do not mix well and nothing of interest emerges out of combining them. A more useful way of building up a theory of sets would be to consider a universal set U, which would include all the items about which we may wish to talk. For instance, such a set may consist of all the people residing in New Delhi at a particular time, if we wish to talk about the inhabitants of that city. Confining ourselves to this universal set U, we may construct a set of all adults. Can we generate another set out of U? Of course we can. Such a set would consist of items of the universal set U *not* included in S. A set like this is known as the complement of S in U and is usually written as S^*. In the example cited, S^* is the set of all the people living in New Delhi who are *not* adults. Now suppose we construct another set, S', consisting of all the literate people living in New Delhi. Out of the two sets S, S' we can construct another more comprehensive set which consists of all the people who are either adult or literate. Some of the adults (members of S) may not be literate (members of S') and some literates may not be adults. No matter; we include in the combined set I all people who are either members of the first, that is, are adult, *or* members of the second, that is, are literate. Such a set I is obviously the sum of the two sets S and S' and is written as $S + S'$. In other words, the sum set I of S and S' consists of all people who are either adult or literate. In symbols, $I = S + S'$.

Another way of constructing a set out of the two given sets S and S' is to take only those items of S and S' which are common to both. In the case under consideration such a set would consist of literate adults, that is, those people who are adults as well as literate and are, therefore, members of both the sets S and S' at the same time. Such a set is known as the product set P of S and S' and is written as $S.S'$. In symbols, $P = S.S'$. See the example in Fig. 36.

With these definitions we can now build up a new algebra—the algebra of sets. For instance, let us try to add a set S and its complement S^*. By definition S^*, the complement of S, consists of all those items of U, the universal set, which are not already included in S. Hence their sum would be the universal set itself.

Or, $$S + S^* = U . \qquad . \qquad . \qquad . \qquad . \quad (1)$$

What is their product? The product of the sets consists of items common

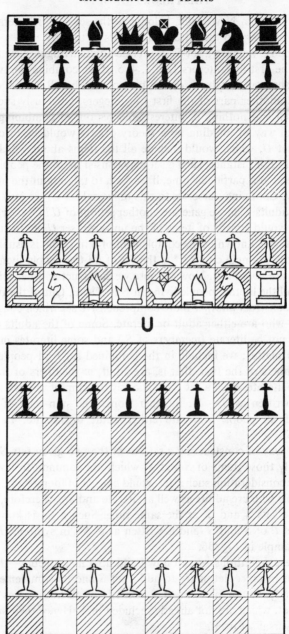

U

S

Fig. 36

S^*

FIG. 36—U (the universal set) = the set of all chess pieces of both players. S = the set of all pawns of both players. S^* (the complement of S) = the set of the pieces of U, excluding the pawns. Note that $S + S^* = U$, and that S and S^* have no common pieces. That is, $S.S^* = 0$.

to both S and S^*. Since S^* consists of only those items of U which do not belong to S, clearly there is no common element between them. A set which contains no element is known as the empty or null set and is denoted by 0. The product set of S and S^* is therefore the empty set 0. In symbols,

$$S.S^* = 0 \qquad . \qquad . \qquad . \qquad . \qquad (2)$$

Equalities like (1) and (2) epitomise in algebraic phraseology two well-known laws of classical logic. Thus (2) is really a statement of the 'logical law of contradiction', *viz.*, an object cannot possess a property and not possess it. (Example: A cannot be both red and not red at the same time.) Likewise (1) is another version of the 'law of excluded middle', *viz.* An object must either possess a given property or not possess it. (Example: A is either red or not red.) It is possible to construct a large number of similar equalities and thus reduce all the laws of classical logic to mere algebraic formulae in the theory of sets. Comparatively recent work by a number of mathematicians like Boole, Bertrand Russell, Whitehead and others has practically turned classical logic into a formal algebraic system that operates almost exclusively with symbols like $\subset, \supset, +$, *etc.* We shall discuss later the value of such attempts. Meanwhile we proceed with the theory of sets.

Hitherto we have considered our sets as mere aggregates of objects. What about their magnitude? If the objects within them are discrete, we can count the number of objects within each. Suppose we have a set S of discrete objects m in number and another set S' of n objects. What is the magnitude of the pooled set, $S + S'$? It might, perhaps, be thought that the number of objects in it would be $m + n$, but that would be the case only if S and S' had no common elements. If they did have some common elements, these would be counted *twice* in the pooled set—if we tried to deduce the number of objects in it by adding together m and n. For instance, if you had a team of eleven cricket players and a team of eleven football players, the combined teams would have 22 players if there were no common players who figured in both the teams. If there were some players, say two, who played in both the teams, the number of players in the pooled team would be only $22 - 2 = 20$. In general, the magnitude of the sum of any number of finite and discrete sets is the sum of the magnitudes of the individual sets provided no two of the sets have any overlapping or common elements. This simple commonplace addition rule will be our main guide in evolving a generalised notion of magnitude applicable to more complicated point sets.

*   *   *   *

Just as a set is defined by specifying some common property or properties of all its items, so also the type or class of a number of different sets may be defined by some property or properties belonging to the class of sets as a whole. Suppose, for instance, our universal set U is the set of all houses in a city. (See Fig. 37.) With this main set we can form a large number of sets consisting of one or more houses. Some of these sets may have the property that the houses included in them are in the same block. This would give us the class of all those sets whose houses are in one block. It is obvious that all sets of houses that we can possibly form are not of this type or class. Only some of the total number of sets that we can form will be of this type. Similarly, other types of sets may be constructed. An important class or type of sets is what is known as an *additive* class of sets. Among other conditions which sets of this type must satisfy, the most important is this: If two or more sets belong to an additive class, then the combined set obtained by adding them must also be a member of the class.

Consider, for instance, the class C_1 of sets of houses in one block. If we have two such sets of houses in each of which the houses lie in one block, it is clear that the pooled set formed by combining them need not have all its houses in one block and will not, therefore, be of the type under consideration. The class C_1 of sets of houses in one block is consequently not an additive class. Consider now an extended class C_2 of sets of houses that lie in one or at most two blocks. If we have two sets of this class in each of which the houses are in one block, the combined set will have its houses in at most two blocks and will therefore belong to the class. But if

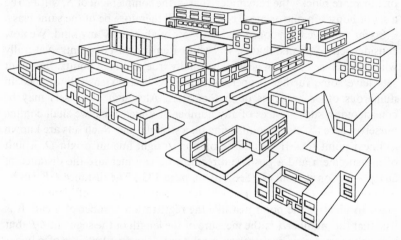

FIG. 37—Additive sets. If our universal set U is a set of houses in a city, a set of houses in one, two or three blocks is not an additive set. On the other hand, the set of houses in one or more blocks is additive.

we add three such sets, we shall produce a set whose houses may lie in more than two blocks. Such a set will not, therefore, belong to the extended class C_2 we have just constructed. This class, too, is therefore not an additive class. We can also construct a still more extended class C_3 of sets of houses whose elements lie in one, two or at most three blocks. A similar argument shows that even this extended class C_3 is not additive.

Now suppose we constructed a class C of sets of houses which lie in one or more blocks. Any number of sets of this class when combined will produce a set which belongs to the class C, as the houses in the pooled set will obviously be in one or more blocks. Such a class C of sets is known as an additive class. In general, an additive class of sets must satisfy three conditions:

1. The universal set U itself belongs to the class.
2. If every one of the sets S_1, S_2, S_3, ... belongs to the class, then the sum set $S_1 + S_2 + S_3 + \ldots$ also belongs to the class.
3. If S belongs to the class, then the complementary set S^* in U also belongs to the class.

It is easy to see that all these conditions are satisfied in the case of class C of sets of houses in one or more blocks cited above. First, the universal set U, the set of all houses in the city, is clearly a set of houses arranged in more than one block. It, therefore, belongs to the class. Secondly, any number of sets of houses in one or more blocks will combine to form a set whose houses are in one or more blocks. The pooled set, therefore, is also of the same type. Finally, if you take out of the universal set a set S of houses in one or more blocks, the remainder, that is, the complement of S, will also be a set of houses in one or more blocks. It will therefore be of the same class.

So far we have been concerned with sets of absolutely any kind. We now consider sets of points, that is, sets whose items consist of points. Naturally whatever is true of sets in general is equally true of sets of points. But point sets have some further properties that have been extensively applied in numerous other branches of mathematics. Although point sets may be considered in super-spaces of any number of dimensions, we shall confine ourselves here to sets of points lying on a straight line. Such sets are known as linear point sets. If we choose on our straight line an origin O, a unit of measurement and a positive direction, we can measure the distance of any point P on it from O. (See Fig 40, page 133.) The distance OP can be represented by a real number x, which can serve as a sort of identification mark to identify the point just like the registration number of a car. It is true that this number x is the measure of the length of the segment OP, but it equally serves as a registration mark for P. The simplest case of a linear point set is an interval, that is, the set of all points of a segment of the line lying between two given points, say O and A. Suppose the length of $OA = 1$, then the identification mark for O is 0 and for A is 1. The interval OA is the set of all points lying between O and A. We also define the same interval analytically, that is, by means of numbers as the point set x, where x is any real number between 0 and 1.

Another instance of a linear point set is the set of points whose identification marks are, say,

$$1, 1/2, 1/3, 1/4, \ldots, 1/10.$$

Such a set is a finite point set as its member points are finite in number, being 10 in all. In fact, we could construct any number of linear point sets by clubbing together any finite or infinite number of points of the line segment OA. Finite point sets are easy to treat but are of limited interest. We make a list of their member points and there is nothing more that we can do with them. Infinite point sets are more interesting. Now, as we saw before, there are infinities and infinities. There is the enumerable or countable infinity of the endless series of integers

$$1, 2, 3, 4, \ldots$$

and there is the much bigger infinity of the non-enumerable set of points on a line. In Chapter 5 we denoted these infinities by a and c respectively. Corresponding to these two types of infinities we have two types of linear infinite point sets. First, there is the enumerably infinite point set, for instance, the set of points consisting of the endless series

$$1/2, 1/3, \ldots, 1/n, \ldots$$

Second, there is the non-enumerably infinite point set consisting of all the points in the interval (0, 1) or any non-vanishing sub-interval thereof.

One common property of both types of infinite point sets is that every *bounded* set has at least one *limiting point*. This statement contains two new terms that have not yet been explained, *viz.* 'bounded' and 'limiting point'. A set is said to be 'bounded' when all its points are situated within finite bounds. That is, there exist two numbers a and b, within which lie all the numbers which stand for the registration marks or co-ordinates of the points of the set. Take, for instance, the infinite set S of points on the line OA, whose registration marks or co-ordinates are the numbers

$$1/2, 1/3, \ldots, 1/n, \ldots$$

This set is 'bounded' because every number belonging to it lies between 0 and 1. Now if every number of S is less than 1, *a fortiori* it will also be less than 2, 2·5, 3 and, in fact, every number greater than 1. As scientists valuing precision we are interested in the lowest number that can serve as an *upper* bound for the numbers of S. Obviously such a lowest number is $1/2$. For, if we took any number less than $1/2$, say, 4/9 it would not be an upper bound of S, because one number of S at least, *viz.* $1/2$, itself exceeds the number (4/9) so selected. At the same time any number greater than $1/2$ would be unnecessarily big as no number of the set exceeds $1/2$. When a set S, as in this case, itself contains a number greater than any and every other number belonging to it, this greatest number of the set will clearly be the lowest number that can serve as an upper bound of S. This greatest number of the set is then known as *the* upper bound of S, and S is said to *attain the upper bound*.

Consider now the lower bound of the set S

$$1/2, 1/3, 1/4, \ldots, 1/n, \ldots$$

Here every number of S obviously exceeds zero and, therefore, *a fortiori*, also -1, -2, ... or, for that matter, any number less than zero. We could thus take any of these numbers 0, $-\frac{1}{2}$, -1, -2, ... as a lower bound of the set. It will serve as an effective lower barrier past which no member of S can go. But, again, as men valuing precision we naturally look for the *greatest* of these numbers, which can serve as a *lower* bound. We scan our set and we find no member in it which we could spot as *the* lowest member of the set. All the numbers of S keep on decreasing without ever coming to an end. Is there then a number which is less than all the members of S. Clearly zero is a number with this property. Moreover, it is also the greatest; there is no number greater than zero with this property. For suppose, if possible, there was another such number, e.g. $\dfrac{1}{1,000,000}$, greater than zero but less than *all* the members of S. Obviously it cannot serve as a lower bound for our set S, as it exceeds the numbers, $\dfrac{1}{1,000,001}$, $\dfrac{1}{1,000,002}$ of the set. The number $\dfrac{1}{1,000,000}$ is therefore not a lower bound of S. The same is true of every other small number greater than zero. Zero is therefore the greatest number which is less than all the members of S. For no matter how small a number greater than zero we may choose we can always discover some member of S which is smaller still. Zero is thus the greatest lower bound of S.

In general, if an infinite set is bounded—that is, if all its numbers lie between two numbers such as a and b—then infinite pairs of other numbers can also serve as bounds for the set. Our problem is to discover *the least* upper bound and the *greatest* lower bound. How do we know that we can always find them? To show that this can always be done, Dedekind, a German mathematician, stated an axiom whose truth, he hoped, would be self-evident to all reasonable persons. If it appears so to you, there will be no difficulty in understanding his 'proof' of the existence of these bounds. If not, you may assume that if a set has one *upper* bound and, therefore, an infinity of them, there will always be one M among these upper bounds which is the least. Using the phraseology of the New Look style of reasoning popularised by the mathematicians of the 'rigorous' school, it has the following property:

Either M belongs to the set and is its greatest number, or if not, given any number ε, however small, there is at least one number of the set which exceeds $M - \varepsilon$ and is less than M.

This statement requires some further elucidation. As we have already seen, in the case of the set S

$$\tfrac{1}{2}, \tfrac{1}{3}, \tfrac{1}{4}, \tfrac{1}{5} \cdots$$

cited above, the least upper bound M is $\tfrac{1}{2}$ and it belongs to S. But M, the least upper bound, need not belong to the set. Consider, for example, the infinite set S'

$$\frac{1}{2}, \frac{2}{3}, \frac{3}{4}, \frac{4}{5}, \frac{5}{6}, \frac{6}{7}, \cdots \frac{1,000,000}{1,000,001}, \frac{1,000,001}{1,000,002} \cdots \frac{n}{n+1}, \cdots$$

This set is clearly bounded, for no matter how far we may go no number of the set (by the very manner of its construction) exceeds 1. As you would have observed, the denominator of every number of the set exceeds the numerator by one. No number of the set can therefore exceed or even equal one. Hence, although the number 1 itself does not belong to the set, it is *an* upper bound of the set. But is it *the* least upper bound? That is, is there any number less than 1 which can also function as an effective barrier past which no number of the set can go?

If possible, let us choose a number less than one but very close to it, as for instance the number, $1 - \dfrac{1}{1,000,000}$, and try if it can function as an upper bound. The answer is no. Because the number $\dfrac{1,000,000}{1,000,001}$ is clearly a member of our set and it exceeds the number chosen, *viz.* $\left(1 - \dfrac{1}{1,000,000}\right)$ $= \dfrac{999,999}{1,000,000}$. It is the same with any other number less than 1 no matter how close to 1. In other words, given any number ε, however small, $1 - \varepsilon$ is *not* an upper bound because there is always at least one number of the set which exceeds $1 - \varepsilon$ but is less than 1.

Hence, the *least* upper bound M of any set can be only one of the two things. *Either* M itself is a member of the set and is therefore its *greatest* number as is the case with the number '$\frac{1}{2}$' *vis-à-vis* the set S; *or*, if not, there is always at least one member of the set greater than $M - \varepsilon$ but less than M, where ε is any arbitrarily small number we choose to nominate. (This is the case with the number '1' *vis-à-vis* the set S'.)

A similar statement is true of the lower bound m of a set but the following changes in phraseology should be noted. *Either* m, the lower bound, is itself a member of the set and is therefore its least number, as, for instance, is the case with the number '$\frac{1}{2}$' *vis-à-vis* the set S'; *or*, if not, there is always at least one member of the set less than $m + \varepsilon$ but greater than m, where ε, as usual, is any arbitrarily small number we choose to nominate. This is the case with the number zero *vis-à-vis* the set S.

To revert to Dedekind's axiom. According to it, if you have found a way of dividing all real numbers into two classes L(eft) and R(ight) such that

(*i*) every real number belongs either to L or to R but *not* to both or to none; and

(*ii*) every member of R exceeds every member of L;

then there exists a dividing number M which forms the frontier of the two classes such that every number greater than M belongs to R and every number smaller than M to L, while the number M itself may belong to L or R. Now suppose we have an infinite set S of numbers with *an* upper bound b. We can divide all real numbers including the members of our set S into two classes, L and R. We place in R all real numbers like b which exceed all numbers of S and in L all real numbers which either belong to S or are less than some or all members of S. This way of partitioning the real numbers in two classes L and R satisfies the two conditions (*i*) and (*ii*) of 'Dedekind's cut', as it has been named. There is therefore a frontier number M and this is the least upper bound of S.

In a similar manner it may be shown that if a lower bound exists there is a number m which is the greatest lower bound.

We now define a limiting point of a set. It is a kind of rallying point or point of condensation round which an infinite number of points belonging to the set cluster. Thus in the case of the infinite point set

$$\tfrac{1}{2}, (\tfrac{1}{2} + \tfrac{1}{2}), \tfrac{1}{3}, (\tfrac{1}{2} + \tfrac{1}{3}), \tfrac{1}{4}, (\tfrac{1}{2} + \tfrac{1}{4}), \ldots 1/n, (\tfrac{1}{2} + 1/n), \ldots$$

the point zero is a limiting point as an infinite number of points of the set tend to congregate round it. A more recondite definition of the limiting point of a set is this: A point x is a limiting point of the set S if at least one point of the set S lies in the interval $(x - \varepsilon, x + \varepsilon)$ no matter how small ε may be. Except for the fact that it uses the fashionable terminology of the new style of reasoning, it is completely equivalent to the one previously given. A limiting point of a set, then, is a privileged sort of point, which contains at least one point of the set in its neighbourhood no matter how small. It itself may or may not belong to the set.

To prove our statement that every bounded infinite set has at least one limiting point, consider a bounded set whose points lie between 0 and 1. If we divided the interval $(0, 1)$

0   1

into two halves, $(0, \tfrac{1}{2})$ and $(\tfrac{1}{2}, 1)$,

                        $\tfrac{1}{2}$

0   1

then one of the two halves at least must contain an infinite number of points of the set, because otherwise the set would be a finite set. Suppose the left half

                        $\tfrac{1}{2}$

0

contained an infinite number of points of the given set. We could split this new interval $(0, \frac{1}{2})$ again into two equal halves.

$$\underset{0}{\rule{0pt}{0pt}} \overset{\frac{1}{4}}{\rule{3cm}{0.4pt}} \underset{\frac{1}{2}}{\rule{0pt}{0pt}}$$

At least one of these two new sub-intervals would again have an infinite number of points of the set. In this way we can go on dividing every interval into two halves, picking up the one that contains an infinite number of points belonging to the set.

$$\underset{0}{\rule{0pt}{0pt}} \overset{\frac{1}{8}}{\rule{3cm}{0.4pt}} \underset{\frac{1}{4}}{\rule{0pt}{0pt}}$$

$$\underset{0}{\rule{0pt}{0pt}} \overset{\frac{1}{16}}{\rule{2cm}{0.4pt}} \underset{\frac{1}{8}}{\rule{0pt}{0pt}}$$

$$\underset{0}{\rule{0pt}{0pt}} \overset{\frac{1}{32}}{\rule{1.5cm}{0.4pt}} \underset{\frac{1}{16}}{\rule{0pt}{0pt}}$$

etc.

In the limit, we shall reach at least one limiting point round which an infinite number of points of the set congregate.

This act of successively breaking an interval or a stick in two halves is a useful artifice which is often employed in mathematics to prove a variety of results. For instance, we may use it to sum up an infinite series that we shall require later. Suppose we have a stick of any length l and we break it into two halves. Take now the right half and break it into two halves and so on indefinitely. We shall have thus broken the stick into an infinite number of pieces of various lengths, each piece being half the size of the preceding as shown in the diagram below:

		l	
First half	$\dfrac{l}{2}$		
0		$\dfrac{l}{4}$	
	Second half		
		$\dfrac{l}{8}$	
	Third half		
		$\dfrac{l}{16}$	
	Fourth half		
	Fifth half	$\dfrac{l}{32}$	

It is plain that the sum of the lengths of all these infinite number of pieces of lengths $\frac{l}{2}, \frac{l}{4}, \frac{l}{8}, \frac{l}{16}, \ldots$ is exactly the same as the length of the original stick. In other words, the sum of the infinite series

$$\frac{l}{2} + \frac{l}{4} + \frac{l}{8} + \frac{l}{16} + \cdots$$

is simply l.

*　　　*　　　*　　　*

The theory of point sets is yet another illustration of the one conspicuous feature of mathematical history that we have stressed repeatedly, *viz.* the intimate mutual interaction between the development of mathematics and its applications or in other words, the close tie-up between 'pure' and 'applied' mathematics. As we remarked before, the theory of point sets originated from studies of heat flow, kinetic theory, molecular motion, thermodynamics, *etc.* These studies were in turn directly inspired by a desire to improve the working and design of the steam engine, which became the chief prime mover in Western Europe during the era of the dark Satanic mills and 'carboniferous' capitalism that descended on it suddenly towards the close of the eighteenth century. The theory of point sets was thus already quite advanced when the 'pure' mathematician began to take it up during the second half of the nineteenth century. Now, as often happens, the 'pure' theory gave rise to certain subtle problems of its own. The chief among these problems that began to claim attention during the last two decades of the nineteenth century was how to measure the magnitude of absolutely discontinuous infinite point sets. As we saw, if we have a finite set of discrete objects such as a bunch of bananas, we can define its 'magnitude' by the number of objects belonging to it. But in the case of infinite point sets this method does not work. We have, therefore, to devise some other way of measuring the magnitude of such sets. If the infinite point set is a continuous interval such as the set of all points lying between two real numbers, say, $1/3$ and $3/4$, the length $3/4 - 1/3 = 5/12$ of the interval can be taken as its 'magnitude'. But it is not clear how we should define an analogue of length when the point set is not a continuous interval but an absolutely discontinuous set such as the set S_1 of *only* irrational numbers lying between 0 and 1. The length of the unit interval $(0, 1)$ cannot *a priori* be taken as a measure of its 'magnitude' as this length also includes an infinity of rational points like $1/2, 2/3, \ldots$ expressly excluded from the set S_1.

The first to suggest a method for assessing the magnitude or 'measure' of absolutely discontinuous infinite point sets like S_1 was the German mathematician, Hankel. He was followed by Harnack, Stolz and Cantor, who further developed his idea during the eighties of the nineteenth century. It was presently superseded by that of the celebrated French mathematician, Émile Borel. In 1902 Henri Lebesgue, a pupil of Borel, extended his master's theory and showed a way of measuring the magnitude or length-analogue of absolutely discontinuous infinite point sets like S_1. Lebesgue's discovery came none too soon, for already discontinuity had begun to invade physics. Planck, during the preceding year, had discovered that a black body radiates energy discontinuously, that is, always in whole numbers of energy-packets or 'quanta', but never in fractions of a quantum. Hitherto it was tacitly assumed that most physical quantities varied continuously. Mathematical theory was therefore mainly concerned with continuous variables, which had no more than a few isolated discontinuities at the most. Now that it began to be realised that the structure of electricity, matter and energy was granular, so that measures of these quantities varied in jumps or discontinuously, studies of point sets, which could serve as replicas of absolute discontinuity, became socially important. Nevertheless, for a time the 'pure' theory of point sets ploughed its lonely furrow without any thought of possible applications and forgot even its past severely practical origin. For, once Lebesgue had shown the way, more and more abstract theories of measuring the 'magnitude' of point sets were created so as to include all the phenomena of continuity, discontinuity, limit, integrability, differentiability, *etc.*, within the orbit of their sweep.

By the late twenties it seemed as though the 'magnitude' or 'measure' theory, as it is generally called, had gone too far in advance of its applications. Fortunately the balance between theory and practice was redressed in the thirties of the present century, when the Russian mathematician Kolmogorov first applied the measure theory of point sets to give a new definition of mathematical probability*—now beginning to dominate quantum physics. Since then the measure theory of point sets has been extensively applied in mathematical statistics, electronics, telephone theory, cybernetics, econometrics, theory of games and economic behaviour, theory of insurance risks, stochastic or random processes such as the growth of biological populations and the fluctuation problem of cosmic ray showers, *etc.*

In all these problems the mathematical model embodying their essential feature is concerned with certain states of affairs denoted by a number of magnitudes. Earlier we saw how the state of a gas consisting of N molecules enclosed in a cylinder could be specified by $6N$ magnitudes. In a like man-

* See Chapter 9.

ner, the state of a biological population at any time may be described by the number of births and deaths of its members at that time. In telephone theory the state of affairs may be indicated by the total number of telephone calls put in and the number that go through. In the problem of cosmic ray showers it is denoted by the number of electrons and photons constituting the shower that disappear or die and the number of new electrons and photons that appear or are born. In mathematical statistics it is the state of the sample on the basis of which we propose to infer the characteristic of the population from which it is drawn.

Now, as we have seen, any such state of affairs can be represented by a 'point' in some imaginary space like the state of gas molecules in a phase space of $6N$ dimensions. A succession of such states of affairs is then represented by a set of 'points' in some such space. These problems thus boil down to a consideration of point sets in imaginary spaces. In many cases the point sets obtained are not continuous sets. To give mathematics a foothold in these cases, it is necessary to devise a way of assigning to such point sets some magnitude or measure which plays the same role with respect to them as length does in the case of continuous intervals. That is why many of these problems cannot be solved without resort to Lebesgue's theory of measure, which attempts to define an analogue of length when the point set is *not* a continuous interval but a complicated set like the set S_1 mentioned above.

To devise a way of ascertaining the 'magnitude', 'length-analogue', or 'measure' of a discontinuous set like S_1, consider first the universal set $(0, 1)$ within which we define any interval, say, $(\frac{1}{3}, \frac{3}{4})$. Let us call it i. (Fig. 38 A.) The length $L(i)$ of the interval i will then be $(\frac{3}{4} - \frac{1}{3}) = \frac{5}{12}$. Now if split a stick in two parts, the lengths of the two parts added together equal that of the original unbroken stick. In exactly the same way, if we divide i into two non-overlapping sub-intervals i_1, i_2 the length of i will be the sum of the lengths of i_1 and i_2. (Fig. 38 B.)

In symbols, $$L(i) = L(i_1) + L(i_2)$$.   .   .   . (1)

Obviously the equality holds even if we divide i into any finite number n of non-overlapping sub-intervals. Thus

$$L(i) = L(i_1) + L(i_2) \ldots L(i_n)$$.   .   . (2)

In fact, it is true even when n becomes infinite. It might seem evident that if the equality (2) is true for any finite n, it is necessarily true when n is infinite. But infinity in mathematics is a tricky business and we have to be on guard when extending relations like (2) that are true for any finite n to

include cases when n becomes infinite. Nevertheless, a 'rigorous' proof of
the equality (2) for the infinite case can be given although it will not be
repeated here. We may thus assume that if an interval is divided into a
finite or an enumerably infinite number of non-overlapping intervals, the
length of the total interval is equal to the sum of the lengths of all its

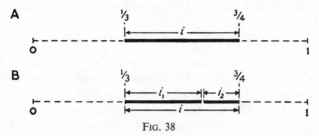

Fig. 38

constituent sub-intervals. The length $L(i)$ of an interval i is thus known as
an additive function of the interval i. However, we can construct sets of
points more complicated than intervals.

Can we define some magnitude or measure, $L(S)$, of a set S which has
the same property with regard to it as the length function $L(i)$ has with
respect to the interval i? In other words, can we devise a function $L(S)$
which could serve as a measure of the density of packing of the points
of the set in the same way as the length of an interval is a measure of the
packing of the points within its fold? If S consists of a single point, say $\frac{1}{2}$,
the measure $L(S)$ of the set will obviously be zero as the point by itself
occupies no length whatever. If S consists of two points, say $\frac{1}{2}$ and $\frac{1}{3}$, the
measure $L(S)$ of the new set will still be zero, as two extensionless points
occupy no more length taken together than singly, just as two penniless
have-nots after pooling their resources remain as penniless as before.
In fact, if S consists of any finite number n of dimensionless points, such
as $\frac{1}{2}, \frac{1}{3}, \frac{1}{4}, \ldots 1/(n + 1)$, the measure, $L(S)$, of S by the same tokens remains
zero.

However, when n becomes infinite it is not clear whether $L(S)$ would
still be zero or different. *A priori*, it cannot be stated that an infinite number of dimensionless points would aggregate to zero length as the infinite
number of points in the interval (0, 1) do produce the non-zero length
unity. The problem, then, is to discover what length would an infinite set
of points occupy, if they were all packed together as tightly as the points
on a line. Now the infinite set may be enumerable or non-enumerable.
Consider first of all an enumerably infinite set of points like

$$\frac{1}{2}, \frac{1}{3}, \frac{1}{4}, \ldots, 1/n, \ldots$$

Let us enclose each of these points in intervals of very small lengths. Since the point itself has no magnitude it can always be so enclosed no matter how small the enclosing piece.

Now think of any number as small as you please, say 0·0001. Take a stick whose length is 0·0001 and break it into two halves. Retain the left half and break the right half again into two halves and so on indefinitely. We have thus broken the stick into an infinite number of successive pieces of lengths—

$$0·0001/2, \ 0·0001/4, \ 0·0001/8, \ 0·0001/16, \ldots$$

each piece being half the length of the preceding. Now enclose the first point, *viz.* $\frac{1}{2}$ in the first piece, the second point, *viz.* $\frac{1}{3}$ in the second, the third, *viz.* $\frac{1}{4}$ in the third, and so on. If we packed all these points, *viz.* $\frac{1}{2}, \frac{1}{3}, \frac{1}{4}, \ldots$ as closely as 'adjacent' points on a line are, obviously they would occupy a length not exceeding the sum of the lengths of all the enclosing pieces. Hence all that the points in the enumerably infinite sequence S

$$\tfrac{1}{2}, \tfrac{1}{3}, \tfrac{1}{4}, \ldots, 1/n, \ldots$$

when packed together would occupy is a length no more than that of the original unbroken stick, *viz.* 0·0001, the arbitrarily small number with which you began the operation. That is, the measure $L(S)$, of the set S, is less than any arbitrarily small number you care to nominate, which is another way of saying that it is zero. The measure of a finite or an enumerably infinite point set is therefore zero.

Now how shall we define $L(S)$, the measure of the extent of inner packing of the points of a set S, when it is a non-enumerably infinite set? If S is a continuous set such as an interval, its length would be the measure $L(S)$ of its inner packing. But what about non-enumerable point sets which are not intervals? It is not very difficult to define such sets. A classical instance is the set of points in the universal set $(0, 1)$, whose co-ordinate is any irrational fraction less than unity. As we know, the interval $(0, 1)$ consists of points the co-ordinate of each of which is any real number, that is, either a rational number or an irrational number lying between 0 and 1. We can, therefore, divide the universal set, the interval $(0, 1)$, into two complementary sets; the set S_1 consisting of only points with irrational co-ordinates and S^*_1 consisting of points with rational co-ordinates.

Now both the sets S_1 and S^*_1 consist of an infinite number of points but the two infinities are of different power. We showed in the last chapter that while the power of the infinite number of real numbers between 0, 1 is c that of the infinite number of integers

$$1, 2, 3, 4, \ldots$$

is only a. It can be shown that the power of the infinite number of rational points between 0, 1 is the same as that of the infinite number of integers.* It follows, therefore, that the infinite point set S^*_1 is enumerable. The set S_1 is therefore non-enumerable. For, if not, let S_1 be enumerable also. The pooled set $S^*_1 + S_1$, that is, the set of points in the interval $(0, 1)$, would also be enumerable because the sum of two enumerable sets is itself enumerable. But we showed in the last chapter that the set of points in the interval $(0, 1)$ is not enumerable. S_1 therefore cannot be enumerable.

What is the measure of the extent of inner packing of the points of a non-enumerable set like S_1? We have agreed to measure the extent of

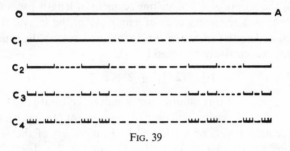

FIG. 39

packing of all the points of the interval $(0, 1)$ by its length, unity. Now all these points fall in two mutually exclusive classes—those belonging to S_1 and those to S^*_1. But the latter set is enumerable so that the measure of the extent of packing of the points of S^*_1 is zero. It follows, therefore, that the measure of the density of packing of the points of the non-enumerable set S_1 is 1.

It might perhaps be imagined that as the power c of non-enumerable infinity is higher than a, the power of the enumerable infinity, the measure of the density of packing of points of a non-enumerable set is always higher than that of the extent of packing of points of an enumerable set. As we have already seen, the latter is always zero. It might therefore appear that the measure of the extent of packing of points in a non-enumerable set is always greater than zero. This, however, is not always true. We can construct non-enumerable sets of points whose measure of the extent of inner packing is zero. Here is one example given by Cantor.

Consider a straight line OA, where O is the origin and OA is a unit length. Remove from it the middle third consisting of all points whose co-ordinate x lies between $\frac{1}{3}$ and $\frac{2}{3}$. (See Fig. 39.) We then obtain two non-overlapping segments or intervals towards each end. Let the set of points in these two end segments be called C_1. Now remove from C_1 the middle third of each of its two segments. This will leave us a set C_2 consisting of four segments, each of length $1/3^2$. Repeat this process of removing the

* For proof see Fig. 30 of Chapter 5.

middle third of each of the four segments of C_2. We thus obtain another set C_3 of eight segments each of length $1/3^3$. By continuing this process *ad infinitum* we can generate successively sets C_4, C_5, C_6, ...

Let C be the set of points on the unit segment OA after all the intervals have been removed so that C is the set of points common to the infinite sequence of sets C_1, C_2, C_3, ... It can be shown by means of Cantor's diagonal process used in the last chapter to prove the non-enumerability of the set of all real numbers in the interval $(0, 1)$ that the set C, like the continuum of real numbers, is non-enumerable. What is the extent of inner packing of the points of C? Since one segment of length $\frac{1}{3}$ was removed at the first step; two segments each of length $(\frac{1}{3})^2$ at the second step; 2^2 segments each of length $(\frac{1}{3})^3$ at the third step; and so on; the total length of the segments successively removed is

$$1 \cdot \tfrac{1}{3} + 2 \cdot (\tfrac{1}{3})^2 + 2^2 \cdot (\tfrac{1}{3})^3 + \dots$$

It can be shown that this infinite sum is unity, the total length of the segment OA. Although the entire length of the segment OA has been removed, there still remain points of the line OA, *viz.* members of the set C. Such, for example, are the points with co-ordinates $1/3$, $2/3$, $1/9$, ..., $2/9$, $7/9$, $8/9$, ... by which the successive segments are trisected. The measure of the extent of inner packing of points of the non-enumerable set C is therefore zero. Thus while the measure of the extent of inner packing of points of an enumerable set is *always* zero that of a non-enumerable set may or may not be zero.

Consider now any non-enumerable discontinuous set S. Obviously a length analogue or measure function $L(S)$ will have to satisfy the requirement that if S is divided into a finite or an enumerably infinite number of sub-sets S_1, S_2, S_3, ... with no common points, then

$$L(S) = L(S_1) + L(S_2) + L(S_3) + \dots . \qquad . \qquad . \quad (3)$$

It will be observed that this is merely a generalisation of the addition rule applicable to finite and discrete sets as also to sets of intervals.

In general, it is not always possible to define a measure function $L(S)$ having this property for every non-enumerable and discontinuous set. But for the class of sets known as additive sets, a measure function $L(S)$ can be defined. Hence the great importance of the additive class of sets in the theory of point sets.

We explained earlier the idea of an additive class of sets by an example of the class of sets of houses in one or more blocks. We now illustrate the idea of an additive class of point sets by another. Let our universal point set be the interval $(0, 1)$ and let us consider the class B_1 of all intervals lying within the main interval. Is this class additive? In order to see this we veiify if it satisfies all the three conditions specified earlier. (See page 119.) To begin with, the universal set U, the interval $(0, 1)$ is itself an interval and therefore belongs to B_1, the class of intervals in $(0, 1)$. The first condition is therefore satisfied. Now take two sets belonging to B_1, say, the interval $(0, \frac{1}{2})$ and the interval $(\frac{3}{4}, 1)$. The combined set of the two intervals is not an interval. (See Fig. 40.) The pooled set is therefore not an interval and does not belong to B_1, the class of intervals. The second condition is thus not satisfied and hence the class B_1 of sets of intervals in $(0, 1)$ is not an additive class.

Consider now the class B_2 of sets which are either intervals or sum of a finite or an enumerably infinite sequence of intervals. Such a class would now satisfy the first and second conditions necessary for an additive class. But what about the third condition? Let us consider the set S^*_1 mentioned above of points whose identification marks or co-ordinates are rational

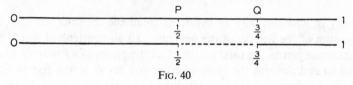

FIG. 40

fractions less than unity. As we saw, S^*_1 is an enumerable set. Each of these points may be considered as an interval of zero length. S^*_1 is thus the sum of an enumerable sequence of intervals each of zero length. It therefore belongs to the class B_2. The complement of S^*_1, viz. S_1, is then the sequence of points in $(0, 1)$ the co-ordinate of each of which is an irrational number less than unity. Now if we try to represent S_1 as a sequence of an infinite number of points (intervals of zero lengths), a non-enumerable infinity of such points will be required. S_1 therefore cannot be represented as the sum of an enumeiably infinite number of intervals. Consequently S_1, the complement of S^*_1, is not a member of class B_2. The third condition is not satisfied so that the class B_2, too, is not additive.

We began with a class B_1 of sets consisting of intervals and found it non-additive. We extended it to include all sets which were intervals or sum of a finite or an enumerably infinite number of intervals. We found that even this extended class B_2 is non-additive. We could go on extending the class of sets in this way and form a class B_3 by including in it sets formed by the sums and products of an enumerable infinity of sets in B_2. We can show that even B_3 is not additive. In fact, we could go on in this way without ever producing an additive class. But if we considered this infinite process of successive extensions giving us classes like B_1, B_2, B_3, ... *ad infinitum* as finally completed, we obtain the totality of all sets ever reached in this way. Such a class is known as the class B of Borel sets in the interval (0, 1), named after the French mathematician, Émile Borel, who first studied them. This class of Borel sets is an *additive* class.

We now revert to the definition of a measure function $L(S)$ of sets which are not intervals. As a first step we consider a class B_2 of sets defined above. Such a class consists of sets which are sums of a finite or an enumerably infinite sequence of intervals. If any of the intervals overlap we can drop the overlapping portion in one of them so that the set may be defined as the sum of a sequence of non-overlapping intervals, like i_1, i_2, i_3, ... with no common points between them. If all the points of such a set were packed together as tightly as they are in a straight line, they would evidently extend to the length of intervals i_1, i_2, ... laid end to end. Its measure function $L(S)$ is therefore merely the sum of the interval lengths i_1, i_2, i_3, ... which go to constitute the set. Thus

$$L(S) = L(i_1) + L(i_2) + L(i_3) + \cdots$$

Obviously in whatever way we divide or subdivide the intervals i_1, i_2, i_3, ... the sum of the lengths of the new intervals so constructed would remain the same just as the total length of all the pieces of a foot-rule when laid end to end remains the same whether you break it into five or fifty pieces. The set function, $L(S)$, is called the Lebesgue measure of the set S, in honour of the French mathematician, Henri Lebesgue.

Let us now consider a class of sets which cannot be represented as the sum of a finite or an enumerably infinite sequence of intervals. One such set, as we have already seen, is the set S_1 of points in the interval (0, 1) the identification mark or co-ordinate of each of which is any positive irrational number less than unity. Many other point sets of similar kind can be defined. How shall we measure the extent of inner packing of points in such sets? Let us first consider the problem in more general terms. We shall revert to the set S_1 consisting only of irrational points later. Take any point set S whatever in the universal set U, that is, the interval (0, 1). Since all the points of the set S lie in the interval (0, 1)—the universal set— the length of this interval, unity, is the upper limit of the measure of the extent of packing of points in S. We could enclose all the points of S within another set I consisting of the sum of one or more intervals. For example, let S be the infinite set

$$\tfrac{1}{2}, \tfrac{1}{4}, \tfrac{1}{8}, \tfrac{1}{16}, \cdots$$

We could enclose all the points of S within a single interval such as $(0, \tfrac{1}{2})$. As we can see, every point of the set S lies within this interval. We could also enclose all of its points within two intervals such as the interval $(0, \tfrac{1}{4})$ and the interval $(\tfrac{1}{2} - \tfrac{1}{100}, \tfrac{1}{2} + \tfrac{1}{100})$. For the first interval $(0, \tfrac{1}{4})$ covers every point of S except the first, *viz.* $\tfrac{1}{2}$, which is covered by the

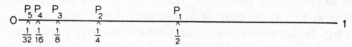

S is the set of points $P_1, P_2, P_3, P_4, P_5, \ldots$ whose distances from O are $\frac{1}{2}, \frac{1}{4}, \frac{1}{8}, \frac{1}{16}, \frac{1}{32}, \ldots$ respectively.

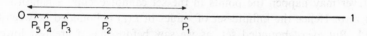

The length OP_1 or the interval $(O, \frac{1}{2})$ includes every point of S.

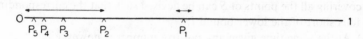

The length OP_2 or the interval $(O, \frac{1}{4})$ includes every point of S except P_1, which is included in the interval $(\frac{1}{2} - \frac{1}{100}, \frac{1}{2} + \frac{1}{100})$.

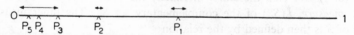

The length OP_3 or the interval $(O, \frac{1}{8})$ includes every point of S except P_1 and P_2, which are included in the intervals $(\frac{1}{2} - \frac{1}{100}, \frac{1}{2} + \frac{1}{100})$ and $(\frac{1}{4} - \frac{1}{1000}, \frac{1}{4} + \frac{1}{1000})$ respectively.

FIG. 41

second interval $(\frac{1}{2} - \frac{1}{100}, \frac{1}{2} + \frac{1}{100})$, and so on. (See Fig. 41.) Thus a set I of intervals enclosing all the points of S can be built up in an infinite number of ways. I may be a single interval like $(0, \frac{1}{2})$, or alternatively a combination of two intervals like $(0, \frac{1}{4})$ and $(\frac{1}{2} - \frac{1}{100}, \frac{1}{2} + \frac{1}{100})$ or three intervals like $(0, \frac{1}{8})$, $(\frac{1}{4} - \frac{1}{1000}, \frac{1}{4} + \frac{1}{1000})$ and $(\frac{1}{2} - \frac{1}{100}, \frac{1}{2} + \frac{1}{100})$, etc. There are infinitely many more complicated ways of manufacturing I but whatever we do we must ensure that each one of these ways of constructing I produces a set of one or more intervals such that every point of S is enclosed by at least one interval of the set I. To each such set I of intervals there corresponds a measure function $L(I)$. We thus have *an infinite set* of values of $L(I)$ corresponding to the infinite number of ways of constructing

I so as to cover every point of *S*. Which one of these infinite values of *L(I)* shall we pick as the measure of the extent of packing of the points in *S*? Clearly, the minimum value of *L(I)*. Now if you were given a finite set of values of a function like *L(I)*, you could write them all out in a tabular form and then look for the minimum. But when the set of values happens to be infinite how shall we even know whether there is any such minimum at all? However, in any case we know that *L(I)* cannot be more than 1, the length of the universal set in which all the points of *S* and all the intervals of the set (*I*) are included. Nor can *L(I)* be less than zero, for whatever may happen the points in the set cannot occupy a length less than zero. Hence the infinite set of values of *L(I)* is bounded between 0 and 1. But every bounded set, as we saw before, has a greatest lower bound and a least upper bound. Let *l* be the greatest lower bound of the set of values of *L(I)*. What it means is that a number *l* with the following two properties can always be found:

(*i*) No *L(I)* of the set is less than *l*. In other words, no set (*I*) of intervals covering all the points of *S* can be devised such that the corresponding *L(I)* falls short of the lower bound *l*;

(*ii*) At the same time given any positive number ε, however small, there always exists some system *I* of intervals covering all the points of *S* such that the corresponding *L(I)* is less than $l + \varepsilon$.

We may denote *l* by $\bar{L}(S)$. It is known as the *outer* measure of the set *S*. But from any given set *S*, we can derive another set *S**, the complement of *S* in (0, 1). We can, therefore, in exactly the same way, also define the *outer measure*, $\bar{L}(S^*)$ of the complementary set *S**. The inner measure $\underline{L}(S)$ of *S* is then defined by the relation:

$$\underline{L}(S) = 1 - \bar{L}(S^*).$$

In general, $\bar{L}(S)$ and $\underline{L}(S)$ are not equal. But when $\bar{L}(S) = \underline{L}(S)$, the set *S* is said to be *measurable* and the common value of the outer and inner measure of *S* is known as the *measure* of *S*. It is the complete analogue of length when *S* is no longer the sum of a finite or an enumerably infinite sequence of intervals. It can be shown that $\bar{L}(S) = \underline{L}(S)$ whenever *S* is a Borel or an additive class set.

Let us now revert to S_1, the set of points in (0, 1) with irrational co-ordinates. There are various ways of constructing a set *I* of intervals such that each point of S_1 is covered by at least one interval of *I*. One of the members of *I* is the set consisting of the single interval (0, 1)—the universal set *U*—in which all the points of S_1 are enclosed. *L(U)* is unity. There are many other ways of constructing *I* but it is not difficult to see that the corresponding *L(I)* would never be less than *L(U)*. In other words,

$L(U) = 1$ is the greatest lower bound of the set $L(I)$. Hence *the outer measure* (l) *of the set* S_1 *is unity*. What is the inner measure of S_1? To find it consider its complement S^*_1, the set of points with rational co-ordinates. S^*_1 is therefore enumerable. But in the case of an enumerable set like S^*_1, we showed that a set I of enclosing intervals can always be chosen such that $L(I)$ is less than any arbitrarily small number. The greatest lower bound of the set of values of $L(I)$ corresponding to the set S^*_1 is therefore zero. In other words, $\bar{L}(S^*_1) = 0$. But $\underline{L}(S_1) = 1 - \bar{L}(S^*_1)$ so that $\underline{L}(S_1) = 1$.

The inner measure of S_1 is therefore unity as is also its outer measure. S_1 is consequently a measurable set and its measure is unity. It may be remarked that the measure of S_1 remains the same as that of the universal set, the interval (0, 1), although S_1 is constructed from it by excluding an enumerable infinity of points with rational co-ordinates from the interval. This is only a particular case of the more general theorem that the exclusion of a set of measure zero from another set has no effect on the measure of the latter. In the case of the discontinuous set S_1, consisting of only irrational points in (0, 1), the extent of inner packing of its points is the same as that of all the points in the continuous interval (0, 1). The exclusion of rational points from the interval (0, 1) no doubt creates an infinite number of gaps in it; but it makes no difference whatever to the density of packing of the remaining points in the interval. The reason is that the excluded points form only an enumerable set whose measure is zero.

* * * *

(This section may be omitted on first reading)

We explained in Chapter 4 how the problem of summing or integrating a function over a continuous interval, area, or space crops up over and over again in the most varied fields of science. For instance, to calculate the distance travelled by a moving body during any interval of time we have to integrate the speed function over the time interval. In the design of irrigation dams, it is necessary to integrate the pressure function over the entire area of the dam surface to derive the whole pressure that the dam face has to endure. In many electrical problems we have to integrate the electric force function over a region of space. In all these problems the value of a function is defined at every point of a continuous interval, area, or region of space, and we have to sum these values corresponding to all the points of the interval, area or region in question.

In more complex problems, the function is *not* defined over a continuous interval, area or region but *only* over a non-continuous set of points lying in such intervals, areas or regions. The solution of these problems requires the summation of such functions over non-continuous sets of points. For example, consider the fundamental problem of mathematical statistics. Here we wish to infer the unknown characteristic of a given population by observing samples of some fixed size drawn from it in a suitable manner.

For the sake of simplicity let the sample size be such that a single item drawn from our population constitutes the sample. The sample is therefore defined by a single magnitude measuring the particular attribute under study of the selected item. If we take on a straight line a point whose distance from a fixed origin is the single magnitude pertaining to the observed sample, we may represent this single-unit sample by such a point. If the entire population consists of a finite number of members, say 10,000, it is obvious that the number of samples that can be drawn from it is also finite. In fact, in this case the number of single-unit samples that can be drawn is the same, *viz*. 10,000. Consequently the totality of all possible samples that we can draw may be represented by a set S consisting of 10,000 points on a straight line, each point representing some possible sample. A sub-class of samples out of this totality of all possible samples, such as, for example, those whose sample magnitudes lie between any two arbitrarily given values will naturally be denoted by some sub-set S' of the main set S.

Now each sample has a probability of its occurrence depending on the method of drawing the sample and the constitution of the population. Hence to each sample point of the set S of the aggregate of all possible sample points there corresponds its probability of selection. In other words, the probability function of the sample is defined only for the non-continuous set S of sample points and no others. But the solution of the sampling problem requires the calculation of the probability that the sample selected belongs to the sub-set S' of some particular sub-class of samples. This means that the probability function has to be summed over the non-continuous set S'. When the population considered is infinite, the corresponding sets S and S' of sample points become infinite point sets which moreover are non-continuous in many cases. That is why it is necessary to generalise the notion of integration so as to permit integration of functions over non-continuous sets.

To arrive at such a generalisation we observe that the 'measure' of a continuous set like an interval i between any two real numbers a, b is the length $(b - a)$ of the interval. We can also interpret the length of this interval as the integral $\int_a^b dx$ or $\int_i dx$ over the interval i. In exactly the same manner, the 'measure' or length-analogue of an infinite non-continuous but measurable set s can also be interpreted as a new kind of integral $\int_s dx$

over the set s. This new type of integral is known as the Lebesgue integral. In other words, the concept of ordinary integral of a function $f(x)$ over an interval i can be generalised to yield that of a Lebesgue integral of a function over a non-continuous but measurable set s such as the set S_1 of only irrational points within the unit interval (0, 1). In fact, the parallel between an ordinary integral and a Lebesgue integral is so close that we can actually derive the latter from the former by merely substituting the measurable set s for the interval i everywhere as follows:

PARALLEL BETWEEN ORDINARY AND LEBESGUE INTEGRAL

Ordinary Integral

Let $f(x)$ be a function defined for every point in an interval i between any two real numbers a, b.

Lebesgue Integral

Let $f(x)$ be a function defined for *some* points of the interval i and let s be the set of these points for which $f(x)$ is defined such as the set of only irrational points of the interval i. We assume that the set s is measurable.

To derive the ordinary integral of $f(x)$ over the interval i, we first divide the interval i into a finite number n of non-overlapping sub-intervals $i_1, i_2 \ldots, i_k, \ldots, i_n$, such that

$$i = i_1 + i_2 + \ldots i_k + \ldots i_n.$$

To derive the Lebesgue integral of $f(x)$ over the set s, we first divide the set s into a finite number n of non-overlapping sub-sets. Let these sub-sets be $s_1, s_2, \ldots, s_k, \ldots, s_n$, such that

$$s = s_1 + s_2 + \ldots s_k + \ldots s_n.$$

Since s is measurable, there is a measure function $L(s)$ analogous to the length of an interval such that

$$L(s) = L(s_1) + L(s_2) + \ldots L(s_k) + \ldots L(s_n).$$

We form the product of the length of each sub-interval i_k and the value $f(x_k)$ of $f(x)$ at any point of that sub-interval and add all these products to obtain the sum

$$S^*_n = f(x_1)i_1 + f(x_2)i_2 + \ldots f(x_k)i_k, \ldots f(x_n)i_n.$$

We form the product of the measure $L(s_k)$ of each sub-set s_k and the value $f(x_k)$ of $f(x)$ at any point of the sub-set s_k and add all these products to obtain the sum

$$Z_n = f(x_1)L(s_1) + f(x_2)L(s_2) + \ldots f(x_k)L(s_k) + \ldots f(s_n)L(s_n).$$

The limit of S^*_n as n is indefinitely increased in such a way that the length of every sub-interval tends to zero is known as the ordinary integral

$$\int_a^b f(x)dx \text{ or } \int_i f(x)dx.$$

The limit of Z_n, as n is indefinitely increased in such a way that the measure of every sub-set tends to zero is known as the Lebesgue integral

$$\int_s f(x)dx \text{ over the set } s.$$

If the limit of S^*_n as well as Z_n as n tends to infinity in the manner described above does not exist, as might happen, $f(x)$ is not integrable neither in the ordinary nor in the Lebesgue sense. But it is quite possible for the limit of Z_n to exist while that of $S*_n$ does not. Consider, for instance, the absolutely discontinuous function $f(x)$ mentioned in Chapter 3 defined within the interval (0, 1) by the formulae

$$f(x) = 0, \text{ for all rational values of } x,$$
$$= 1, \text{ for all irrational values of } x.$$

It can be shown that the ordinary integral $\int_0^1 f(x)dx$ over the interval (0, 1) does *not* exist in this case, whereas the Lebesgue integral of $\int_{S_I} f(x)dx$ over the set S_1 of all *irrational* values of x in the interval (0, 1) does exist. The reason is that the value of $f(x)$ is uniform all over the set S_1 whereas it continually oscillates between 0 and 1 in the unit interval (0, 1). On the other hand, whenever the ordinary integral of a function over an interval does exist, then so does its Lebesgue integral over the set of *all points in that interval*. In fact, the two are identical as the Lebesgue measure of a sub-set that is a sub-interval is merely the length of that sub-interval. That is why Lebesgue integral is a tremendous generalisation of ordinary integral.

The goddess of mathematics, when in a giving mood, is often embarrassingly bountiful. If she allows one gift she allows an infinity. For more than twenty years mathematicians looked round for a suitable measure of the extent of inner packing of points in a set without much success. At last, when Lebesgue found one, they discovered an infinite number of other ways of doing the same thing. Suppose, for instance, we have a measurable set S, whose Lebesgue measure is $L(S)$. Let $f(x)$ be a fixed non-negative function integrable over any finite interval. If we have a measurable set S in the interval, we can define a new function $P(S)$ of the set S by the equation

$$P(S) = \int_S f(x)dx,$$

whenever $f(x)$ is integrable over S. If not, we simply equate $P(S)$ to infinity. This new function $P(S)$ has the additive property similar to that of the Lebesgue measure $L(S)$. In other words, if

$$S = S_1 + S_2 + S_3 \text{ - - -,}$$

then,

$$P(S) = P(S_1) + P(S_2) + P(S_3) + \text{ - - -.}$$

We could, therefore, take $P(S)$ also as a measure of the density of packing of the points in S. The only difference is that when S is a continuous interval like (a, b), $L(S)$ is merely the length $(b - a)$ of the interval, whereas $P(S)$ becomes the ordinary integral

$$\int_a^b f(x)dx.$$

The difference between $P(S)$, or P-measure, and $L(S)$, the Lebesgue measure, of a set S may be illustrated by means of an analogy. Suppose we have a straight thin metal rod. If its density is uniform, we can express the weight of any portion of its length by the length of that portion. Suppose we want the weight of its length lying between two points A and B distant a and b respectively from one end (see Fig. 42). To deduce the actual weight of AB, we have only to multiply its length $(b - a)$ by its density which is

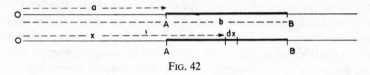

FIG. 42

uniform all over. We may, therefore, take $(b - a)$ itself as a measure of its weight. Suppose now that the density of the rod varies from point to point and depends on the distance of the point from one of its ends. Obviously, we cannot now take the bare length $(b - a)$ of the portion AB as a measure of its weight. We have to 'weight' each portion of its length by an appropriate factor to take account of the variable density.

If the density at a point is a function, $f(x)$, of its distance x from one extremity, the weight of a small portion dx of its length will be $f(x)dx$ and, therefore, that of a finite portion of the length AB will be the integral $\int_a^b f(x)dx$. Thus, whereas in the former case the length $(b - a)$ could be taken as the measure of the concentration of the material of the rod in the length AB, in the latter case it has to be $\int_a^b f(x)dx$. The former case is only a special case of the latter when $f(x)$ happens to be unity throughout. In the same way P-measure of a set S is a generalisation of its L-measure or Lebesgue measure. P- measure reduces to L-measure when $f(x)$ assumed for the definition of $P(S)$ is put equal to unity everywhere in S.

The introduction of P-measure has been of great service as with its help it has been possible to define a yet more general class of integrals that have been widely applied in statistics, diffusion problems, population theory, cosmic ray research, *etc.*

7

THE THEORY OF GROUPS

THERE are in many men two conflicting impulses—one urging them towards science and the other towards mysticism. The former springs from the belief that the universe is potentially explicable by rational enquiry and sensuous observation. The latter originates from faith in a way of knowledge which does not depend on sense, reason or analysis but on revelation, insight or intuition. In the last analysis mysticism postulates that the laws of the universe are fundamentally unknowable and no 'true' understanding of the ways of the world is really possible without invoking God, spirit, élan, nisus, entelechy, or some similar mystical principle. The conflict between these two antipodal beliefs is not new. It existed (as it exists today) in an acute form in antiquity almost since the very inception of the scientific method. Sometimes a compromise is attempted by letting God or the gods continue their rule in an Olympian world of their own and virtually forbidding them from infringing upon the workings of the mundane world. Epicurus did this in the ancient world, and in the modern world Newton did the same when he assumed, at least in some passages of the *Principia*, that after its initial creation by God the world has been quite independent of Him for its continued existence and motion.

Nevertheless, it seems to me that these two world-views are essentially irreconcilable, although there are many cases where men's minds have been exercised deeply by both views at the same time. For mysticism, in spite of its appeal to intuition or insight, often fascinates men of the finest intellect, who are lured into it by an unwarranted extension of some indubitable scientific fact. Thus, the fact that the ratio of the lengths of two vibrating strings emitting two different notes of the musical scale can be expressed in terms of natural numbers, suggested to Pythagoras—the Greek number mystic—and also to his followers that number is God, and led them even to compose prayers such as—'Bless us, divine number, thou who generatest gods and men,' *etc*. In our own day recent developments in nuclear physics have led some eminent contemporary scientists to the view that, alongside the physical world which science probes, there lies a 'spiritual' world which a scientist may recognise but cannot explore by the scientific method. Their reason is as follows.

Till about fifty years ago the physicists tried to explain and correlate observable phenomena of the physical world by means of models, which were supposed to function like ordinary objects of everyday experience. For instance, the kinetic theory of gases explained their behaviour on the hypothesis that gas molecules were miniature billiard balls, and the wave theory of light conceived light to be a form of wave motion in a cosmic ocean of jelly called the ether. But as they began to delve deeper into the nature of atomic and sub-atomic phenomena, the attempt to explain the microscopic world of atoms and electrons by means of concepts drawn from the level of everyday experience broke down.

Since what actually happens inside the atoms is unobservable, some scientists have come to the conclusion that the world of physics is a scheme of abstract metrical symbols connected by mathematical equations, which in some mysterious way 'shadows' the familiar world of our consciousness but can never penetrate it because its real nature is unknowable. For handling this 'shadow-world' of abstract metrical symbols they have recourse to a more recondite algebra than the ordinary high-school algebra. In ordinary algebra we encounter unknown quantities x, y, z, ... Although the quantities are unknown, we are ultimately able to find their values because, by subjecting them to known operations like addition, multiplication, subtraction, squaring, cubing, *etc.*, we obtain known results. But that is a defect in the eyes of a mystic, who is dominated by the faith that the universe is essentially unknowable. For him an instrument that tells us so much is thereby almost disqualified for 'treating a universe which is the theatre of unknowable actions and operations'. So Eddington argues that 'we need a super-mathematics in which the operations are as unknown as the quantities they operate on, and a super-mathematician, who does not know what he is doing when he performs these operations. Such a super-mathematics is the Theory of Groups.'

Now what is this 'super-mathematics', the Theory of Groups? It is simply an algebraic theory that came into being during the early decades of the nineteenth century in order to clear up certain questions in the theory of equations. Mathematicians had dealt with algebraic equations since remote antiquity, although they did not write them in the way we do now. For example, the Rhind papyrus, written by the scribe Ahmes sometime before 1700 B.C. and based on a text that may be at least two centuries older, mentions the equation: 'Heap; its 1/7, its whole; it makes 19', which in our language would be written as $x/7 + x = 19$. With the very dawn of civilised life in cities, whether in ancient Egypt, Sumeria, China, India, or elsewhere, men were faced with problems in astronomy, architecture, calendar making, barter, interest, discount, partnership, *etc.*, which required at least a rudimentary knowledge of forming equations and solving them. When the slave basis of these ancient civilisations gave birth to a firmly entrenched leisure class, the same device was used to construct and solve numerical puzzles or riddles which became a favourite social pastime.

The Palatine Anthology, for instance, cited by Cajori, included about fifty arithmetical epigrams (like Euclid's riddle quoted below), which implied knowledge of algebraic equations: 'A mule and donkey were walking along laden with corn. The mule says to the donkey, "If you gave me one measure I should carry twice as much as you. If I gave you one, we should both carry equal burdens." Tell me their burdens, O most learned master of geometry.' Bhaskara's great work, *Lilavati*, is full of several such epigrams or puzzles written in florid poetical language.

Although equations of varying degrees of complexity had been handled since remote antiquity, it took a long time before there was any clear understanding of their general theory. This was only natural as the number system with which the algebraists operated was incomplete. It did not include even negative numbers, not to speak of the imaginary numbers. Even as late as the twelfth century, the 'learned Brahmins of Hindustan', who, according to Hankel, were the real inventors of algebra and who had already recognised the existence of negative numbers by means of ideas of 'debt' and 'possession', rejected negative roots of equations. Thus Bhaskara, who gave $x = 50$ and $x = -5$ for the roots of $x^2 - 45x = 250$, discarded the second value, 'for it is inadequate, and people do not approve of negative roots'. It was not till the close of the eighteenth century that the number field was fully extended by the explicit inclusion of integers, fractions, irrationals, negatives and imaginaries in one number field, the field of complex numbers. Only then was it possible even to state the fundamental theorem of the theory of equations, *viz.* every equation of nth degree

$$a_0 x^n + a_1 x^{n-1} + a_2, x^{n-2} + \ldots a_{n-1} x + a_n = 0$$

has exactly n roots, no more and no less.

As long as there was no explicit recognition of imaginary and negative numbers, such roots were ignored and it was not clear how many solutions an equation ought to have. However, by the time Gauss proved this fundamental theorem for the first time towards the close of the eighteenth century, earlier investigators had already given general methods for the solution of equations of second, third and fourth degrees, that is, equations in which the highest power of x is 2, 3 and 4 respectively. In other words, rules or formulae were given so that by the adding, multiplying, subtracting, dividing, extracting square or cube roots, *etc.*, of the co-efficients of the unknown x in the equation, all the solutions (or roots) of the equation could be found. For instance, the general form of a quadratic equation, in which the highest power of x is 2, is

$$ax^2 + bx + c = 0,$$

and by our fundamental theorem, it will have just two roots, say x_1 and x_2. The solution of the equation means the explicit statement of a rule by means of which we can calculate both x_1 and x_2 from a, b, c, the coefficients of the equation. In this particular case, the rule is comparatively simple:

$$x_1 = \frac{-b + \sqrt{b^2 - 4ac}}{2a},$$

and

$$x_2 = \frac{-b - \sqrt{b^2 - 4ac}}{2a}.$$

Naturally the corresponding rules for equations of the third and fourth degrees are more complicated but they were discovered by the sixteenth century. During the succeeding two centuries, many unsuccessful attempts were made to give similar general rules for solving equations of the fifth and higher degrees. But as no such formulae could be found, men began to wonder whether it was at all possible to frame them. In mathematics it often happens that after repeated failures to solve a problem someone comes along and *proves* that the solution sought for is impossible. Thus, for centuries mathematicians tried to square a circle, trisect an angle, or prove the parallel postulate till it was shown that none of these feats was possible.

The same thing happened with the problem of solving generally equations of the fifth and higher degrees. Abel, Wantzel and Galois showed conclusively that such equations could not be solved generally. In other words, it was impossible to give formulae involving the sums, differences, products, quotients, squares, cubes or other roots of the coefficients of the equation whereby all its roots could be calculated—at least in principle. This, of course, does not mean that the equations have no roots. They have them; only we cannot trap them in algebraic formulae of the sort described above.

Now, lack of a general solution of equations of higher degree than the fourth is no serious handicap. For, short-cut methods of finding the approximate values of the roots of any given numerical equation of any degree are available. Even in the case of equations of the third and fourth degrees, we often prefer not to use the complicated algebraic general formulae and we work out their approximate solutions by simpler methods of numerical computation such as Horner's or Newton's methods or even by a calculating machine. Nevertheless, the algebraic investigations, which Abel and Galois conducted to prove the impossibility of a general solution in the case of equations of degree higher than four, had fruitful consequences.

Out of them grew the Theory of Groups and other revolutionary developments in modern algebra.

Now what is a group? First of all a group, as in ordinary language, is a collection, set, or aggregate. The elements of the group or the set may be objects of any kind, such as quantities, numbers, points, planes, curves, surfaces, displacements, rotations or other operations of any kind whatever. In elementary algebra we use the mystery symbol x to denote some unknown number, but when we want a 'super-algebra' we no longer restrict the mystery symbol x to represent a number. We use it to denote an 'indefinable'. That is to say, x is just a symbol concerning which nothing is assumed except that it obeys certain fundamental laws of algebra. We then start with a set of 'indefinables' denoted by the mystery symbol x. If we want to distinguish between the various elements of this set, we may represent them by $x_1, x_2, \ldots$.

Now at the outset, it might be objected that to call the mystery symbol x an 'indefinable' is an unwarranted piece of sophistication, for we require it to obey the fundamental laws of algebra and what else other than a number could obey those laws? But it will be recalled that there are sets of elements which are not numbers, though they behave like numbers in some respects. The set of symbols O, I, denoting the truth values of propositions introduced in Chapter 2, is a case in point. It is easy to construct examples of sets of elements other than numbers which obey algebraic laws.

Take, for instance, three playing cards—the ace, two, and three—out of a pack (Fig. 43). From any arrangement of these three cards you could produce several others by various ways of shuffling them. Let us recount these ways. First of all, we may leave the cards undisturbed. This is the way of no shuffle. We get three distinct ways of shuffling by interchange of the first and second, the first and third, and the second and third cards. The fifth way of shuffling is produced by interchanging the first and second followed by an interchange of the second and third. The sixth way of shuffling is given by interchanging the first and second followed by an interchange of the first and third. There is no other way, as we shall show presently. Let us now consider a set whose 'indefinable' member x is one or other of the aforementioned six ways of shuffling three cards. Let us denote them by the symbols $x_1, x_2, x_3, x_4, x_5, x_6$. The following scheme then defines the 'indefinable' operations $x_1, x_2, \ldots, x_6$ composing the set.

x_1: The way of no shuffle or leaving the cards as they are.
x_2: The way of interchanging the first and second cards.
x_3: The way of interchanging the first and third cards.
x_4: The way of interchanging the second and third cards.

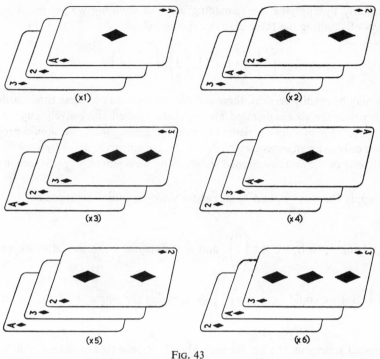

Fig. 43

x_5: The way of interchanging the first and second *followed* by an interchange of the second and third cards.

x_6: The way of interchanging the first and second *followed* by an interchange of the first and third cards.

Now suppose you started with any arrangement of cards, say $\begin{pmatrix}1\\2\\3\end{pmatrix}$,

which means ace on top, two in the middle and three at the bottom. If we apply the shuffle denoted by the symbol x_1, we have the same arrangement as x_1 is the way of no shuffle or leaving the original arrangement intact. If

we apply the shuffle x_2 to it, we get the arrangement $\begin{pmatrix}2\\1\\3\end{pmatrix}$. We may denote

the result symbolically by the equation,

$$x_2\begin{pmatrix}1\\2\\3\end{pmatrix} = \begin{pmatrix}2\\1\\3\end{pmatrix}$$

Similarly by applying the remaining ways of shuffling we get in all six ways of shuffling the three cards as shown below:

$$\begin{pmatrix}1\\2\\3\end{pmatrix}, \begin{pmatrix}2\\1\\3\end{pmatrix}, \begin{pmatrix}3\\2\\1\end{pmatrix}, \begin{pmatrix}1\\3\\2\end{pmatrix}, \begin{pmatrix}2\\3\\1\end{pmatrix}, \begin{pmatrix}3\\1\\2\end{pmatrix}.$$

As may be readily verified, there are no arrangements of these three cards other than the six enumerated above. If we applied successively any two or more ways of shuffling cards to any given arrangement, we should produce only an arrangement which we could have got straightway by applying only one of the aforementioned six shuffling ways singly. For instance, let us apply the ways x_2 and x_3 successively to our initial arrangement $\begin{pmatrix}1\\2\\3\end{pmatrix}$.

Applying x_2 to it, we get $\begin{pmatrix}2\\1\\3\end{pmatrix}$; and *then* applying x_3 to the latter we get $\begin{pmatrix}3\\1\\2\end{pmatrix}$. But we could have got $\begin{pmatrix}3\\1\\2\end{pmatrix}$ by applying the single shuffle x_6 to our original arrangement $\begin{pmatrix}1\\2\\3\end{pmatrix}$. We may, therefore, take two successive shuffles x_2, x_3 in this order as equivalent to a single shuffle x_6. Symbolically,

$$x_3 . x_2 \begin{pmatrix}1\\2\\3\end{pmatrix} = \begin{pmatrix}3\\1\\2\end{pmatrix},$$

$$\text{and } x_6 \begin{pmatrix}1\\2\\3\end{pmatrix} = \begin{pmatrix}3\\1\\2\end{pmatrix}.$$

Hence $x_3 . x_2 = x_6$.

Let us now apply the shuffle x_4 after we have applied the shuffles x_2, x_3 in succession. As we saw, the latter two shuffles produce the arrangement $\begin{pmatrix}3\\1\\2\end{pmatrix}$, to which we now apply x_4. This gives $\begin{pmatrix}3\\2\\1\end{pmatrix}$, which we could have obtained from our original arrangement $\begin{pmatrix}1\\2\\3\end{pmatrix}$ by applying to it the single shuffle x_3.

Symbolically, $\quad x_4 . x_3 . x_2 \begin{pmatrix} 1 \\ 2 \\ 3 \end{pmatrix} = \begin{pmatrix} 3 \\ 2 \\ 1 \end{pmatrix}$,

and $\quad\quad\quad\quad\quad x_3 \begin{pmatrix} 1 \\ 2 \\ 3 \end{pmatrix} = \begin{pmatrix} 3 \\ 2 \\ 1 \end{pmatrix}$.

Therefore, $\quad\quad\quad\quad x_4 . x_3 . x_2 = x_3$.

We could apply the same operation successively. For instance, $x_2 \begin{pmatrix} 1 \\ 2 \\ 3 \end{pmatrix}$ gives

$\begin{pmatrix} 2 \\ 1 \\ 3 \end{pmatrix}$. If we apply to this result the same shuffle a second time we get

$\begin{pmatrix} 1 \\ 2 \\ 3 \end{pmatrix}$. We could get the same shuffle by a single application of shuffle x_1.

Hence, $\quad\quad x_2 . x_2 \begin{pmatrix} 1 \\ 2 \\ 3 \end{pmatrix} = \begin{pmatrix} 1 \\ 2 \\ 3 \end{pmatrix}$, and $x_1 \begin{pmatrix} 1 \\ 2 \\ 3 \end{pmatrix} = \begin{pmatrix} 1 \\ 2 \\ 3 \end{pmatrix}$

or, $\quad\quad\quad\quad\quad\quad x_2 . x_2 = x_1$.

If we apply x_2 a third time to $\begin{pmatrix} 1 \\ 2 \\ 3 \end{pmatrix}$, we get again $\begin{pmatrix} 2 \\ 1 \\ 3 \end{pmatrix}$, which we can obtain

from $\begin{pmatrix} 1 \\ 2 \\ 3 \end{pmatrix}$ by a single application of x_2. In other words, three successive

applications of x_2 are equivalent to a single application of x_2. Symbolically $x_2 . x_2 . x_2 = x_2{}^3 = x_2$.

We thus see that the resultant effect of applying any numbers of shuffles $x_1, x_2, \ldots, x_6$ *any* number of times to any given arrangement of the cards is the same as that of a *single* application of some one or other of these six shuffles. As we have seen, two successive applications of x_2 are equivalent to x_1, three to x_2, and so on. In fact, we can give a table showing the single shuffle to which a combination of any two of the six shuffles specified above are equivalent. The table is reproduced below and may be verified by applying any two of the shuffles successively. To read it proceed as follows:

Suppose we want the result of applying two successive shuffles, first x_3 and then x_4. Look in the column headed by the *first* shuffle, *viz.* x_3, and the row headed by the *second* shuffle, *viz.* x_4. The result is easily read as x_6. And so on for other pairs.

Columns

		x_1	x_2	x_3	x_4	x_5	x_6
	x_1	x_1	x_2	x_3	x_4	x_5	x_6
	x_2	x_2	x_1	x_5	x_6	x_3	x_4
Rows	x_3	x_3	x_6	x_1	x_5	x_4	x_2
	x_4	x_4	x_5	x_6	x_1	x_2	x_3
	x_5	x_5	x_4	x_2	x_3	x_6	x_1
	x_6	x_6	x_3	x_4	x_2	x_1	x_5

A set or collection of 'indefinables' x, like the set of six shuffles described above, whose elements combine in such a way that any two or more of them are equivalent to some single item of the set, is known as a *group*. The essential point is that the elements of the group may be combined according to a law and any combination of them produces an element belonging to the set itself. It is, therefore, completely self-contained. It is obvious that every set or aggregate cannot be a *group*. For instance, the set of all males aged thirty in India, or the set of all concentric circles in a plane, is not a group as there is no way specified in which any two elements of the set may be combined. Even if we specified a way of combining the elements of a set, it may still not form a group as the combination of two elements of the set may form an element not belonging to the group. Take, for instance, the set of the first ten integers, 1, 2, 3, . . ., 10; we could combine any two of its elements by adding any two numbers of the set. If we apply the ordinary addition rule for combining elements in this set, we may or may not get an element belonging to the set. Thus, if we combine say 3 and 5, or 4 and 2 by addition, we get elements belonging to the set; but if we combine two elements like 9 and 8 of the set, we get an element 17 which does not belong to the set. The set of the first ten integers is, therefore, not a group. *For a set to be a group it is necessary that there should be a rule for combining any two of its elements* and *that the resultant element so produced is itself a member of the set.* Although this definition of a group will suffice for our purposes, we shall give below a technically more precise definition, just to show how the group is the most characteristic concept of modern mathematics. You may skip over it and the rest of the following paragraph if you find it too difficult.

A group is a system of a finite or infinite number of objects within which an operation is defined which generates from any two elements a, b

another object ab also belonging to the system, subject only to two conditions:

(*i*) The operation according to which the elements of the system are combined obeys the associative law $a(bc) = (ab)c$;

(*ii*) If a, b are any two elements of the system, there also exist two elements x, y of the system such that $ax = b$ and $ya = b$.

It is really a wonder that from these two insignificant-looking assumptions there springs an abundance of profound relations tying up in the single framework of an axiom system an astounding variety of seemingly unrelated branches. For instance, an *integral domain*, which is a set for which two operations, *viz.* addition and multiplication, are defined, is a group under addition alone. A *field*, on the other hand, is a set which is not only a group under addition but also under multiplication provided we exclude the zero element. If a set is a group and if we take any of its elements, say a and further also take b identical with a, by hypothesis (*ii*) above, there exists an element x_0 such that $ax_0 = a$. In other words, the group also contains x_0 which is an identity element* of the group, that is, an element which leaves unaltered any element with which it is combined. It can be shown that there is one and only one such identity element in a group. You may easily verify by means of the table given in the text that the set G of six shuffling operations x_1, x_2, x_3, x_4, x_5 and x_6 satisfies the aforementioned two conditions of a group. Can you figure out its identity element?

In the foregoing example of the group G of six shuffles $x_1, x_2, \ldots, x_6$, the sub-set G' of three ways of shuffling, *viz.* x_1, x_5, x_6 is by itself a group. For, consider this sub-set (G') of the three shuffles, x_1, x_5, x_6. The rule for combining shuffles given in the aforementioned table shows that $x_1.x_5 = x_5$, $x_1.x_6 = x_6$, $x_5.x_6 = x_1$, $x_1.x_1 = x_1$, $x_5.x_5 = x_6$, $x_6.x_6 = x_5$. Thus any repetition or successive applications of the three shuffles x_1, x_5, x_6 produce an arrangement which could be obtained by a single application of some one of these three shuffles. Hence the sub-set G' of the three shuffles x_1, x_5, x_6 is itself a group. A sub-set like G' which, being a part of a larger group G, itself possesses the group property in its own right, is known as a *sub-group* of the original group. Another example of a sub-group is the set x_5, x_5^2, $x_5^3 = x_1$, for any combination of x's in the sub-group leads back to one or other element of itself. This may also be verified easily from the table reproduced above.

A group like the above, which is obtained by permuting cards, letters, symbols, *etc.*, is called a *permutation* group. The number of cards, or letters permuted, is called the degree of the group. In the aforementioned

---

* It is exactly what zero is in an integral domain.

example we produced our group of 6 x's by permuting three cards. Its degree is therefore three. The number of all possible arrangements or permutations of 3 cards is $3.2.1 = 6$, and is known as the *order* of the group. The group of six x's is called a symmetric permutation group of degree 3 and order 6. The sub-group consisting of x_1, x_5, x_6 is of *degree* three and order three. The sub-group consisting of $x_5, x_5{}^2, x_5{}^3 = x_1$, whose elements are generated by the powers of a single element like x_5, is known as a *cyclic* group.

A group may consist of a finite or infinite number of elements. So far we have given only examples of finite groups. As an example of an infinite group, consider the set of positive integers. We can combine any two elements of the set in two ways by adding them or multiplying them. Let us choose the ordinary addition (which should also include its inverse subtraction) as the rule for combining the elements of this set. With this broader concept of 'addition' which includes subtraction, the 'sum' of two positive integers is not always a positive integer. Hence the set of all positive integers is *not* a group under 'addition'. On the other hand, the set of all positive and negative integers including zero is a group under 'addition'. For the sum (or difference) of any two integers of the set would belong to the set itself. To say that the set of positive and negative integers is a group under addition is merely another way of paraphrasing the statement made in Chapter 2 that such a set is closed under addition and subtraction.

If instead of generalised addition we adopted 'generalised' multiplication, which includes its inverse division as well, the set of positive and negative integers—including zero—is not a group, for the division of one integer by another is not always exact. On the other hand, the set of all real numbers (excluding zero) forms a group if the rule of combination is generalised multiplication.

Now we may have two groups consisting of widely different elements but with identical group structure. What this means may be illustrated by an example. Suppose we have a group G of four numbers,

$$1, \sqrt{-1}, -1, -\sqrt{-1}.$$

If we adopt the usual multiplication rule as the rule of combination, it may easily be verified that the four elements form a group. For instance, the product of $\sqrt{-1}$ and $-\sqrt{-1}$ is $+1$, which is an element of the set, and so on for other products in pairs. We can also form another set of four elements, a group G' of rotations of a line OA in a plane through angles of $0°, 90°, 180°, 270°$. Any two successive rotations of the set are equivalent to some single rotation belonging to the set. Thus, suppose we first rotate the line OA through $270°$ and then follow it up by another rotation through

180°. The final position is the same as if it had been subjected to a single rotation through 90° (see Fig. 44). The set G' of four rotations is therefore a group. We thus have two groups G, G' with four elements in each. Now suppose we postulate the following correspondence between the elements of G and G':

$$\begin{array}{llll}
1 & \text{corresponds to a rotation through} & 0° \\
\sqrt{-1} & \text{,, \quad ,, ,, \quad ,,} & \quad\quad 90° \\
-1 & \text{,, \quad ,, ,, \quad ,,} & \quad\quad 180° \\
-\sqrt{-1} & \text{,, \quad ,, ,, \quad ,,} & \quad\quad 270°
\end{array}$$

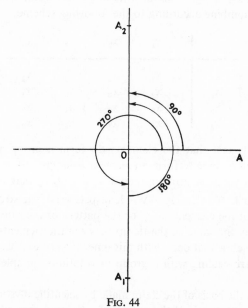

Fig. 44

If we combine *any* two numbers of G and any two *corresponding* rotations of G' according to their respective laws of combination, the resultant elements of G and G' are also corresponding elements. Suppose we combine $\sqrt{-1}$ and -1. The product is the element $-\sqrt{-1}$ of G. The elements in G' corresponding to $\sqrt{-1}$ and -1 in G are rotations through 90° and 180° respectively. Two such successive rotations are equivalent to a single rotation through 270° which, as shown above, corresponds to $-\sqrt{-1}$. This is true of any pairs of elements of G or G'. Thus the two groups G and G' have the following properties:

(*i*) Both have the same number of elements;

(*ii*) To every element of G there corresponds just one element of G' and vice versa;

(*iii*) If we produce an element x of G and an element x' of G' by combining *any* two given elements of G with corresponding elements in G', then the element x of G will correspond to the element x' of G'.

Such groups are equivalent to one another as far as their group structure, or the pattern of inner interlocking between their respective elements, is concerned. The group structure of both is therefore the same although the nature of the elements of the two groups differs radically from one another in other respects. Now suppose we construct another group g of four 'indefinables', x_1, x_2, x_3, x_4 about which we know nothing except that any two of them combine according to the following scheme:

	x_1	x_2	x_3	x_4
x_1	x_1	x_2	x_3	x_4
x_2	x_2	x_3	x_4	x_1
x_3	x_3	x_4	x_1	x_2
x_4	x_4	x_1	x_2	x_3

We may easily test that the elements do form a group as any two or more of them combine to form an element belonging to the set itself. We may also verify that the group g is equivalent to G by making x_1, x_2, x_3, x_4 correspond to $1, \sqrt{-1}, -1, -\sqrt{-1}$, respectively. The structure of inner interlocking of the elements of g, or the pattern of their internal relationship, is exactly the same as the tie-up between the elements of G (or G'). The 'super-algebra' that deals with this inner structure is therefore the same whether we are dealing with a group of rotations, complex numbers, or 'indefinables'.

This fact is the basis of the claim made by scientific mystics like Eddington that 'the universe is of the nature of a thought or sensation in a universal Mind.' Eddington admits that this statement 'is open to criticism' and 'requires more guarded expression if it is to be accepted as a truth transcending forms of thought.' Nevertheless, he considers it 'true in the sense that it is a logical consequence of the form of thought which formulates our knowledge as a description of a Universe.' However that may be, it seems to me that its truth, in whatever sense it may be said to be true, is not suggested, much less established, by the fact that mathematical physics nowadays is a scheme of symbols interlocked by the existence of abstract structure. To understand their argument in favour of this claim and the fallacy underlying it, we may further illustrate how these ideas have been applied by mathematicians to the concept of space. The simplest case mathematically is that of spherical space.

To save ourselves turning somersaults in four-dimensional space, consider a spherical space like the surface of a globe such as is used to represent the earth in geography. Let us consider the 'space' formed by its surface only, that is, the set of points lying on the globular shell excluding those inside or outside the globe. Any point on its surface can be changed into any other by a rotation of the globe round its North–South axis. Thus you could shift a point marked, say Paris, from a position just in front of you to any other position. Starting from any orientation of the globe you could obtain an infinite number of others by rotating it through different angles from 0° to 360°.

All such rotations form a group because any two rotations of the globe round its axis would give you a position which you could obtain by a single rotation belonging to the group itself. Thus if you first applied a rotation through 90° and then another through 60°, the resultant of these two rotations is equivalent to a single rotation through 150°. Two successive rotations through 270° and 180° degrees are equivalent to a single rotation through 90°, and so on. In fact, if you could rotate your globe round *any* axis instead of only one—the North–South axis—you could generate a much bigger set, the set of all possible rotations of the globe round any axis whatever. This set of rotations also possesses the group property, and the group of rotations round the North–South axis is only a sub-group of this larger group. Now, mathematically, the structure of the 'space' given by the surface of the globe is the same as that of any group of 'indefinables' equivalent to this group of rotations. Hence, argues Eddington,

'When we introduce spherical space in physics, we refer to something— we know not what—which has this structure. . . . The general concept, which attempts to describe space as it appears in familiar apprehension— what it looks like, what it feels like, its negativeness as compared with matter, its "thereness"—is an embellishment of the bare structural description. So far as physical knowledge is concerned, this embellishment is an unauthorised addition.' In other words, all that the 'super-algebra' used by physics for describing the structure of space requires is some group of 'indefinables' equivalent to the group of rotations.

In the same way, Eddington considers that although we actually experience sensations with all the welter of pictorial details accompanying them, the starting point of physical science is knowledge of only the *group structure* of a set of sensations in a consciousness. Eddington is, of course, aware that to get at such knowledge of group structure we have to visualize rotations in actual space and experience sensations in the ordinary way and that this is impossible without the aid of what he calls 'embellishments' and 'unauthorised additions'. But when he says that the starting point of physical science is knowledge of the group structure of a set of sensations, he does not mean the historical starting point but the logical starting point.

In this way he makes a distinction between the genesis of this knowledge and its subsequent logical formulation, which no longer requires the concrete pictorial details out of which it has actually grown. This is fair enough, for in mathematics the distinction between an abstract logical scheme and its actual realisation is well understood. But he certainly goes too far when, on the basis of the fact that for applying group algebra to certain aspects of physical phenomena all that we require is the abstract and logically pure concept of structure without the pictorial details, he claims that physical theory, concerned as it is with only this abstract structure of a group of metrical symbols, can tell us nothing of the external world which is of an inscrutable nature.

The fact that the group theory deals only with the inner structure of the elements of equivalent groups and pays no heed to their other characteristics is no reason for claiming that the 'real' nature of the external world is inscrutable. For the abstraction introduced by group theory is of a piece with the abstractions of ordinary algebra and is in keeping with the general trend of modern mathematics towards greater abstraction. Even elementary algebra is in its own way abstract enough. If we have two sets or aggregates, say a bunch of bananas and a heap of wheat sacks, whose elements can be matched with the fingers of one hand, we represent the plurality of both by the same symbol, the number 5. For purposes of calculation the difference between bananas and wheat sacks is of no importance and what is common to both, *viz.* their plurality, is expressed by one and the same symbol—the number 5. On this ground, we do not claim that elementary algebra justifies the view that the 'true' nature of bananas and wheat sacks is unknowable and that all that we can know about them is the possibility of matching them on an equivalent set of 'indefinables'.

Why then should we accept Eddington's claim that the 'super-algebra' of Group Theory justifies the view that physical science can give us no communicable knowledge of space, sensations, *etc.*, other than that of the inner structure of a group of 'indefinables' equivalent to the group of rotations, sensations, etc.? Take, for instance, the concept of the two-dimensional spherical space discussed above. All the qualities of this space, its 'thereness', 'curvature', *etc.*, are necessary before we can even define the group of rotations and the groundwork of their internal relationship, *i.e.* their internal structure. *After* we have done this, we may construct an equivalent group of 'indefinables' with the same structure, just as we may construct an equivalent set of 'indefinables' with the same plurality to count concrete sets like bunches of bananas and wheat sacks; but that gives us no right to claim that physical science tells us no more about space than the fact that its group structure is the same as that of a group of 'indefinables'—'we know not what'. On the other hand, we actually use all the qualities of space, which a logician may subsequently dismiss as an 'unauthorised embellishment' in his formulations, just to discover that its group structure is that of a group of 'indefinables' which we literally create afterwards on the pattern of the structure of the group of rotations. It seems that confusion is caused by the use of the word 'indefinable' by 'super-algebraists' of the group theory. 'Indefinable' of the 'super-algebraists' is not the ineffable of the mystics—a mysterious something that is too deep for words and cannot be defined. It is simply the *undefined* that is not pinned down to anything concrete in order to create an instrument of analysis that may be as widely applicable as possible. In other words, we leave the mystery element x of our group deliberately *undefined* in order to secure the widest possible generality in our theory. But when we apply it to any particular case as, for instance, to space, we do define rotations with all the 'embellishments' of space before we can even know that they form a group. Having ascertained this, we neglect these 'embellishments' and apply the algebra of group theory to develop a mathematical theory of spherical space.

It happens that the application gives us a 'picture' of space very much like a navigator's chart which is a *representation* of the actual course of his ship through the oceanic wastes. The mystic is so fascinated with the picture he creates that like Narcissus he can do nothing but gaze at it. His tragedy is that in his state of self-hypnotism he cannot 'kiss the lips he so ardently desires because they are his own'. It seems that in this crisis-ridden world of today Narcissism is one way of escape for sensitive souls anxious to avoid the raging conflicts that threaten to engulf them. Eddington's mysticism is an expression of this tendency* in science like the work of the symbolists in art and literature. In fixing his gaze on the symbolic world of physics, the abstract structure of the group of sensations, Eddington reminds one of the literary Narcissus, André Gide, who, in his *Les Nourritures Terrestres*, and other works is so busy looking at himself that he himself becomes just vision: 'Et même moi, je n'y suis rien que vision.'

*   *   *   *

Another variety of mysticism with a strong flavour of modern science sprang up during the first two decades of the twentieth century. Following the success of Einstein's theory of Relativity, Minkowski put forward the metaphysical view that henceforth space and time were to have no separate

* Eddington is fully aware of this link between himself and the artist. He says that the artist may partly understand his statement that our account of 'the external world is a "Jabberwocky" of unknown actors executing unknown actions.' In fact, this is also Eddington's explanation of the 'Jabberwockies' that we see hung in Art galleries. *New Pathways in Science*, page 256.

significance, but were to be only shadows of a 'higher' unity transcending both. Basing himself on the Minkowski dictum of space-time unity, Alexander evolved a hierarchy of categories beginning with Existence, Relation, Order, *etc.*, and ending ultimately with Deity as the final end-product of the universe and not the origin, as is usual in orthodox theologies. In spite of magnificent work, the scientific support that Alexander* sought from Relativity physics for his metaphysical views on matter, mind, Deity, *etc.*, was gratuitous. A. N. Alexander had several imitators of lesser calibre, who, exploiting the great difficulty involved in understanding Relativity mathematics, let loose a deluge of mystical theories of God, World-Will, World-Intelligence, *etc.*, all claiming to emanate from Relativity equations. In order to show how unwarranted is the support claimed for these views from relativity physics, we shall explain the ideas underlying the mathematical apparatus used by Einstein in formulating his Relativity Theory.

Einstein's theory is geometrisation of physics. Now, geometry deals with 'space', and space, as we saw earlier, may be of one, two, three or more dimensions. It is true that our actual physical space is three-dimensional, but 'pure' geometry may deal with space of any number of dimensions. During the nineteenth century mathematicians developed a generalised 'pure' geometry of n-dimensional 'space'. In formulating his special Relativity Theory, Einstein used this geometry for the special case $n = 4$. As we are concerned here only with the ideas underlying this geometry, we shall confine ourselves to two-dimensional space such as the plane of paper or the surface of a spherical globe. The four-dimensional geometry actually used by Einstein is no doubt more complicated, but all the essential ideas underlying it can be more simply explained in terms of the two-dimensional case.

Let us consider a two-dimensional space such as the plane of paper or the surface of a spherical globe. If we wish to develop its geometry, we shall naturally have to talk about points which, in their totality, make up that space. Now, if we desire to talk about the individuals of any aggregate, such as the employees of a mill, we give each one of them a name. For certain purposes, we also assign them numbers such as ticket numbers, provident fund numbers, ration card numbers, *etc.* Each individual then has a name and a set of numbers each one of which identifies him and thus serves as a label for him. To have complete information about our employees, we require a table showing against each individual his respective ticket number, provident fund number, ration card number, *etc.* Knowing any *one* of these particulars, we can ascertain every other particular of the individual in question by consulting our table. This procedure is, however,

* A. N. Alexander: *Space, Time and Deity*, 1928.

workable only if the number of individuals in our aggregate is finite. But if the number of individuals in the aggregate is infinite and if we want complete information about every one of them, it is of no help. For, obviously, we cannot construct a table with an infinite number of entries even if we could find the time to scan its never-ending columns. If we want to know all the other particulars of an individual, given any one of them, we shall have to replace our table of individual entries by a set of general rules. For instance, we may lay down the rule that a ration card number of an employee shall always be twice the number of his ticket and thrice

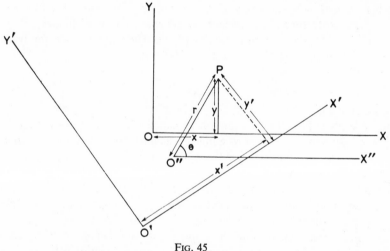

FIG. 45

his provident fund number. Knowing these rules, we can deduce all other relevant numbers pertaining to any individual provided we are given any *one* of his specification numbers.

The need for such general rules in geometry is all the greater as each one of the elements in the aggregate, that is, the points in space, can have an infinity of registration numbers, the counterparts of ticket numbers or ration card numbers of our illustration. For example, for certain purposes we measure the distances x, y of a point P from two perpendicular lines in a plane and take them as a label or registration mark for the point P. For other purposes, we may measure similar distances x', y' from *another* set of axes and take the new numbers x', y' as another registration mark for P. For still other purposes, we may take the distance r of P from a fixed origin O'' and the angle θ which $O''P$ makes with a given direction. This gives yet another label or registration mark for the same point P and so on *ad infinitum*. (See Fig. 45.) In such cases, we need a neat method of deriving all kinds of registration marks of P, given any one of them

Obviously we cannot use the tabular method; we have to specify general rules whereby we can derive any kind of registration mark, say (x', y') of P from a knowledge of any other, say (x, y). Such general rules are known as rules of transformation as they enable us to transform one kind of label or registration mark into those of another kind. The most general rule of transformation of the co-ordinates (x, y) of P into the co-ordinates x', y' of another kind is

$$x' = f(x, y),$$
$$y' = \varphi(x, y),$$

where f, φ are any functions of x, y whatever. If we know the functions f, φ, the above formulae enable us to calculate the co-ordinates (x', y') of any point, given its co-ordinates (x, y). The simplest rule for transforming x, y into x', y' is given by the formulae,

$$x' = a_1x + a_2y$$
$$y' = b_1x + b_2y$$ . . . . (1)

where the coefficients of x, y, *viz.* the numbers a_1, a_2, b_1, b_2 are presumed to be known. This pair of equations is called a linear transformation.

We have shown above how a linear transformation may arise when we try to express the co-ordinates (x', y') of a point P with respect to a new set of axes in terms of its co-ordinates (x, y) with respect to the old axes. In other words, we transform the co-ordinate axes but keep the point fixed to obtain a linear transformation. However, we could also generate a linear transformation by reversing the procedure, that is, by keeping the axes fixed and transforming the point P into another point P' by some process. A simple way of producing a linear transformation is by means of reflection. Here we keep the axes fixed and transform each point P into its image P' reflected in a mirror placed along any given line through the origin such as the x-axis. Since the image of any point lies as far behind the mirror as the point is in front, a glance at Fig. 46 shows that if the co-ordinates of any point P are (x, y), then those of its image P', *viz.* (x', y') are given by the equations:

$$x' = x = 1.x + 0.y$$
$$y' = -y = 0.x - 1.y$$

Obviously, this transformation is only a particular case of the more general linear transformation (1), when $a_1 = 1$, $a_2 = 0$, $b_1 = 0$ and $b_2 = -1$.

Likewise, we may transform each point on the surface of an elastic material, such as India rubber or steel plate, by stretching, bending, *etc.* In many cases the transformations produced in this manner are also linear. In whatever manner we may evolve a linear transformation—whether by transforming the co-ordinate axes or by reflection, stretching, *etc.*—the important point is that it is expressed by equations of the same form as (1).

So far we have confined ourselves to points in two-dimensional spaces such as the plane of paper in which we have drawn our diagram. But we can easily extend the notion of linear transformation to three-dimensional space and even to hyper-spaces of more than three dimensions. In order to

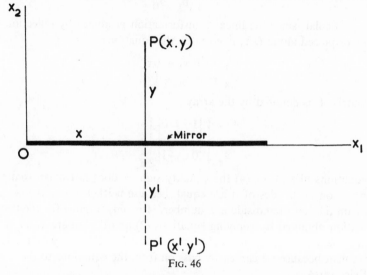

Fig. 46

be able to point out more easily the corresponding extensions to hyper-spaces of three, four, ... and n dimensions, we shall change our notation and denote the co-ordinates x, y of our point P by the symbols x_1, x_2. The co-ordinates of a 'point' in n-dimensional space would then be x_1, x_2, x_3, ..., x_n. Reverting now to linear transformations in two-dimensional space we may write the formula for transforming x_1, x_2 into x'_1, x'_2 as

$$x'_1 = a_1 x_1 + a_2 x_2$$
$$x'_2 = b_1 x_1 + b_2 x_2 \qquad \cdot \qquad \cdot \qquad \cdot \qquad \cdot \qquad (2)$$

This linear transformation is completely determined if we know the four coefficients or numbers, a_1, a_2, b_1, b_2. We may denote it by the symbol A or by the combination (a_1, a_2, b_1, b_2) so long as we know where to put each coefficient in writing out the equations of transformation (2). This notation is, however, inconvenient when we deal with transformations in n-dimensional space. So we write it in the form of a square array that we obtain by erasing every symbol in the equations (2) except the coefficients. This gives us the square array

$$\left\| \begin{array}{cc} a_1 & a_2 \\ b_1 & b_2 \end{array} \right\|.$$

This square array is known as a *matrix* and completely defines the transformation A. We express this by writing

$$A = \left\lVert \begin{matrix} a_1 & a_2 \\ b_1 & b_2 \end{matrix} \right\rVert \qquad \cdot \qquad \cdot \qquad \cdot \qquad \cdot \quad (3)$$

In particular, since the linear transformation produced by reflection in a mirror placed along OX_1 is given by the equations:

$$x'_1 = 1 . x_1 + 0x_2,$$
$$x'_2 = 0 . x_1 - 1x_2,$$

its matrix A' is denoted by the array

$$A' = \left\lVert \begin{matrix} 1 & 0 \\ 0 & -1 \end{matrix} \right\rVert . \qquad \cdot \qquad \cdot \qquad \cdot \qquad \cdot \quad (4)$$

In equations like (3) or (4) the equality sign $=$ does not mean that the numbers on both sides of it are equal, because neither A nor the square array on the right-hand side is a number. A is only a name for the transformation obtained by applying equations (2) and the square array is an arrangement of writing the coefficients involved in these equations. It is convenient because we can easily pass on from the equations to the array and vice versa.

Now suppose we transform the co-ordinates from x_1, x_2 to x'_1, x'_2 by a linear transformation A given by the matrix (3), and transform x'_1, x'_2 into x''_1, x''_2 by another transformation A' given by the equations

$$x''_1 = a'_1 x'_1 + a'_2 x'_2$$
$$x''_2 = b'_1 x'_1 + b'_2 x'_2 \qquad \cdot \qquad \cdot \qquad \cdot \qquad \cdot \quad (5)$$

The matrix of the new transformation A' is

$$A' = \left\lVert \begin{matrix} a'_1 & a'_2 \\ b'_1 & b'_2 \end{matrix} \right\rVert \qquad \cdot \qquad \cdot \qquad \cdot \qquad \cdot \quad (6)$$

What is the matrix of the transformation A'' which converts x_1, x_2 directly into x''_1, x''_2 instead of through the intermediary of A'? The resultant A'' of two successive transformations A *followed* by A' is called the product of A and A' and is symbolically written as

$$A'' = A' . A.$$

Here again A'', A', A are *not* numbers but mere names for the three transformations specified above. By the product of any two transformations A', A, we understand the final effect of applying *first* A and then A'.

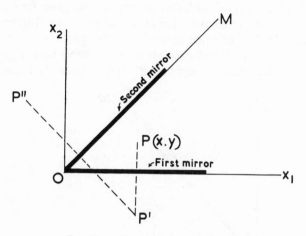

FIG. 47A—A'. A transforms P into P''.

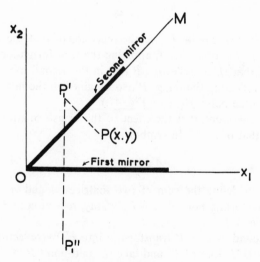

FIG. 47B—$A.A'$ transforms P into P_1''.

In general, the product of any two transformations such as A and A' depends on the order in which they are applied. Suppose, for instance, we place two mirrors, one along OX_1 and the other along any other line through O such as OM. Let reflection in the former be the transformation A and that in the latter the transformation A'. If we apply to any point P the transformation A first, we get its image P'. Applying now the transformation

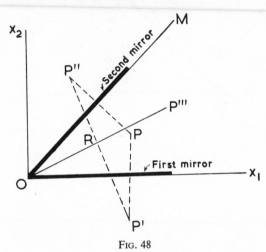

FIG. 48

A' to P', we obtain the point P'', *viz.* the reflection of P' in the second mirror along OM. On the other hand, applying the transformation A' first and then A to P, that is, by reflecting P first in the second mirror along OM and then by reflecting the image P'_1 so obtained in the first mirror along OX_1, leads to the point P''_1. (See Fig. 47.)

It follows, therefore, that the effect of the transformation $A'.A$ is *not* the same as that of $A.A'$. In symbols,

$$A'.A \neq A.A'.$$

We can also define the sum of two matrices A and A'. Let the first matrix A transform a point P into P' by, say, reflection in a mirror along OX_1. (See Fig. 48).

Let the second matrix A' transform P into P'' by reflection in another mirror along OM. Join P', P'' and take the mid-point R of $P'P''$. Join OR and produce it to P''' so that $RP''' = OR$. It can be shown that the transformation which changes P into P''' directly is also linear. Its matrix is known as the sum of the two matrices A and A'. It is obvious that we reach the same point P''' whether we reflect P first in the mirror along OX_1 and then in that along OM or vice versa. In other words, the sum (unlike the product) of the two matrices A, A' does *not* depend on the order in which they are added. In symbols,

$$A + A' = A' + A.$$

With these definitions of the sum and product of matrices we can develop an algebra of matrices in much the same way as that of sets or truth values of propositions. This algebra has manifold applications from the theory of elastic strains, games and economic behaviour to that of splitting atoms. The reason is that matrix is a neat way of mathematising operations of the most complex kind or writing several sets of magnitudes. For instance, the railway engineer has, to ensure the safety of trains, to figure out under what strains rails fracture. This leads him to the study of stresses and

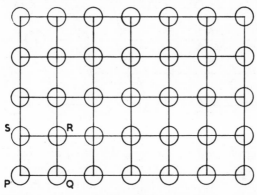

Fig. 49

strains in material bodies. Now the strains set up in a steel rail when a train passes over it can be specified by a matrix. What actually happens is that in its free unloaded state, the molecules of the material of the rail are arranged in small squares very much as shown in Fig. 49. When the rail is strained by the passage of a train the square arrangement of the molecules is disturbed. The little squares such as $PQRS$ in Fig. 49 become little parallelograms like $P'Q'R'S'$ as in Fig. 50.

If we keep the co-ordinate axes fixed, the transformation of points like P, Q, *etc.* into P', Q', *etc.*, may be denoted by a pair of linear equations like (1). (See page 160.) In other words, the state of strain produced by the moving train may be denoted by the matrix of a linear transformation.

Matrix algebra has also been used to rescue the classical atomic theory from the cul-de-sac into which it found itself about thirty years ago. According to the classical theory an atom is a miniature planetary system in which a number of electrons continually dance round a central nucleus. On the basis of such a model and with the help of certain *ad hoc* modifications of the classical laws of electrodynamics, it was possible to explain some of the experimental observations. But the theory soon reached the limits of its achievements and proved powerless to account for a number of other observations. This failure of the theory led some physicists to the conclusion that the trouble was perhaps due to the fact that the theory took too much for granted.

The theory operated with such concepts as the position and velocity of electrons, their orbits and so forth and then applied its conclusions to interpret experimental observations. And yet what is actually observed is

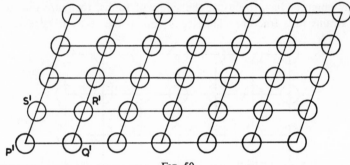

Fig. 50

never the electrons and their orbits, velocities, *etc.*, but the radiations emitted by a whole aggregate of atoms. These radiations are nothing but light waves, even though many of them cannot be seen in the ordinary way. Like light waves of different colours—or radio waves for that matter— each kind of radiation has its own characteristic wave length and intensity. They are actually observed by examining the radiation by means of a spectroscope.

It therefore happens that the classical theory has recourse to the concepts of motion, position and velocities of electrons, while what experimental observations yield is wave lengths and intensities of the radiations emitted by manifolds of excited atoms. This would seem to show that the basic concepts of the theory are of no account for the final result. One may therefore be tempted to examine how far one could go if the classical model were to be abandoned. Now if one were to construct a theory of atomic behaviour without invoking the assumed but unobservable electronic motions within the atom, one could hardly do without matrices and matrix algebra for the following reason.

When we examine the pattern of radiation emitted by an aggregate of atoms—and this is all that observation can do—we may conceive of each radiation of one wave length as emanating from some single atom of the aggregate at any one time. The pattern of radiations emitted by an aggregate of atoms, and through it indirectly the state of the entire aggregate of atoms, may thus be described by a succession of numbers giving the characteristic wave length and intensity of each kind of radiation that the aggregate may possibly emit. Such a succession of numbers may be arranged as a matrix. A matrix therefore is one way of describing the state of affairs prevailing within a manifold of atoms, if we wish to avoid reference to quantities which, like the position and velocity of an electron, are in principle unobservable. That is why matrices and matrix algebra have helped advance the classical atomic theory beyond the range of its original validity.

This success of matrix algebra in solving a number of problems of quantum physics has been used to justify the view that the world of physics is merely a 'shadow-world' of metrical symbols. But with the progress of time these new theories, too, have got into as great—though quite different —difficulties as the older quantum theory. New phenomena revealed by the study of the atomic nucleus have forced a multitude of *ad hoc* amendments of the theory, with the result that it has again become a mere medley of assumptions without any internal consistency. Consequently the 'scientific' basis of the mystical superstructure that was being raised on it is spurious.

*   *   *   *

Suppose now we have two equations

$$
\begin{aligned}
x'_1 &= a_1x_1 + a_2x_2 \\
x'_2 &= b_1x_1 + b_2x_2
\end{aligned}
\qquad \cdot \quad \cdot \quad \cdot \quad \cdot \quad (7)
$$

Can we express x_1, x_2 in terms of x'_1, x'_2? Under certain conditions, we can. For we may now treat these equations inversely as two simultaneous equations in two unknowns x_1, x_2 and solve them in terms of a's, b's and x'_1, x'_2. This gives

$$
x_1 = \frac{b_2x'_1 - a_2x'_2}{a_1b_2 - a_2b_1}, \; x_2 = \frac{a_1x'_2 - b_1x'_1}{a_1b_2 - a_2b_1}.
$$

With two equations the algebra is not complicated but if we have three or four equations, it begins to become increasingly tedious. With a still larger number of equations, it is completely unmanageable. If, therefore, we wish to transform co-ordinates in a generalised space of n dimensions, we must devise a neater way of solving simultaneous equations in n variables. This neater way is by the use of *determinants*. In elementary arithmetic if we are given a number like $(2 \times 7 - 3 \times 4)$, we simplify it by actually carrying out the multiplications and other arithmetical operations involved in the expression and write the result as 2. In higher algebra, we often find it more convenient not to perform these operations or at least to postpone them till the very end. A determinant is an expression of this kind in which the numbers involved are to be combined in a certain way if we want its exact value but may with advantage be left untouched. The above expression can be written in a determinant form

$$\begin{vmatrix} 2 & 3 \\ 4 & 7 \end{vmatrix}$$

in which the arrows link up the numbers to be multiplied. This is only another way of writing the number 2, which is derived from it by multiplying the first number, 2, by its diagonally opposite number, 7, and subtracting from it the product of the other two diagonally opposite numbers, viz. 3 and 4. In other words,

$$\begin{vmatrix} 2 & 3 \\ 4 & 7 \end{vmatrix} = (2 \times 7 - 3 \times 4) = 2.$$

In practice, the arrows are omitted.

In general,

$$\begin{vmatrix} a_1 & a_2 \\ b_1 & b_2 \end{vmatrix} = (a_1 b_2 - a_2 b_1)$$

Another way of performing the same calculation is this. Fix your attention on the numbers in any row, such as the numbers 2 and 3 in the first row. Take first the number 2 in the first row. Equally it occurs in the first column also. If you delete all the numbers in the first row *and* first column in which the number 2 occurs, you are left with only one number, *viz.* 7

$$\begin{vmatrix} 2 & 3 \\ 4 & 7 \end{vmatrix}$$

Form the product 2×7. It gives 14.

Now take the second number in the first row, *viz.* 3. As before, erase all the numbers in the first row *and* second column, that is, the row and column in which the number 3 occurs.

$$\begin{vmatrix} 2 & 3 \\ 4 & 7 \end{vmatrix}$$

This leaves the number 4. Again form the product 3×4. It gives 12. The determinant then is the difference between the two products

$$14 - 12 = 2.$$

The advantage of this apparently involved way of performing the calculation over that previously given is that it is applicable to higher order determinants. For example, a third-order determinant is a square array with three rows and three columns such as

$$\begin{vmatrix} 2 & 1 & 3 \\ 3 & 4 & 2 \\ 1 & 1 & 2 \end{vmatrix}$$

This means that the numbers like 2, 1, 3, 3, *etc.*, written above, are to be combined according to a certain rule. This rule is a simple extension of that for second-order determinants already described.

The great utility of these fanciful expressions is that they enable us to give a neat solution of simultaneous linear equations, or at least a neat way of writing it out even if we do not get the actual numerical values in any given case without a lot more tedious computations. For instance, in the case of equations (7), page 167, *viz.*

$$x'_1 = a_1 x_1 + a_2 x_2,$$
$$x'_2 = b_1 x_1 + b_2 x_2,$$

we found that

$$x_1 = \frac{b_2 x'_1 - a_2 x'_2}{a_1 b_2 - a_2 b'_1},$$

$$x_2 = \frac{a_1 x'_2 - b_1 x'_1}{a_1 b_2 - a_2 b_1}.$$

We may also write this solution in terms of determinants as under

$$x_1 = \frac{\begin{vmatrix} x'_1 & a_2 \\ x'_2 & b_2 \end{vmatrix}}{\begin{vmatrix} a_1 & a_2 \\ b_1 & b_2 \end{vmatrix}}, \; x_2 = \frac{\begin{vmatrix} a_1 & x'_1 \\ b_1 & x'_2 \end{vmatrix}}{\begin{vmatrix} a_1 & a_2 \\ b_1 & b_2 \end{vmatrix}}.$$

When we have to solve linear simultaneous equations in three or more unknowns it is possible to write the solution in determinant form by a simple extension of the procedure adopted in the case of two unknowns. The problem may therefore be regarded as completely solved thereby. But this solution is a mere figure of speech when the number of simultaneous equations to be solved is really large, say fifty. For the actual values of fifty unknowns of a set of fifty linear equations are buried as deep under the dead-weight of fifty-order determinants as in the original fifty equations. The reason is that it is not a simple matter to evaluate a determinant of order fifty and you would have to evaluate fifty-one of them before you could get at the actual values of the fifty unknowns in question. And yet fifty unknowns is no large number.

As we saw in Chapter 4, Southwell's relaxation methods reduce the problem of solving a differential equation to that of solving a set of algebraic equations involving the unknown values of the wanted function at a number of its nodal points. But a function could hardly be said to be known unless we knew its values at fifty or more nodal points. Consequently, algebraic equations involving fifty or more unknowns are of frequent occurrence nowadays in the solution of various problems in physics and engineering. A theoretical solution in determinant form is not of much help here and a practical method for pencil and paper work is provided by Southwell's relaxation technique.

Although it would be possible to design a special machine to carry out this process, Hartree considers that Southwell's technique is not very suitable for a general-purpose machine. He has therefore suggested another which is more suitable for the general-purpose calculating machines already constructed or under construction. The idea underlying Hartree's method is very simple. Taking the two equations (7) in two unknowns x_1, x_2 we have

$$a_1 x_1 + a_2 x_2 - x'_1 = 0,$$
$$b_2 x_1 + b_2 x_2 - x'_2 = 0.$$

If we square and add the expressions on the left-hand sides we obtain a quadratic expression

$$(a_1 x_1 + a_2 x_2 - x'_1)^2 + (b_2 x_1 + b_2 x_2 - x'_2)^2 = 0$$

which vanishes for just those values of x_1 and x_2 which are solutions of the equations (7). But since the quadratic expression is a sum of two square terms it cannot be negative. Consequently the *minimum* value of this quadratic expression can only be zero, which it assumes for just those values of x_1, x_2 which are solutions of (7). The problem of solving equations (7) is therefore reduced to that of minimising the value of the aforementioned quadratic expression which is more suitable for a general-purpose calculating machine than Southwell's relaxation technique.

This is one instance of the way in which modern methods of computation as developed by Hartree, Southwell and others are leading to a new revolution in mathematics to which we alluded in Chapter 1. The older point of view was to regard the minimising of a quadratic form *as reducing to the* solution of a set of simultaneous algebraic equations. But in view of the computing facilities now available it is more feasible to think the other way round, that is, to regard the solution of the simultaneous equations being reduced to the minimising of a quadratic form. In a similar manner the older method of solving a differential equation was to approach the solution through a sequence of functions each of which satisfies the equation but *not* all the initial or boundary conditions. Again, in view of the computing facilities now provided by the automatic calculating machines, the final solution is approached through a sequence of functions each of which satisfies all the boundary conditions but *not* the differential equation. The reason is that many forms which appear very sophisticated according to the older point of view are very suitable for the new methods of numerical computation by calculating machines, many of which can perform about a million multiplications an hour.

*   *   *   *

The theory of matrices and determinants is a highly developed branch of algebra, and has been extensively applied to the theory of equations, geometry and, as we have seen, more recently to quantum theory of the atomic nucleus. It has also been made to carry grist to the mystic mill. The mystic tries to find support for his views in the occurrence of certain invariant expressions in physical theory. An invariant, as the name indicates, is something that does not vary. In a world of change and flux, 'something' that does not vary or remains 'eternal' is apt to be equated to a godhead; but mathematical 'invariants', as we shall see, are by no means such exalted beings. Without a mystic's intuition they would at best look rather like stuffed images of 'immutability'.

The fact of the matter is that when we apply algebra to geometry by assigning co-ordinates to various points of the space or surface under study, we can also express certain inherent or intrinsic properties of configurations of its points by means of algebraic expressions involving their co-ordinates. For instance, suppose we have two points P, Q in a plane with co-ordinates (x, y) and (x', y') with respect to two rectangular axes. The distance, s, between P and Q can be easily expressed in terms of the co-ordinates of P and Q. If we apply Pythagoras's theorem to the right-angled triangle PQL (see Fig. 51), we obtain the equation

$$S^2 = PQ^2 = PL^2 + LQ^2$$
$$= (x' - x)^2 + (y' - y)^2,$$

or, $\qquad S^2 = (x' - x)^2 + (y' - y)^2$.   .   . (1)

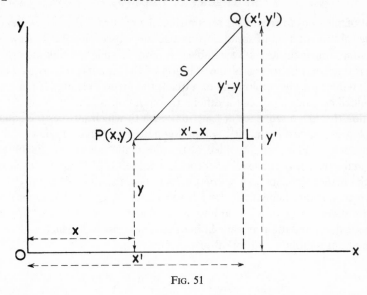

FIG. 51

The distance formula (1) is thus merely the well-known Pythagorean theorem translated into the language of algebra. But since the distance PQ is an intrinsic property of the points themselves and does not in any way depend on the co-ordinate axes used to assign co-ordinates to P and Q, it is obvious that we shall get the same value for S if we use any other system of rectangular axes. Suppose we use another set of such axes and suppose the new co-ordinates of P and Q are (x_1, y_1) and (x'_1, y'_1). They are, of course, quite different from the old co-ordinates (x, y) and (x', y'); but the value of the expression $(x'_1 - x_1)^2 + (y'_1 - y_1)^2$ is exactly the same as that of the old expression $(x' - x)^2 + (y' - y)^2$, as both measure the square of the same distance PQ. Such an expression is called an *invariant* as it retains the same value so long as we use any system of rectangular co-ordinates.

In this case the expression remains invariant only for a rather special class of co-ordinates—the rectangular co-ordinates—but we can also derive similar algebraic expressions which retain the same value however we may change the co-ordinate axes. The idea may be illustrated by an analogy. To express a thought or an idea, we need a language but there are many kinds of languages. Although every expression of an idea in a language is often charged with linguistic and emotional overtones peculiar to that language, it is, nevertheless, possible to say that the idea common to a number of translations of an expression is the 'invariant' kernel that remains however we may alter its linguistic outfit. In the same way analytical geometry formulates algebraic expressions in terms of co-ordinates for various geometrical magnitudes, such as distances between points, angles between lines, areas of figures, *etc*. However we may alter the co-ordinates, the final value of the expression remains invariant, as the magnitude expressed by it is independent of the co-ordinate system in much the same way as the idea underlying its linguistic expression in any language is independent of that language.

The discovery of 'invariants' was made almost as soon as algebra began to be applied to geometry. It caused no excitement then or afterwards. But the trouble arose during the first decade of the twentieth century when algebraic geometry began to be applied to physics. Physics is mainly concerned with events, such as the arrival of a light ray at a particular point, the explosion of an atom, the emergence of a nova, *etc*. To be able to talk about events in the same way as we talk about points in geometry, we need a method for assigning registration marks to these events. This is easy, for every event must take place somewhere and sometime. So, if we took the co-ordinates of the point of its occurrence and watched the instant of time at which it occurred, we would have a complete set of specification marks to identify the event. Every point or 'point-event', to use geometric terminology, has therefore a set of four numbers x_1, x_2, x_3, x_4 to identify it.

We may thus conceive of our eternally changing world of events as a static geometric manifold of 'point-events'. In dealing with this imaginary static manifold of 'point-events' by the methods of analytical geometry we encounter certain algebraic expressions, which, no doubt, involve the co-ordinates of point-events but nevertheless retain the same value however we may alter the co-ordinate system. This fact enables us to express certain physical laws, *e.g.* the law of gravitation, in invariant forms, that is, forms which stand like a rock among the shifting quicksands of co-ordinate transformations. These invariant forms of point-events are in principle no different from the invariant expressions for distances, angles, areas, *etc*., of the ordinary three or two-dimensional space, though they have quite unjustifiably been used to support mystical theories claiming to 'prove' 'mathematically' the existence of God, Deity, World-Will, World-Intelligence, *etc*.

*   *   *   *

An interesting application of the theory of sets, vectors, matrices and groups is the mathematical theory of games created by Neumann and Morgenstern in an attempt to provide a new approach to economic questions as yet unsettled. As they rightly point out, classical economics left out of consideration a vital element of political economy, that is, group rivalries and clash of interests. Of old, it has been known that whereas some economic policies may benefit everybody in more or less degree, certain policies would benefit one or more groups only at the cost of others. In spite of this conflict of various group interests, classical economics was dominated by the idea that if an individual were free to pursue his own good, he in some mysterious manner promoted the good of everyone else at the same time. Hence its advocacy of the Benthamite formula—the promotion of 'the greatest good of the greatest number'.

In an economy of the isolated Robinson Crusoe type it is, no doubt, possible for an individual (that is, Crusoe) to implement this formula, at least in principle, and direct his economic effort to maximise his own good. But even a Crusoe begins to get the creeps when he sees any sign of another will intruding on his domain as, in fact, Defoe's Crusoe did when he first saw the footsteps of the cannibals. Consequently, the Benthamite formula is practical economics for an isolated Crusoe but not for a participant in a real economy. For, while the former faces merely 'dead data' or 'the unalterable physical background of the situation', the latter has to face, in addition 'live data'—that is, the actions of other participants which he can no longer control and whose interests may even run counter to his own.

This consideration introduces a difficulty hitherto disregarded by classical economy. It is with a view to overcoming it that Neumann and Morgenstern have developed the mathematical theory of 'games of strategy', their object being the creation of theoretical models designed to play the same role in economic theory as the various geometrico-mathematical models have played successfully in physical theory.

Now, what is a game? In ordinary parlance it is a contest between a number of players, played for fun or forfeit according to some predetermined rules and decided by skill, strength or chance. Although a game may be played for mere fun or some non-monetary forfeit, such as Cupid played with Lyly's Campaspe,* for most econometric purposes it would do if it were assumed to be played for money or some such thing as 'utility' which we may suppose to be measurable. In most games played for monetary stakes such as Bridge, Poker, *etc.*, the algebraic sum of the gains and losses of all the players is zero. Such a game is called a *zero-sum* game. In other words, most ordinary games are zero-sum games wherein the play does not add a single penny to the total wealth of all players. It merely results in a new distribution of their old possessions.

A game theory of economic behaviour can therefore deal only with a

---

* John Lyly, *Campaspe*, III, v:

> Cupid and my Campaspe play'd
> At cards for kisses, Cupid paid;
> He stakes his quiver, bow, and arrows;
> His mother's doves, and team of sparrows;
> Loses them too; then, down he throws . . .

pure problem of distribution or imputation, that is, a problem wherein an economic group of Peters could only be paid by robbing a group of Pauls. In any real economy the actual situation is different. In fact, most economically significant schemes cannot be treated as zero-sum games at all for the sum of all the payments—the total social product—is in general not zero. It does not even remain a constant.

To take account of this important feature of social economy, Neumann and Morgenstern broaden the concept of a game wherein the sum of the total proceeds* of all players is *not* zero. This is done by proving that a non-zero sum game played by $(n - 1)$ persons is very closely related to a zero-sum game played by n persons. Consequently the case of a zero-sum n-person game is sufficiently broad to cover the general problem of social economy, *viz.* the problem of imputation with or without the creation of 'utility' during the process of play. We shall, therefore, confine our description to the case of a zero-sum game only.

The simplest case of such a game is the case of a single player. As he faces no opponent his task is the extremely simple one of maximising his own 'good', 'satisfaction' or 'utility'. At least it is so in theory. This case corresponds to a rigidly established dictatorship in which one unalterable scheme of distribution prevails and the interests of all the members of the society are assumed to be identical with those of the dictator. There is nothing further that the game theory can tell us in the solution of the imputation problem in this case.

Next in order of complexity is the case of a zero-sum two-person game. This case corresponds to a market wherein a single buyer 'bargains' with a single seller. The game becomes more complex when there are three or more players. The reason is this. While in a two-person game there is always a total clash of interests, in a three-person game there occurs a partial mitigation of this total clash. The mere existence of a third player opens up possibilities of coalitions and alliances by any two of them against the third. This case corresponds to dupolistic market wherein a buyer faces two producers of the same commodity. The possibilities of coalitions and alliances increase enormously as the number (n) of players increases.

Nevertheless, it is often possible to reduce a general zero-sum n-person game to the simpler case of a zero-sum two-person game. For consider a game played by n players denoted, for brevity, by the numbers 1, 2, 3, . . ., n. Let $I = (1, 2, 3, . . ., n)$ be the set of all these players. Let S be a sub-set of any players who decide to form a coalition and co-operate fully against the rest. Let S^* be the complement of S in I and let it be further assumed that the players in S^*, too, decide to play as a coalition in self-defence. The result is merely a zero-sum two-person game between the two coalitions

* These may be measured in terms of either money or utility.

S and S^*. That is why the case of a zero-sum two-person game is quite fundamental in the whole theory, and we shall now consider it in greater detail.

For the sake of simplicity we may visualise a zero-sum two-person game as a sequence of only two moves: a first move by player 1, followed by a second move by player 2. Let player 1's move consist in throwing a die which may turn up with any of the following six integers uppermost:

$$1, 2, 3, 4, 5, 6.$$

Let player 2's move consist in throwing a coin which has integers 1, 2 stamped on its two faces instead of the usual head and tail. Player 1 then chooses an integer i from the set of six integers 1, 2, 3, 4, 5, 6, by a throw of the die, and player 2 integer j from the set of two integers 1, 2, by a throw of the coin. In other words, i can assume any one of the six values 1, 2, 3, 4, 5, 6, and j any one of the two values 1, 2. Any actual play will give rise to an i and a j, the individual choices of the two players. It is clear that there will be in all $6 \times 2 = 12$ such pairs of i's and j's, each one of which is an outcome of some particular play. The entire set of 12 pairs of i's and j's covering the entire totality of all possible plays defines the game G as distinct from any particular play thereof. Now the rules of the game must also prescribe the payment functions K_1, K_2, that each player has to make to the other as a result of a play. K_1, K_2 can only be functions of the choices i, j that the two players make in the two moves permitted to them by the rules of the game. We may therefore write $K_1(j, i)$ as the amount that player 1 gets if the play results in a choice i by player 1 and a choice j by player 2. Likewise $K_2(j, i)$ is the sum obtained by player 2 for the same choice i, j. Since the game is zero-sum,

$$K_1(j, i) = -K_2(j, i).$$

We may express this by writing

$$K_1(j, i) = K(j, i), \quad K_2(j, i) = -K(j, i).$$

The course of the play will be determined by the desire of the first player to maximise K_1 or K and that of the second player to maximise K_2 or, what comes to the same thing, minimise K. Thus both concentrate on the same function K, one with the intention of maximising it and the other with that of minimising it. However, neither of the players is in a position to do anything in the matter as the moves of both are *chance* moves decided by the throw of a die and a coin.

To give some scope to the play of free will of the participants let us now assume that player 1 makes a *personal* move instead of a *chance* move. In other words, he selects a number i out of the six members 1, 2,

3, 4, 5, 6 by an arbitrary act of free choice instead of by a throw of a die. Likewise let the second player also select a number j out of the two numbers 1, 2 by a similar act of free choice. We have now to define the payment function $K(j, i)$ for each pair of (j, i)'s which specifies a particular outcome of the game. For instance, let us suppose that if player 1 chooses 5 and player 2 chooses 1, the former receives 2 coins, whereas if these choices are 5 and 2 respectively he pays 2 coins. In other words, we assume

$$K(1, 5) = 2 \text{ and } K(2, 5) = -2.$$

In the same way the values of $K(j, i)$ for another ten pairs of possible choices j, i can be assumed. We exhibit these assumed values of K function in the form of a rectangular matrix as follows:

j \ i	1	2	3	4	5	6
1	$K(1, 1)$ $= 1$	$K(1, 2)$ $= -8$	$K(1, 3)$ $= -3$	$K(1, 4)$ $= -9$	$K(1, 5)$ $= 2$	$K(1, 6)$ $= 7$
2	$K(2, 1)$ $= 3$	$K(2, 2)$ $= 9$	$K(2, 3)$ $= -1$	$K(2, 4)$ $= -7$	$K(2, 5)$ $= -2$	$K(2, 6)$ $= 8$

Now if player 1 moves first and makes his choice of an i, player 2, who now makes his move in full knowledge of his opponent's move, can so select his j as to minimise K. Thus suppose player 1 selects $i = 5$. Once he has chosen $i = 5$, he can receive only one of the two figures under the column headed 5 in the above matrix, *viz.* 2 or -2. Player 2, who knows that 5 has been chosen, will naturally choose $j = 2$ so as to allow him the minimum of these two figures. In other words, although player 1 is free to choose any i out of the six integers

$$1, 2, 3, 4, 5, 6,$$

he can expect to get only the minimum of the two figures shown in each of the six columns headed successively 1, 2, 3, 4, 5, 6. As a glance at the matrix would show, these six column minima are

$$1, -8, -3, -9, -2, 7.$$

Player 1 will therefore choose that column which contains the *maximum* of these minima. This is 7, which is the value of K when $i = 6$ and $j = 1$. In other words, a rational way of behaving in this case would be for player 1 to choose $i = 6$ and player 2 to choose $j = 1$. This will lead to player 1's gaining 7 coins at the cost of player 2.

Now it might seem unfair to let player 1 make the first move and player 2 the second move. To redress the balance between the two let us reverse their roles and see what happens. Player 2 has now to make the first move, that is, choose a j, and player 1 chooses an i in full knowledge of his opponent's move. If player 2 chooses $j = 1$, he restricts the value of player 1's gains to the figures in the first row of the K-matrix. He may then expect player 1 to choose an i which corresponds to the maximum figure in the first row. This maximum of the first row is 7. If he chooses $j = 2$, player 1 will choose the maximum of the second row which is 9. Consequently, it would be rational for player 2 to choose $j = 1$, in which case player 1 chooses $i = 6$. This again leads to player 1's gaining 7 coins (the matrix element corresponding to $i = 6$, $j = 1$) at the cost of player 2.

It would thus be seen that in this case it is immaterial who makes the first move. There is only one solution of the game which corresponds to a choice of $i = 6$, $j = 1$ and $K(1, 6) = 7$. In other words, in the case of figures tabulated in the aforementioned K-matrix, whether we first choose the *minima* of the columns and then choose the *maximum* of these *column-minima* or we choose the *maxima* of the rows and then choose the *minimum* of these *row-maxima*, the end-product is the same. When this is the case the game is said to be *strictly determined* as there is only one possible outcome with a definite value of K which one player has to pay the other. But such strictly determined games are exceptions. In the example cited above we artificially constructed the K-matrix so as to ensure that the maximum of the column-minima equals the minimum of the row-maxima. In other words,

Max. (Column-minima) = Min. (Row-maxima).

This is not the case in general. Consider, for instance, the K function defined by the following rectangular matrix.

i j	1	2	3	4	5	6
1	$K(1, 1)$ $= 1$	$K(1, 2)$ $= -1$	$K(1, 3)$ $= 1$	$K(1, 4)$ $= -1$	$K(1, 5)$ $= 1$	$K(1, 6)$ $= -1$
2	$K(2, 1)$ $= -1$	$K(2, 2)$ $= 1$	$K(2, 3)$ $= -1$	$K(2, 4)$ $= 1$	$K(2, 5)$ $= -1$	$K(2, 6)$ $= 1$

This means that if i, j are either *both* odd or both even, player 1 gains 1 coin from player 2, otherwise he loses the same amount to him. In this case the minimum of the first column is -1. In fact, the minimum of all the six columns is the same number -1. Hence the maximum of the column minima is -1. In other words,

$$\text{Max. (Column-minima)} = -1.$$

Likewise, the maximum of the first row is 1 and so also of the second row. Consequently the minimum of the row maxima is 1.

Or, $\text{Min. (Row-maxima)} = 1.$

Consequently, Max. (Column-minima) $\neq$ Min. (Row-maxima).

Here, priority in the first move is a very material factor in the outcome of a play. If it is player 1, he knows whatever i he may choose, his opponent will choose a j to make loss a certainty. Suppose he chooses $i = 3$, in that case player 2 chooses $j = 2$ and $K(2, 3) = -1$. Similarly, if he chooses $i = 2$, player 2 will choose $j = 1$ and $K(1, 2)$ is again -1. In other words, the mere fact that player 1 makes the first move ensures that $K = -1$, which means that he loses 1 coin.

The situation is altogether changed if player 2 has to make a first move. It is easy to see that whatever row number (j) he may choose player 1 can choose a suitable i to make $K = 1$. In other words, he is now certain to gain 1 coin. But in the abstract game G in which neither player knows the move of his opponent there is no strictly determined outcome and the payment function K will continually oscillate between the two values -1 and 1. In general, a zero-sum two-persons game is not strictly determined and the value of a play will vary within a range V_1 to V_2. In our example, V_1 was -1 and V_2, 1 and the value of the play oscillated between these two numbers.

A typical application of the theory of two-person games is the case of a two-person market which is equivalent to the simplest form of the classical problem of bilateral monopoly. Here let us equate player $1 =$ seller, player $2 =$ buyer.

First move by player 1: Choose a price p for his commodity A and offer it to player 2.
Second move by player 2: Accept or reject the offer.

In order that the transaction have any sense it has also to be assumed that the value (u) of the possession of A to 1 is less than the value v of its possession to 2. If the buyer accepts the price p offered, the amounts the two players get are

$$\begin{aligned} \text{Player } 1 &= p, \\ \text{Player } 2 &= v - p \end{aligned}$$

Since player 1 agrees to the sale at price p, p must exceed u, the pre-sale value of the commodity to him. Since the buyer agrees to buy, the value v must exceed the price p paid by him. In other words,

$$p \geqslant u \text{ and } v \geqslant p,$$

or
$$u \leqslant p \leqslant v.$$

Here u, v correspond to V_1 and V_2 of the general theory of the two-person game described above. The actual value of the play, that is the price p, oscillates between these two values. In this model we have assumed that the game is played in two moves giving rise to a single bid which is either accepted or rejected. But in actual practice a good deal of bargaining, haggling and negotiating takes place. Consequently, a satisfactory theory of this oversimplified model will have to leave the entire interval (u, v) free for p to move in.

This result could, no doubt, have been reached by common sense straightway without the long theoretical preamble preceding it. But the value of a theory is not judged by its treatment of simple and trivial cases but rather by that of complicated cases where common-sense considerations fail to lead anywhere. Neumann and Morgenstern believe that the mathematical theory of games is valuable in that it provides a possible theoretical foundation for econometrics without which it will remain no better than a morass of empirical or semi-empirical formulae or worse still, a pseudo-science of vague generalisations and half-truths.

But a serious limitation of the theory in view of sociological applications lies in the fact that the mathematical method suggested is not quite suited to the field to which it is being applied. It seems impossible to devise sufficiently realistic assumptions which can provide a basis for the mathematical super-structure. Even the basic assumption of the game theory that all players in the game are equally rational or intelligent is not really true in economics. If the struggle for livelihood is a game, it is too much like that of herding a majority of sheep by a minority of shepherds, to make Neumann's assumption realistic. An added source of complication is the fact that while some shepherds sometimes behave like sheep some sheep exhibit unusual intelligence. It is therefore likely that Neumann's theory may remain a mere mathematician's delight or at best be applicable to actual societies to a very limited extent.*

*   *   *   *

One of the most spectacular successes of group theory is in the field of differential equations which, as we have seen, is the heart of almost all

---

* For an interesting application of the theory of games to industrial statistics, see page 253.

applied mathematics. Whether we study the motion of pendulum bobs or cannon balls, of planets or galaxies, of fluids in pipes or river beds, of winds in laboratory tunnels or over aircraft wings in the sky, of tiny solid particles like sand and gravel suspended in moving water in river and harbour models, or of underwater missiles like depth charges, we have, in the last analysis, to solve a set of one or more differential equations. In many cases the situation is so complex that it may not be possible even to frame the appropriate differential equation or set of equations. But our ability to frame it by no means implies our ability to solve it. In fact, given a set of differential equations, the odds are heavy against its being amenable to any known treatment. That is why any new way of treating them is so valuable. Group theory is valuable as it is the master key that solves a large class of equations that can be solved in no other way.

In order to see this we shall first remark that a group may consist of a finite or infinite number of elements. As an illustration of a finite group we may cite the case of six shuffling operations x_1, x_2, x_3, x_4, x_5, x_6 by which we introduced the idea of group. As an illustration of an infinite group we have the group of all rotations of a sphere about any diameter by which we defined the two-dimensional space that is the surface of its spherical shell. Another instance of an infinite group is the infinite aggregate of scales on which we may measure any physical magnitude such as distance. Obviously any given length may be measured in any unit—miles, yards, feet, inches, kilometres, metres, centimetres, kos, versts, lis, leagues, light-years, parsecs, or any of their infinite aliquot parts, fractions or multiples. We can easily transform a length measured in any unit into any other by multiplying it by an appropriate conversion factor. Mathematically we can express the entire aggregate of operations whereby we convert distance in any unit into any other as follows:

Let S be the transformation of a length expressed in any unit, say yards, to another, say centimetres. If x is its measure in the former and x' in the latter and a the conversion factor,* then algebraically S, the operation of converting yards into centimetres is the transformation $x' = ax$. If now we considered another transformation S' of the length (x') expressed in centimetres to another x'' expressed in, say leagues, S' is the transformation $x'' = a'x'$ where a' is the new conversion factor appropriate for converting centimetres into leagues.† From x'' in leagues we could derive another transformation S'' which gives the measure x''' in some other units, say versts. It is clear that we could construct an infinity of such transformations $S, S', S'', S''', \ldots$ corresponding to an infinite number of values $a, a', a'', a''', \ldots$ of the conversion factor. The infinite set of transformations S, S',

---

* That is, one yard = a centimetres.
† That is, one centimetre = a' leagues.

S'', S''', ... whereby we are able to convert a measure in one unit to any other, possesses the group property as we shall now show.

Algebraically, all these transformations S, S', S'', ... may be subsumed under a single equation of the type $x' = ax$, where a is *any* real number other than zero. In other words, S is the transformation $x' = ax$, S' the transformation $x'' = a'x'$, S'' the transformation $x''' = a''x''$, S''' the transformation $x^{\text{iv}} = a'''x'''$, and so on. Now if we perform in succession any two transformations belonging to this aggregate, the end-product is again a transformation belonging to the aggregate. For example, let us combine the transformations S and S', then SS' is evidently the transformation $x'' = a'x' = a'(ax) = aa'x = \alpha x$, where α, the product of the conversion factors a, a', is again a real number. Consequently SS' is the transformation $x'' = \alpha x$ which by definition belongs to the same aggregate. The number a appearing in the transformation $x' = ax$ is called the *parameter* of the group. The transformation $x' = ax$ is therefore an *infinite, continuous one-parameter group.*

Now the link between group theory and differential equations is provided by an extension of an old Greek idea, the idea of geometrical similarity, ratio and proportion. For example, two similar triangles are identical except for the scale or conversion factor. One is a miniature replica of the other. We could derive all the dimensions of one from those of the other if we knew the value of this conversion factor. But what is true of similar triangles is true of every physical quantity. If we think systematically about the conversion factors needed to convert physical quantities from one system of units to another, we find that every physical quantity has certain 'dimensions' to be written as exponents. Take, for instance, velocity. It is simply the quotient of distance divided by time. If we represent the dimension of distance or length by L, and the dimension of time by T, the dimensions of velocity are L/T which may also be written as LT^{-1}. In other words, if we change the unit of length by the conversion factor a and of time by the conversion factor b, then the measure of velocity will have to be changed by the conversion factor ab^{-1}. Similarly acceleration, which is rate of change of velocity, that is, the quotient of increase in velocity divided by time, will have the dimensions of velocity, *viz.* (LT^{-1}), divided by T. Its dimensions will therefore be LT^{-2}. Consequently if we make the same change of units of length and time as before, the measure of acceleration will have to be changed by the conversion factor ab^{-2}.

Fourier was the first to show that by simple change of units it is easy to treat problems involving heat conduction in small spheres and in the earth by the same analytical formulae. For after all the earth is only a small sphere magnified by an appropriate conversion factor. Even though

Fourier's deductions went badly astray,* due to the neglect of the effect of the then unknown radioactive materials in the earth's core, Fourier's idea was basically right and suggested fundamental laws in the hands of Stokes, Savart, Reynolds, Rayleigh, Froude and others. The idea has now been vastly generalised and provides the basis for what is called dimensional analysis.

The essence of dimensional analysis is that any physical equation—differential or algebraic—that is, any equation involving physical magnitudes, must be dimensionally homogeneous. In other words, any such equation remains valid no matter in what units we may decide to measure the physical quantities appearing therein. In the language of group algebra the same thing is expressed more precisely by saying that every physical equation is invariant under the group of transformations employed for converting magnitudes from one set of units to another. Take, for instance, the differential equation of the motion of a pendulum bob we encountered in Chapter 3.

$$d^2x/dt^2 = -gx/l \qquad . \qquad . \qquad . \qquad . \quad (1)$$

If we use the transformations $x' = ax$ and $t' = bt$ to express the length x and time t in new units by means of the conversion factors a and b, the new equation involving new magnitudes x' and t' in new units will be exactly of the same form as the original equation in terms of x and t. Making the substitutions given above we may easily verify that

$$\frac{d^2x'}{dt'^2} = \frac{a}{b^2}\frac{d^2x}{dt^2}, \quad \text{or} \quad \frac{d^2x}{dt^2} = \frac{b^2}{a}\frac{d^2x'}{dt'^2} = \frac{1}{ab^{-2}}\frac{d^2x'}{dt'^2}.$$

To convert g, which is acceleration of gravity, into new units, we shall have to multiply it by the conversion factor ab^{-2}. Hence the value g' of gravity in new units will be $(ab^{-2})g$. Likewise x' and l' in new units will be ax and al respectively. The original equation (1) in old units x, t therefore becomes

$$\frac{1}{ab^{-2}}\frac{d^2x'}{dt'^2} = -\frac{g'}{ab^{-2}}\frac{x'}{a}\frac{a}{l'},$$

or
$$d^2x'/dt'^2 = -g'x'/l' \qquad . \qquad . \qquad . \qquad . \quad (2)$$

The new equation (2) is thus exactly of the same form as the old equation (1) except that the new magnitudes x', t', g', l' are now expressed in new units derived from the old by the conversion factors a and b. This simple fact—the invariance of physical equations under the group of

---

* This again shows the hidden pitfalls of 'pure' deduction. Even the most plausible deduction may go over the rocks due to the neglect of some unforeseen factor.

transformations of units—has far-reaching consequences. For if a differential equation is invariant under a group G of transformations then so must its solution be under G. This fact may be combined with other considerations, such as those of geometrical or physical symmetry, to solve differential equations. For example, if we could solve the differential equation (1) we could calculate the period p of a short swing of the pendulum bob.

But the group-theoretic approach just described enables us to make this calculation directly. For clearly p can depend only on g and l, the only physical quantities appearing in our equation. We may assume that it is a simple function of these magnitudes such as

$$p = Cg^m l^n \qquad . \qquad . \qquad . \qquad . \qquad (3)$$

where C is some constant, a pure number. If this solution is to remain invariant under the transformation $x' = ax$ and $t' = bt$, then the various magnitudes in new units denoted by a dash will be

$$p' = bp, \; g' = \frac{a}{b^2}g, \; l' = al.$$

Making this substitution in (3), we have

$$\frac{p'}{b} = C\left(\frac{b^2 g'}{a}\right)^m \left(\frac{l'}{a}\right)^n,$$

or

$$p' = Cg'^m l'^n \frac{b^{2m+1}}{a^{m+n}}.$$

But the invariance of the solution implies

$$p' = Cg'^m l'^n.$$

This means that $\dfrac{b^{2m+1}}{a^{m+n}}$ must be unity for all values of the conversion factors a and b. This can happen only if the indices of both a and b are zero. In other words, we must have

$$2m + 1 = 0 \text{ and } m + n = 0,$$

or

$$m = -n = -\frac{1}{2}.$$

Hence the period of the swing is $C(g^{-\frac{1}{2}} l^{\frac{1}{2}})$ or $C\sqrt{\dfrac{l}{g}}$, where C is some constant which we may determine by experiment. We may, for example, measure the period of a short swing of the bob when l is some known magnitude. The same value of C will then hold for every other length of the pendulum.

We have described in some detail the solution of the differential equation (1) by means of the group-theoretic approach, although it can also be solved in a more direct way which has the additional merit of determining the constant C as well. But the group-theoretic approach applies in more complicated cases where no other method is available. For instance, the differential equation of heat diffusion from a point in a medium of constant thermal diffusivity or the differential equations of stationary and non-stationary flows of compressible non-viscous fluids are amenable to group-theoretic treatment, while other methods fail or only partially succeed.

Group-theoretic approach owes its great power to the fact that equations of fluid motion remain invariant not only under the group of transformations of units considered above but also in many cases under wider groups of transformations such as the group of rotations and translations. This, however, is not all. The value of the group-theoretic approach in fluid mechanics does not only depend on its usefulness in solving the differential equations of fluid motions: it is also the unifying principle in innumerable questions of fluid mechanics—as indeed it has already proved to be in other branches of physics. For instance, it is the very core of modelling analysis whereby we use river, harbour, aerodynamical and other models to study actual fluid behaviour experimentally to bridge the gap between hydrodynamical theory and experience.

The logical foundations of dimensional and modelling analysis are already a very intricate study. Their fundamental assumptions have been explicitly stated by Bridgman and searchingly examined by Birkhoff. Some of these assumptions are still under debate, but beyond all dispute is the hypothesis underlying the group-theoretic approach, *viz.* that if the premises of a theory are invariant under a group G of transformations, then so are its conclusions. This axiom of group theory is really a mathematical re-formulation of the old philosophical principle that there is a hidden order in Nature. The mystics find in this a new confirmation of their belief that intuition is often more 'reliable' than intellect. But the hidden order that intellect discovers in nature only by prolonged and patient search is indeed a far cry from the vague visions of the unity of nature conjured by mystic intuition. The former is knowledge acquired after a patient unravelling of the complexity of the real facts of nature. This knowledge is thus quite literally power, that is, it confers power to change and tame raw nature to our service. On the other hand, the 'unity' of nature that a mystic sees in a flash of inspiration, ends in inane and passive contemplation of nature which on its own bases is not even communicable, or is almost incommunicable.

8

SPACE AND TIME

W HEN, in the time of the Pharaohs, the Nile annually overflowed its banks, there arose the social problem of re-marking the obliterated land boundaries after the flood waters receded. Thus originated geometry, the science of measuring land, and many of its rules were merely the result of empirical observations made by the ancient Egyptian land surveyors. A similar development took place more or less independently in almost all other ancient civilisations such as the Babylonian, the Chinese and the Indian, as a result of the activities of land surveyors and architects engaged in building palaces and altars. In time this empirical knowledge percolated to Greece, where it underwent a remarkable transformation. From being a haphazard collection of empirically derived rules of thumb, it became a deductive science of great aesthetic appeal. For it was discovered that all these rules, and many more besides, could be deduced by a process of logical deduction from a few simple propositions. Although this work of logical deduction was actually done piecemeal by several thinkers over a pretty long period, it was finally systematised by one man—Euclid. In his *Elements*, Euclid showed in detail how all the geometrical theorems could be logically deduced from a dozen odd axioms or postulates. If the truth of his postulates were granted, that of his theorems would automatically follow.

Now, there was no difficulty in accepting the truth of almost all his postulates because they appeared 'self-evident'. For instance, one of these postulates was that if equal magnitudes are deducted from equals, the remainders are equal among themselves. No one can seriously deny the truth of this proposition, and other postulates of Euclid were equally plausible and self-evident. But there was one exception which ultimately upset the entire Euclidean apple-cart. This was Euclid's fifth postulate about the behaviour of parallel lines, which signally failed to appear self-evident to his successors. For over 2,000 years after his day, mathematicians tried in vain to prove its truth, till about 120 years ago the Russian Lobachevsky and the Hungarian Bólyai showed indisputably that such a demonstration could not be given. They pointed out that no *a priori* grounds exist justifying belief in the disputed postulate and that an equally consistent system of geometry could be constructed by replacing it by its contrary.

Now what is Euclid's parallel postulate, which has been discussed so much by mathematicians for over two millennia? It may most simply be explained by means of a diagram (Fig. 52). Let AB be a straight line and P a point outside it. Naturally we believe that the straight line AB, if produced, would go on for ever without coming to an end. In other words, it is infinite. If we draw straight lines joining P to various points of AB, we

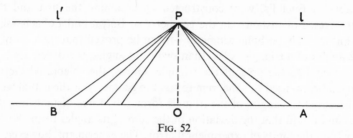

FIG. 52

get a pencil of lines radiating from P to points more and more remote from O. In the limit when one of the lines, say l, meets the line OA at infinity towards the right, we say l is parallel to OA. Similarly if we consider the points of the line towards the left, there will be another line l' through P which would meet the line OB at infinity towards the left.

Now Euclid assumed that both the lines l and l' would in effect be in one and the same straight line. In other words, he assumed that the angle between these two lines l, l' through P would be 180°. But Bólyai and Lobachevsky argued that it was not *logically* necessary that the limiting positions of the right and left parallels l, l' must be in the same straight line. The only reason for assuming it was that in ordinary diagrams such as we draw on paper the two parallels do appear to lie in one straight line. But suppose we drew a diagram on a really cosmic scale by taking P as the centre of Sirius and AB as the line joining, say, the Sun and Polaris. Will the left and right parallels in this case be in the same straight line, that is, enclose an angle of 180°? They may or may not do so. There is no *a priori* reason why they should. The argument of Bólyai's opponents, based on the behaviour of parallels in accurately drawn paper diagrams, is no more applicable than the argument that the surface of a continent must be flat merely because to all intents and purposes a football field appears flat.

Both Bólyai and Lobachevsky believed that a straight line in space extends to infinity in both directions. About twenty-five years later Riemann suggested that there is no logical reason why a straight line should necessarily be of infinite length. The character of the space in which we live might very well be that all straight lines return to themselves and are of the same length like the meridians on the surface of the earth. If we reject the usual assumption that the length of a straight line is infinite, we can draw no parallels to it from P as all straight lines drawn from P will intersect it at a finite point. Logically speaking, therefore, there are three alternatives. We may be able to draw none, one or two parallels to any given line from a point outside it.

Euclid assumed the second, Bólyai and Lobachevsky suggested the third and Riemann the first alternative. New geometries, logically as impeccable as Euclid's, were constructed by assuming the first and third alternatives. For instance, in the geometry of Bólyai and Lobachevsky—also known as hyperbolic geometry—it can be proved that the sum of the three angles of a triangle is less than two right angles, the deficiency being proportional to the area of the triangle. Gauss, the celebrated German mathematician, used this theorem to test the geometry of the actual world. He measured the angles of the triangle formed by three distant mountain peaks and found that the deviation of the sum of its angles from 180° was well within the limits of experimental error. The experiment, however, was inconclusive as the size of Gauss's triangle, though large compared to triangles drawn on paper, was small compared to the dimensions of the universe. But Lobachevsky concluded from the very small value of the parallaxes of the stars that the actual space could differ from Euclidean space by an extremely small amount. We shall revert later to the fundamental question, whether it is possible to determine by observation what system of geometry is valid for the actual space we live in. Meanwhile, we pursue in greater detail the further development of these new ideas in geometry.

That these developments took place at all was due in a large measure to the work of a French philosopher-mathematician, René Descartes, who liberated geometry from the slavery of diagrams. Before Descartes there was no way of developing a geometric argument except by drawing a figure, as every student of school geometry knows. But Descartes showed that geometry could be reduced to algebra. Instead of denoting points by dots and crosses, as in a geometrical diagram, we could designate them by their co-ordinates, that is, distances from a set of mutually perpendicular reference lines or co-ordinate axes. This algebraicisation of geometry by the introduction of co-ordinates showed that many inherent or intrinsic properties of geometric configurations could be expressed by invariant algebraic expressions. For instance, as we saw in the previous chapter, the distance formula

$$s^2 = (x' - x)^2 + (y' - y)^2 \qquad . \qquad . \qquad . \quad (1)$$

expresses the distance s between two points P and Q (see Fig. 53) in terms of their co-ordinates (x, y) and (x', y'). Since the distance is an intrinsic property of the points P and Q, irrespective of the choice of the co-ordinate system, this formula yields the same value for all rectangular co-ordinate axes with respect to which the co-ordinates of P and Q may be measured. Such an expression, which embodies the intrinsic property of a figure and is independent of the accidental selection of co-ordinate system, is known as an *invariant*.

The distinction between the intrinsic and extrinsic properties of a geometrical configuration may be explained by means of an analogy. If I say in English 'he brags' and in French 'il se vante' I express the same thought though the words used are different. But if I say 'brag' is a word whose letters when reversed form the word 'garb', I make a statement which will

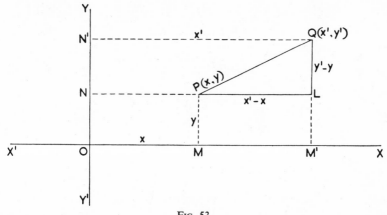

Fig. 53

not be true when translated into another language. Just as in any one language we have statements which are no longer valid when translated into another, so also we have in geometry statements or formulae which may be true in one co-ordinate system but not in others. For instance, the measure of the projection of PQ on the x-axis is not the same in different co-ordinate systems, though the distance PQ is. Naturally, therefore, invariant formulations, that is, formulae which embody the intrinsic properties of geometric configurations and consequently are independent of the accidental co-ordinate system chosen, are more important than those which depend on the particular system of co-ordinates arbitrarily chosen.

At this stage the question may be raised whether there is a large class of invariant properties of geometric configurations, such as curves, surfaces and spaces of three or more dimensions. The answer is that there are. Gauss and Riemann gave a general method for discovering them. This method depends on the use of a new system of co-ordinates since known as Gaussian co-ordinates. Till the beginning of the nineteenth century, the co-ordinates of a point on a curve, surface or space were its distances from the co-ordinate axes, or distances from arbitrary origins and angles with arbitrary reference lines. The inconvenience in the use of such co-ordinates is that, in order to get them, we have to pass out of the surface or curve on which our point lies. Very often we do not wish to, or cannot, move out of the surface under study, and it is desirable to have a method of generating co-ordinates without leaving the surface or the curve. In other words, we require a method whereby we can derive the co-ordinates of any point on our surface by means of measurements carried out without leaving the surface and moving into a third dimension or plenum into which the surface is embedded. Such, for instance, would be the case if we wished

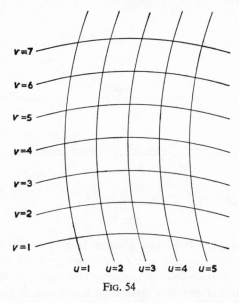

FIG. 54

to discover the nature of the earth's surface by measurements carried out thereon and not by digging into its interior or flying out into the atmosphere.

As a first step towards understanding Gaussian co-ordinates, we remark that, unlike Cartesian co-ordinates, the idea of distance is not essential to them. For instance, if someone enquired of you in Oxford Street, London, the location of Selfridges, you could indicate it by saying that it was at a distance of x yards from where you were. Or, alternatively, you could also specify it as the nth building from the one opposite which you stood. In the former case, the co-ordinate x of the building is its distance from a given point; in the latter, the co-ordinate n is only a label or identification mark like the registration number of a car or a building and does not stand for any magnitude.

A similar labelling device can be constructed in the case of a geometrical surface. Suppose we have a surface such as a sphere, an ellipsoid, or any other figure whatever. We could cover it with a system of curves such that no member of the system cuts any other. (See Fig. 54.) Let this system be called the *u*-system. Beginning with any curve we could consecutively label the *u*-curves $u = 1$, $u = 2$, $u = 3$, ... Similarly, we can draw another system of curves *v*, labelled $v = 1$, $v = 2$, $v = 3$, . . . such that no *v*-curve cuts any other of the system though every one of them cuts all the curves of the *u*-system. The *u*, *v*, systems of curves thus cover the surface with a

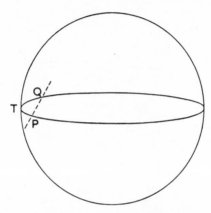

Fig. 55—The distance between any two points *P*, *Q* on a sphere may be measured in two ways. One is the shortest distance between *PQ* along the straight line joining *P*, *Q* and piercing the sphere at *P* and *Q*; the other is along the great circle arc *PTQ* on the surface of the sphere. We can assume the two distances to be identical when *P* and *Q* are infinitely close together.

network and each mesh of the net is labelled by the labelling numbers of the *u*- and *v*-curves which enclose it. These numbers do not represent distances, angles, or any other measurable magnitudes but are merely registration numbers or identification marks of the curves.

But such labelling does not suffice, and we have now to introduce some measure relation. Suppose we have a very *small* rigid measuring rod which we can use to measure distance *on* or along the surface. This is an important point and may be further elaborated. Suppose, for instance, we have a point *P* on a sphere. Let *Q* be another point close to it. (See Fig. 55.) The shortest distance between *P* and *Q* is the distance along the straight line *PQ* which pierces the sphere at *P* and emerges at *Q*. To measure it, we have to leave the surface of the sphere and *bore* into it. But if we wish to explore its intrinsic geometry by measurements carried out without leaving the surface we shall have to measure the distance *PQ* along the circular arc lying on its surface.* If the points P, Q are sufficiently close together, the two distances along the straight line PQ and arc PQ are equal at least approximately. By successively carrying out measurements in infinitely small regions of the surface, we can calculate this distance between any two points on the surface.

We have now the meshes of the network of the u- and v-systems of curves and a very small rigid measuring rod. We can use the measuring rod to measure the small meshes one after another and make a map similar in structure to the region of our surface. Suppose, for instance, we con-

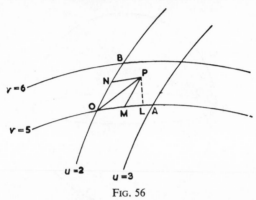

FIG. 56

sider a mesh of the network bounded by the u-curves Nos. 2 and 3 and v-curves Nos. 5 and 6 (Fig. 56). Let P be any point within the mesh and s be its distance from O, the corner of the mesh with co-ordinates $u = 2$, $v = 5$. Let PN and PM be lines drawn from P parallel to the mesh lines OA, OB and M, N be their intersections with them. The points M and N can also be given their registration numbers or Gaussian co-ordinates.

The co-ordinates of M may be determined by measuring the side OA of the parallelogram on which M lies and the distance of M from O. The ratio of the two lengths $OM:OA$ varies from zero to 1 as the point M travels from O to A along OA. This ratio itself can then be regarded as the *increase* of the u-co-ordinate of M as M moves away from O. If this ratio is du, the u-co-ordinate of M will be $2 + du$, 2 being the u-co-ordinate of the corner O of our mesh. As M coincides with A, the other corner of the mesh, $du = 1$, and the u-co-ordinate of A becomes 3 as it ought to. Similarly, we determine the v-co-ordinate of N, the intersection of PN and the mesh line OB. If the ratio of the lengths ON and OB is dv, then we can take dv

---

* It may be of interest to notice that this circular arc PQ is the exact analogue of a 'straight' line for two-dimensional beings unable to move out of the spherical surface. That is, it is the direction which a string assumes when it is fully stretched between P and Q along the surface.

as the increase of the v-co-ordinate of N over that of O. The v-co-ordinate of N is thus $5 + dv$, 5 being the v-co-ordinate of O. As N advances towards B, dv becomes unity and the v-co-ordinate of B becomes 6 as we assumed to start with. The Gaussian co-ordinates of P are then $2 + du$, $5 + dv$. As P roams within the mesh du, dv vary between 0 and 1.

As du, dv are ratios, they do not give us the lengths of OM and ON. If we wanted these lengths, we have to discover the scales on which OA and OB are drawn. Suppose, for instance, a is the scale number for OA. Then any distance OM in the mesh line OA is given by multiplying the ratio

$$\frac{OM}{OA} = du \text{ by } a, \text{ the scale number, that is, the actual length } OA \text{ to be found}$$

by actual measurement. du will, of course, vary as M moves along OA, but the scale number a, which converts the ratio du into the actual distance OM, will not change as long as we remain within the mesh under consideration. Similarly, the scale number b, which converts the ratio dv into distance along the mesh line OB, is also a definite number and remains constant within the mesh. Now if we wanted the distance ds of OP, we have by a well-known extension of Pythagoras's theorem,

$$OP^2 = OM^2 + PM^2 + 2OM.ML \quad . \quad . \quad . \quad (2)$$

Here $OP = ds$, $OM = a\, du$, $PM = ON = b.dv$. ML is the projection of $MP = b.dv$ on OA; it also has a fixed ratio to MP.* Let c be the fixed number that converts the ratio dv into ML, the projection of MP on OA, whence $ML = c.dv$. Substituting these values in (2), we obtain

$$ds^2 = a^2\, du^2 + b^2\, dv^2 + 2a\, du.c\, dv$$
$$= a^2\, du^2 + b^2\, dv^2 + 2ac\, du\, dv \quad . \quad . \quad . \quad (3)$$

This formula gives the distance ds between any point P within the mesh and the mesh corner O in terms of the increments du, dv of its Gaussian co-ordinates. The important point to note is that the numbers a, b, c are *fixed* numbers and do not change *so long as we remain within the same mesh*, although they may and often do have different values from one mesh to another.†

The *internal* geometry of the surface is then known precisely if we know the values of a, b, c for every mesh by means of measurements carried out by always remaining on the surface and without ever going out of it. We could even represent it on a plane map although the surface may be a

---

* This may also be seen by use of elementary trigonometry. If the angle between the mesh lines OA, OB is α then $ML = bdv \cos \alpha = b \cos \alpha$, $dv = cdv$. Hence a fixed number c exists which converts the ratio dv into ML, the projection of MP on OA.

† The scale members a, b, c vary from point to point, just as in a Mercator's map the scale varies from point to point. It is impossible to make a Mercator's map of the earth or even a continent or country on a uniform scale.

complicated surface like that of the earth. For this purpose, we draw two straight lines and take them as our u and v axes (Fig. 57). Although the u, v lines are actually curved on our surface, yet they are represented on our plane map by straight lines, just as the circular lines of latitude and longitude on earth's surface become straight lines on a Mercator's map.

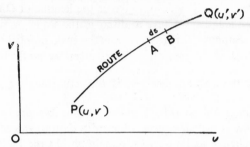

FIG. 57—The distance between P and Q depends on the route we follow. Taking an infinitesimal bit ds, (AB), of the route S, the distance between P, Q along the route S is given by the sum of an infinite number of such infinitesimal bits. In other words, it is the integral $\int_s ds$.

If we want the distance between two points P (u, v) and Q (u', v') we shall have to specify the route by which we travel from P to Q. The shortest distance route PQ as *appearing* on the map will not in actual fact be the shortest route on the surface as all those who have ever handled navigational charts know very well. If we wish to find the distance PQ along a specified route S, we shall have to divide the route into an infinitely large number of bits. Take any one such bit, say AB. The distance, ds, between A and B is given by

$$ds^2 = a^2\,du^2 + 2\,ac\,du\,dv + b^2\,dv^2 \qquad . \qquad . \quad (3)$$

where u, v are the co-ordinates of A and $u + du$, $v + dv$ those of its infinitely close neighbour B. We must also know the values of a, b, c at A which will in general be different for every point of the route. The distance s between P and Q along any given route S is then the integral $\int_s ds$ over the route S, ds being given by (3).

In this way, we are able to measure the distance between any two points P, Q of the surface along any given route S without ever moving out of the surface.

The functions a, b, ac are often written as g_{11}, g_{22}, g_{12} respectively and the co-ordinates u, v as x, y. If we adopt the usual notation, our formula (3) becomes

$$ds^2 = g_{11}dx^2 + 2g_{12}dxdy + g_{22}dy^2 \qquad . \qquad . \quad (4)$$

The g's with different labels are exactly the same things as the ratios a, b, ac used formerly and are called the factors of measure *determination*. The expression (4) is called the *metrical groundform*, or simply the metric of the surface. It is the open sesame of the entire internal or intrinsic geometry of the surface, for a knowledge of the values of g at every point of the region or the surface enables us to determine all *intrinsic* geometrical properties of the surface. We pause here to glance at the main ideas underlying this *internal* theory of surfaces which Gauss developed without making any reference to the plenum in which they are embedded.

First, a knowledge of distance is not required for defining Gaussian coordinates of a point on a surface. The distances are then calculated by assuming that Pythagoras's theorem holds in any infinitely small region of the surface. By means of actual measurements carried out in this small region, we derive the three fixed ratios g_{11}, g_{12}, g_{22}, which convert the increments of Gaussian co-ordinates in the small region into distances. The values of the g's do not remain the same from one region to another but we build up their values for all regions (by actual measurements) carried out in them. The main-spring of the theory is, therefore, the principle of gaining knowledge of the external world from the behaviour of its infinitesimal parts. Secondly, by this device, the role of the accidental co-ordinates is minimised and invariant expressions that remain unchanged for all co-ordinate systems as, for instance, the expression (1) for the distance PQ, are obtained for intrinsic properties of the surface. We shall refer to one such invariant expression as it is of great importance not only in the theory of surfaces themselves but also in its applications elsewhere.

This expression is related to a property of the surface called its curvature. In ordinary speech we understand perfectly well what we mean by the curvature of a curve such as the path of a meandering stream flowing through a valley. For scientific purposes, such as building railway lines and roads, we need a more exact definition of curvature. In Chapter 3 we defined curvature of a plane curve as the limit of the ratio

$$\frac{\text{total bend}}{\text{length of the arc } PQ}$$

when the distance PQ on the curved line shrinks to zero. In other words, it is the rate at which the tangential directions of the curve diverge as we travel along it. But the rate at which the tangential directions diverge is also precisely the rate at which lines drawn from any point at right angles to the tangents diverge.

Suppose from any point O we draw a unit length Op perpendicular to the tangent PT to the curve at any arbitrary point P. (See Fig. 58.) We thus obtain a point p corresponding to any point P of the curve. As P

moves along the curve, its counterpart p moves on a circle of radius unity, op being always 1. Consequently as P describes a small arc ds of the curve, the corresponding point p traces a small arc $d\sigma$ of the unit circle. This arc is also a measure of the plane angle through which the tangent to the curve has diverged in describing the arc ds from P to P'. Its rate of change is the limiting value of $\dfrac{d\sigma}{ds}$ as ds tends to zero.

Gauss extended this idea of the curvature of lines in a plane to surfaces in space. He defined it as the measure of the rate at which lines at right angles to tangent planes diverge in an exactly analogous manner. In place

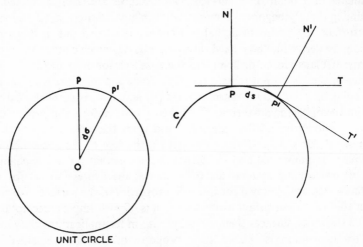

UNIT CIRCLE

FIG. 58—The curvature of the curve C at P is the rate at which the perpendicular line PN at P to the tangent PT diverges as P travels *along* the curve from P to P'. At P the direction of this perpendicular line is parallel to op and at P' to op'. The direction of the perpendicular changes from op to op' as P moves along the curve from P to P'. If the angle pop' is $d\sigma$, the rate of change of the direction of the normal at P per unit length of the curve is $\dfrac{d\sigma}{ds}$.

of the unit circle we now trace a unit sphere by drawing unit vectors from O perpendicular to the tangent planes to the surface. If a small area $d\omega$ of this sphere corresponds to a small area ds of the surface, then $d\omega$ is the solid angle formed by the perpendicular lines erected at the points of ds on the surface. The limiting value of the ratio $\dfrac{d\omega}{ds}$ as ds shrinks to zero is the Gaussian measure of curvature. As both $d\omega$ and ds are areas, their ratio, as also its limiting value, is a pure number. Gauss also showed that this pure number, the curvature, is determined by the inner measure relations of the surface alone.

More precisely, for calculating the curvature of a surface he discovered a formula involving only the coefficient g's of the metrical groundform and their differential coefficients such that its value remains the same for any two groundforms that arise from each other by a transformation of co-ordinates. What it means is this:

If we are working with any co-ordinate system x, y we can, by actual measurements on the surface, discover its metrical groundform, that is, the ratio numbers g which enable us to calculate distances and lengths on the surface in any infinitesimally small mesh of the surface. Now suppose we switch over to another co-ordinate system x', y', we can then find formulae which transform the co-ordinates x, y of a point P to x', y', its new co-ordinates in another system. The g's previously discovered were functions of x, y, and if we substitute their values in terms of x', y', we transform the g's into g''s, the new functions of x', y', which are the coefficients of the new metrical groundform appropriate to the new co-ordinate system x', y'. Gauss's formula for the curvature of the surface gives us the *same* value whether we use the original g's and x, y or the new g''s and the new co-ordinates x', y'. In other words, it is independent of the particular co-ordinate system in exactly the same way as the distance formula, $s^2 = (x - x_1)^2 + (y - y_1)^2$ gives the distance between two points whose co-ordinates are (x, y) and (x_1, y_1), whether we use one or another system of rectangular co-ordinates. The *form* of the formula remains unchanged in spite of all the changes that co-ordinate system itself may undergo. Of course, the word 'all' here must not be understood too literally but must be taken for a fairly wide and general class of transformations to which the co-ordinate system may reasonably be subjected.

Gauss's theory provided the model, on which his pupil, Riemann, built up a more general theory of the internal geometry of spaces of three and more dimensions without referring to hyper-spaces, in which they may be imagined as immersed in much the same way as the surface of a sphere is immersed in three-dimensional space. Riemann thus created the heavy armour by means of which geometry in the hands of Einstein was able to revolutionise physics and, half a century later, completely occupy some fairly wide regions of physics. For a time it seemed imminent that mechanics, optics, electrodynamics, and other branches of physics would all fall victims, but the latest developments in quantum physics have so far resisted all attempts to geometrise this branch. While the majority of physicists are now turning towards a statistical interpretation of funda-mental physical processes, Einstein was hopeful to the end that all physics would one day be reduced to a minor detail in a purely geometrical theory of surfaces and hyper-spaces.

Riemann considered a three-dimensional space in lieu of a two-dimensional surface. There would thus be three sets of numbers x_1, x_2, x_3 instead of two to define the Gaussian co-ordinates of a point. Instead of imposing on a surface an arbitrary network of meshes defined by a system of u and v curves, Riemann imposed on his space a network of cells defined by three families of surfaces labelled

$$x_1 = 1, 2, 3, \ldots$$
$$x_2 = 1, 2, 3, \ldots$$
$$x_3 = 1, 2, 3, \ldots$$

Just as the distance ds^2 of a point P from the corner of a mesh in a two-dimensional surface was given by the metrical groundform $g_{11}dx^2 + 2g_{12}dxdy + g_{22}dy^2$, so also the distance ds^2 of a point P in any cell, from its corner, is given by the metrical groundform:

$$g_{11}dx_1^2 + g_{22}dx_2^2 + g_{33}dx_3^2 + 2g_{12}dx_1dx_2 + 2g_{13}dx_1dx_3 + 2g_{23}dx_2dx_3.$$

As before, the six g's in this form are ratios or pure numbers and have the same value in the same cell though they vary from one cell to another. Their real values for each region or cell have to be discovered by actual measurements carried out in each region by means of an infinitesimally small rigid rod. Once we have written the metrical groundform for three-dimensional space, there is no difficulty in writing it for hyper-spaces of four or more dimensions. Take, for instance, a four-dimensional space, in which we specify a point by its four Gaussian co-ordinates (x_1, x_2, x_3, x_4). We can again divide the hyper-space by a network of cells by four families of three-dimensional spaces given by

$$x_1 = 1, 2, 3, \ldots,$$
$$x_2 = 1, 2, 3, \ldots,$$
$$x_3 = 1, 2, 3, \ldots,$$
and $$x_4 = 1, 2, 3, \ldots,$$

although we have no visual notion as to what our 'space-cell' that we thus carve out would look like. A homogeneous expression of second degree in dx_1, dx_2, dx_3, dx_1 would thus have four squares of terms like dx^2, viz. dx_1^2, dx_2^2, dx_3^2, dx_4^2 and six products of these four differentials taken two at a time, viz. dx_1dx_2, dx_1dx_3, dx_1dx_4, dx_2dx_3, dx_2dx_4, dx_3dx_4. The complete groundform would thus be

$$g_{11}dx_1^2 + g_{22}dx_2^2 + g_{33}dx_3^2 + g_{44}dx_4^2 + 2g_{12}dx_1dx_2 + 2g_{13}dx_1dx_3$$
$$+ 2g_{14}dx_1dx_4 + 2g_{23}dx_2dx_3 + 2g_{24}dx_2dx_4 + 2g_{34}dx_3dx_4.$$

In other words, in a four-dimensional space we need to know the values of ten g's in each space cell by actual exploration therein with a small measuring rod. As in the case of two- and three-dimensional spaces, the values of the g's change from cell to cell though they remain the same in each cell. A knowledge of the values of these ten g's suffices for a complete description of all the intrinsic geometrical properties of our hyper-space. No doubt we cannot conceive what 'distance', 'curvature' or 'straight line' means or looks like in such an *imaginary* space, but we can at least set up suitable extensions of the corresponding algebraic formulae or equations in two or three dimensions of our perceptual space. Riemann succeeded in extending the conception of curvature to spaces of three and more dimensions. Gauss had found that the curvature of a surface was a pure number like temperature or a ratio. Riemann found that the curvature of a space of three or more dimensions is no longer a pure number but a tensor.

What is a tensor? We saw before that the study of mechanics had led to a generalisation of number from complex numbers to space vectors. In order to express mechanical entities like force, velocity, acceleration, *etc.*, mathematicians had been forced to invent vectors which required two things for their specification, a pure number to indicate their magnitude and a line to indicate the direction of their action. As mechanics developed further and led to a study of elastic bodies, it was discovered that even vectors do not suffice to express the stresses and strains that are set up within elastic bodies or viscous fluids. Such entities could be expressed only by many numbers—tensors—which thus represent a further generalisation of vectors.

*   *   *   *

In his studies of abstract geometry, Riemann turned to good account this generalisation that had been made by the physicists. So far, in our study of the intrinsic geometry of spaces, we have assumed that the set of g's entering in the metrical groundform is known by actual measurements carried out in each small region, mesh or cell of our space. Now if we consider the actual three-dimensional space in which we live, and not the purely hypothetical spaces of a mathematician's imagination, the question arises as to how we should discover its metrical groundform in any given co-ordinate system. Obviously, all regions of our actual space are not accessible to us and we cannot always explore them with our measuring rods in order to discover the values of g's in those regions. To obtain the g's, we therefore actually proceed in a different way.

One way would be to *assume* any arbitrary set of functions to define values of g's and then work out the geometric properties of our space such as its curvature, *etc.*, at various points. We thus obtain a rich abundance of geometries and we may choose out of them the one that fits our actual space the best. Another way is to look around and make some *plausible* assumptions regarding the geometric properties of our actual space and then see to what values of g's they lead us. Surprising as it may seem, just one simple assumption that appears eminently reasonable regarding the character of our actual space ensures that g's can have only a very restricted set of values and no others. All the rich abundance of geometries that can be invented by giving the g's arbitrary values thus disappears at one blow and we are left with only a few types of geometries that are applicable to our actual space.

Now we know that a rigid body may be transferred from one place to any other and put in any arbitrary direction without altering its form and content or metrical conditions. This means that space is everywhere homogeneous in all directions, that is, one chunk of space is as good as any other. But if space is homogeneous, its curvature must be the same at all points. If we make this assumption, the g's in our metrical groundform can take only certain restricted values. In fact, Riemann showed that in a homogeneous n-dimensional space of constant curvature a, the factors g_{11}, g_{22}, ... of its metrical groundform are simple functions of the coordinates x_1, x_2, etc. and the constant curvature a. In other words, a knowledge that the curvature of space is a given constant everywhere enables us to derive all the factors of its metrical groundform, which, as we saw before, suffice for a complete description of all the intrinsic geometrical properties of our space.

This fact may also be appreciated in a more intuitive manner by considering the case of a one-dimensional 'space' of constant curvature. The set of points on the circumference of a circle as distinct from other points of the plane in which it is drawn can be taken as an instance of a one-dimensional 'space' of constant curvature. Here we can easily see that the curvature is constant because the circle bends uniformly everywhere in the two-dimensional plane in which it is embedded. A knowledge of its curvature (or radius) tells us all there is to know about the internal geometry of this one-dimensional 'space' that is our circle. A sphere—that is, the set of points constituting its surface as distinct from those lying within or without it in the three-dimensional space in which it is immersed—is the exact analogue of a two-dimensional space of constant curvature. Here, too, we can see that the spherical surface bends uniformly everywhere in the three-dimensional space in which it is embedded.

Again, a knowledge of the curvature of this spherical surface tells us all that there is to know about its internal geometry. When we come to a three-dimensional space of uniform curvature we have to view it as imbedded in a super-space of four dimensions before we can 'see' it bend uniformly all over. But that is what we, three-dimensional beings, cannot do. So we have to content ourselves with an extrapolation and conclude that what is true of one- and two-dimensional spaces of uniform curvature is equally true of a three-dimensional space of uniform curvature where, too, a knowledge of its curvature suffices to give a complete knowledge of its intrinsic geometry.

Now if our actual space around us is homogeneous with constant curvature *a*, there are just three possibilities according as *a* is zero, negative or positive, corresponding to the three possibilities we enumerated earlier concerning the behaviour of parallels drawn from a point *P* to any given line *AOB* (Fig. 52). We saw on page 188 that logically there were just three possibilities; there may be only one, two or no parallels from *P* to *AOB*. If we assume with Euclid that there is only one parallel through *P*, we have Euclidean geometry and this corresponds to zero value of *a*, the space curvature. That is why Euclidean space is also known as 'flat' space. If we assume with Bólyai and Lobachevsky that there are two parallels through *P*, we have hyperbolic geometry and this corresponds to a negative value of *a*, the space curvature. The last case, in which there is no parallel through *P*, gives rise to spherical geometry and corresponds to a positive value of *a*, the space curvature.

What then is the geometry of the actual space we live in? We saw that Gauss tried to ascertain it by measuring the angles of a triangle formed by three distant mountain peaks, and Lobachevsky tried a guess at it by an examination of the values of stellar parallaxes (see page 188). These attempts proved nothing. Now, if Gauss's experiment proved inconclusive only because the size of the triangle he took, though large compared with paper diagrams of the text-books, was small compared with the dimensions of the universe, obviously the next step is to repeat the test with a larger triangle—say one whose vertices are formed by three distant stars or nebulae. But in this case the measurement of angles would depend on the observation of light rays and consequently on the physical laws governing the propagation of light through space. The experiment would therefore tell us more about the behaviour of light rays during their voyage through interstellar space than about the nature of physical space itself—and we might be able to interpret the result in any of several different ways.

This fact is the basis for the positivist assertion that the question whether physical space is Euclidean or not is meaningless. The great French mathematician, Poincaré, for instance, held that the science of geometry cannot tell us what kind of space we actually live in, because geometrical laws are only arbitrary conventions and we could adopt any set of conventions we liked. In his *Science and Hypothesis* he wrote that the question whether Euclidean geometry is true or false is as senseless as to ask whether the metric system of measurement is true and the F.P.S. system false. The answer is that both the systems are conventional ways of measuring quantities and it matters little whether you measure, say, a distance in kilometres or miles. Likewise, the various systems of geometries—Euclidean and non-Euclidean—are merely conventional ways of representing our space and we are free to make any choice we please provided we adopt physical laws appropriate to each mode of representation. What this means may be elucidated by the example of Mercator's map.

We all know that the earth is a large sphere and we have developed a geometry of the sphere, that is, a way of measuring distances on its surface. For certain purposes, we also represent the surface of this sphere on a piece of paper, as in navigation, when we draw a map of the world, *e.g.* by Mercator's projection. Mercator's map is quite a good substitute for the surface of the globe because it represents all essential terrestrial locations in relation to one another. In other words, you could assume that the surface of the earth is 'flat' like that of a piece of paper on which you draw its map, *provided* you adopted a different system of geometrical laws. For instance, on Mercator's chart the ordinary notion of distance does not apply. Thus, it is impossible to specify a scale that will hold throughout the chart. A distance of one inch near the equator may represent 100 miles, while the same distance farther north or south—say in Greenland—may represent only 10 miles. In consequence, Greenland would appear on the map far bigger than, say, Greece, although a traveller may experience the same amount of fatigue in covering both the countries from one end to the other.

Now Poincaré's thesis was that the global and Mercator's way of representing terrestrial distances and positional relations of continents, countries, oceans, *etc.*, are equally valid, and the facts of our experience can be fitted under both the schemes provided we adopt suitable geometrical laws appropriate to each way of representation. Hence the question whether the earth's surface is 'really' round or flat is meaningless—it is only a matter of convention. Similarly, we could assume that our space is Euclidean; then we should have to devise a more or less complicated system of physical laws to describe the behaviour of light rays and moving particles. Or, alternatively, we could assume a more complicated space and simplify our physical laws, as, in fact, Einstein has done, by giving up the idea of force and the law of gravitation but by assuming a more complicated space. It is again a matter only of convention. However, it will perhaps be agreed that it is stretching facts a bit too far to say that the question whether the earth is round or flat is 'meaningless' or a matter of mere convention simply because mathematicians can find a way of representing points of a sphere on a plane!

*   *   *   *

The question of the nature of our actual space (as also of time) is at the root of Einstein's Generalised Theory of Relativity. One consequence of this theory is the emergence of cosmology, that is, scientific speculation regarding the origin and structure of the universe as a whole. In cosmology, too, the very first question that arises is the nature of our space and time. First, take space. Our common-sense view of space is that it is some sort of an immovable container, of boundless extent, which is the scene of everything that happens. Even that which happens in sheer imagination, such as Dante's inferno, is often believed to be located somewhere in this Brobdingnagian box. The size of this cosmic container is infinite, that is to say, there is nowhere where we could locate its walls. Its character is uniform throughout, which means merely that any chunk of its hollow is as good as any other. This common-sense view of space was also the view of classical physcists and mainly due to the influence of Newton. How does it square with the facts of our actual universe? That depends on what we believe would be the outcome of an imaginary experiment which I will now describe.

Suppose we imagine ourselves moving outwards in a straight line past the stars, galaxies or nebulae and beyond them. There are only three possibilities. First, we may ultimately, find ourselves in an infinite, empty space with no stars and nebulae. This means that our stellar universe may be no more than a small oasis in an infinite desert of empty space. Second, we may go on and on for ever and ever meeting new stars and nebulae. In this case there would be an infinite number of stars and nebulae in an infinite space. Third, we may be able to return ultimately to our starting point having as it were circumnavigated the universe like a tourist returning home after a round-the-world trip. If we reject the last possibility as utterly fantastic, we have to do some pretty tough thinking to get over the difficulties which the other two possibilities raise.

A finite stellar universe in an infinite space would either condense into a single large mass due to gravitation, or gradually dissipate, because, if gravitation fails to produce condensation, the stellar radiation, as also the stars themselves would pass out into infinite empty space never to return, and even without ever again coming into interaction with other objects of nature. We find the second possibility—an infinite number of galaxies in infinite space—no more satisfactory. It leads to the rather ridiculous conclusion that gravitational force is everywhere infinite. Every body would find itself attracted by an infinite pull—infinite not in the metaphorical sense of something very large but in the mathematical sense of being larger than any figure we care to name. We could hardly survive such a terrific pull.

It is true that we could avoid this difficulty by an *ad hoc* amendment of

Newton's law but there is a strong feeling against tinkering with well-established physical laws every time we meet an objection. Besides, the amendment would be of no avail because there is another objection which could still be raised. Imagine that space is really infinite with an infinite number of stars and nebulae therein, then the night sky would be a blaze of light with no dark patches between the stars. To understand the reason for this, suppose you were in a big sparsely sown forest. No matter how sparsely sown it might be, provided only it was big enough, the horizon would be completely blotted out of view by the trunks of the trees.

In the same way, however sparse the nebular distribution in space, provided space was infinite, the entire sky would appear to be studded with stars and nebulae leaving not a single point uncovered. We could, of course, explain the darkness of the night sky by assuming that the light of some of the distant nebulae is absorbed *en route*, or that the nebular density becomes progressively less and less as we recede more and more from our present position, or that the vacant spaces of the sky are occupied by dark stars or nebulae that we cannot see. But all these assumptions would be rather arbitrary because all our present experience is to the contrary.

Now about time: Here, too, Newton—and following him the classical physicists—adopted the common-sense view. This view is based on the fact that every one of us can arrange the events that we perceive in an orderly sequence. That is to say, we can tell which of the two events perceived occurred earlier and which later. By means of physical appliances, such as a watch, we can even say how much earlier or later. Suppose I observe the occurrence of two events: (*A*) such as the birth of a new star in the constellation Hercules and another event (*B*) such as a lunar eclipse. Seeing them I could tell whether *A* occurred before, after or simultaneously with *B*. I could even measure the time interval between their occurrence by means of a clock.

Similarly, another observer, say, Voltaire's Micromegas looking from his native star Sirius, could also observe the same two events. He, too, would place them in a time order. Now Newton adopted the common-sense view that the temporal order in which I would place the events *A* and *B* would be precisely the same as that of Micromegas, the Sirian observer. More, my reckoning of the time interval between the occurrence of the two events by means of my watch would be identical with the time reckoning of the Sirian, measured, of course, with a Sirian watch, there being no import of Swiss watches into Sirius.

Unfortunately, this assumption would be valid only if the time at which an event is observed were simultaneous with its occurrence. That would be the case if light travelled, as our ancestors believed, with infinite velocity so that the news of any event, *e.g.* the birth of a nova, could be flashed instantaneously everywhere. There would thus be no time lag between its occurrence at one place and its observance from another, however far. We know now that this is not the case, and instead of instantaneous propagation of light we find that light takes millions of years to come to us from some parts of the universe.

When we allow for this time lag between the occurrence and observance of an event due to finite velocity of light, we find that my reckoning of time between the events A and B is not the same as that of our imaginary Sirian. It is possible for A and B to be simultaneous according to my reckoning and yet the Sirian may find that A occurred before B. It is not that the Sirian watchmakers are in any way inferior to their terrestrial counterparts of Rolex fame. It is because the velocity of light is found to be constant with respect to any moving observer. Normally if you chased a car moving at 30 miles per hour on a cycle moving at 10 miles per hour, the car's velocity relative to you would be only 20 miles per hour. Not so if you were chasing a ray of light. Its velocity relative to you would still remain the same, *viz.* 186,000 miles a second, which you would have found if you stood still on earth. In fact, no matter whether you moved on a cycle at 10 miles per hour or in a supersonic plane at 2000 miles per hour or in a jet rocket at 20,000 miles per hour, light would still continue to elude you at the same even pace, 186,000 miles to a second. It is a result of this peculiarity of the velocity of light—its refusal to mix with other speeds in the normal understandable high-school fashion—that there can be no unique universally valid temporal order in which events can be placed.

A way out of these difficulties of classical physics was shown by Einstein. At the mention of Einstein brows start sweating, but there is no need to be alarmed. Fortunately, like the teeth of an elephant, expositions of Einstein's theory are of two sorts: one sort for display and the other for grinding. I shall describe here only the display sort. You may then make your own guess what the grinding sort is like.

Einstein abandoned the idea of a container type of space and an absolute time, existing in their own right independently of matter and radiation as a sort of theatre for matter and light to enact their drama. He said, in effect, that you do not need the theatre unless there are actors in it, too— or rather, that if you do away with the actors (matter and radiation) you *ipso facto* wipe out of existence the theatre, *viz.* the space-time framework, as well. You may recall the old fable of the giant and the parrot whose lives were bound together. You could kill both by killing either one of the two. We have Einstein's word that it is the same with space-time on the one hand and matter and radiation on the other.

The whole universe, including the space-time framework in which the material events take place, is then a closely bound nexus, one part completely determining the other. Given the distribution of matter and radiation in any region, the character of the theatre—that is, of the space-time framework in which they play their part—is already determined, and vice versa. This device enabled Einstein to do without the Newtonian idea of gravitation as a force in order to account for the motions of natural bodies. He explained planetary motions as a consequence of the very nature of the space-time framework in their neighbourhood. In this way Einstein tried to reduce physics, that is, the theory of the motion of large bodies, to geometry or the theory of space-time and its curvature.

To do so Einstein showed that we could consider the universe to be a four-dimensional continuum in the sense that every event in it requires four specification numbers to identify it—three to indicate the place and one the time of its occurrence. Physical theory—that is to say, the behaviour of matter and radiation—could then be reduced to the geometry of this space-time manifold. In particular, the curvature of this manifold at any point, in Einstein's theory, is simply related to the density of matter in its neighbourhood.* Hence if the mean density of matter in the universe is greater than a fixed number, no matter how small, the curvature of the universe as a whole will be everywhere positive. But a positive curvature of space implies that it must close in upon itself and thus be finite, as we shall presently see.

A straight line in a plane is said to be straight or without curvature, while a circle has curvature which is inversely proportional to its radius. Thus, if the curvature of a circle with one foot radius is taken as unity, that of a circle with a yard as radius will be one-third. That of a circle as large as the earth's circumference will be about one part in 21 million. In the last case the curvature is so slight that we should have difficulty in distinguishing from a straight line any arc of such a circle. Nevertheless, while the straight line extends to infinity with its two ends never meeting, the arc of a great circle, if prolonged indefinitely with the same curvature, will end by closing in on itself. While, therefore, a circle with positive curvature, however small, is always finite, the straight line is infinite. The conception of curvature in the case of curved surfaces and spaces is more complicated, but the essential property remains, namely, that if the curvature of a surface or space is not zero and exceeds an arbitrarily small positive number, the surface or the volume must close in on itself and thus be finite. And as, according to Einstein, the average density of the universe and therefore its curvature is everywhere necessarily non-zero, the universe must be finite with a finite quantity of matter. This is how Einstein proposed to get over the difficulties raised by the idea of infinite space that we mentioned earlier. As will be recalled, we stated

* For a critique of this theory see Appendix I.

that there were only three possibilities if we imagined ourselves moving outwards in a straight line past the stars and nebulae. Einstein's solution of these difficulties is that the universe is not infinite and that we should ultimately return to our starting point if we moved on and on in a straight line. At first sight this may seem incredible, but the theory of curved space-time enabled Einstein to explain a number of puzzling phenomena and even to predict some new. This inspired confidence in the theory, so he proceeded to build a cosmological model of the entire universe on the basis of these ideas. Naturally he had to make a few further assumptions to make a move on.

The simplest assumption that could be made is that the universe is static and uniform all over. That is to say, it stays put in one condition for ever. Of course, we know that the universe is neither static nor homogeneous in all its parts. But the assumption is made to render the problem simple and mathematically tractable so as to obtain a first approximation to the actual state of affairs. Assuming, then, that the universe is static and homogeneous, it can be proved that there are only two possible models to which it can conform—the Einstein model and the de Sitter model, named after their inventors. In a way, these two models may be said to represent the opposite poles of a possible evolutionary tendency in the universe. For in the former, the universe contains as much matter as it possibly could without bursting the relativity equations, while in the latter it is completely empty, permitting neither matter nor radiation. Moreover, the Einstein universe can be proved to be unstable. In other words, any deviation from this condition would tend to increase continually. Some cosmologists, therefore, believe that the actual universe grew out of an initial state nearly—but not exactly—corresponding to the Einstein model. This state being unstable the universe would have to progress towards a final state of extinction corresponding to the empty de Sitter universe. If so, we are at the moment in some intermediate state between these two extremes.

If, on the other hand, we make the more realistic assumption that the universe is not static, though uniform all through, we have, on the basis of relativity mechanics and thermodynamics, only two main possibilities. Either the universe is expanding or it is oscillating between two extremes—expanding in one phase and contracting in the other. As models for the actual universe, the expanding types have two defects. First, they have the disadvantage of spending only an inappreciable fraction of their total existence in a condition comparable to that in which we find ourselves. Second, they lead to the strange conclusion that the universe started from the explosion of a giant atom. In other words, the initial state of the universe was a peculiar condition in which the whole universe was packed within a pin-point and from which it started on its ever-expanding course.

Now you may wonder whether such a state of origin is any more comprehensible than 'the darkness on the face of the deep' of which the Bible speaks. The authors themselves are no less bewildered. That is why they call it a 'singular' state which merely means that it is peculiar or queer. However, the singularity can be got over. We can construct an expanding model in which the universe starts expanding from an initial non-singular state of non-zero radius, and hence the existence of point singularity is not an irremediable defect. Nevertheless, if we believe that the past state of our universe is not a mere flash in the pan in the history of cosmos, or believe that there ought to be something before the initial explosion, we have no choice but to adopt an oscillating model.

In the oscillating model, the universe expands from an initial state to a maximum radius, and on reaching the maximum radius the direction of motion reverses. The contraction thus initiated then continues until expansion begins again on reaching the initial state. Such an oscillating model has the advantage of spending all its life in a condition where there is a finite density of matter such as we find at present. But it has the disadvantage of starting its career from a singular state of zero radius involving an unimaginably dense concentration of all the matter of the universe in the space of a pin-point. Unlike the expanding models, there is no way of constructing an oscillating model which starts from an initial non-singular state of non-zero radius.

At this stage you would, no doubt, want me to stop. For this account of an expanding or oscillating universe, as though it were a rubber balloon that was being blown in and out, would appear too incredible to be sober scientific speculation. But many competent authorities do not think so. They believe that observation confirms the theory. Some years ago the American astronomer, the late Edwin P. Hubble, discovered that the distant nebulae are all receding from us. In fact, the farther off the nebula, the greater the velocity of its recession from us. His observations showed that this velocity of recession is directly proportional to its distance. It is true that the evidence in support of this recession is indirect and that is natural. We cannot measure the velocity of a nebula as we do that of a train or a bullet. But similar evidence—a Doppler shift of spectral lines towards the red end in its spectrum—has hitherto always been found to indicate that the object under observation is moving away from the observer in the line of his sight. Unless the red shifts in the nebular spectra are due to some other cause, Hubble's observations would indicate that every galaxy is running away from every other. In other words, the universe is expanding literally like the surface of a balloon that is being continually inflated. This is taken as a confirmation of the theory underlying the cosmological models described above.

We may mention here some consequences of this nebular recession or expanding universe. At a distance of about 6 million light years the velocity of nebular recession is about 300 miles per second. Since the velocity of recession of a nebula or galaxy is proportional to its distance, a galaxy at a distance of 150 million light years would be receding at a rate of about 7,500 miles per second and a galaxy at 1,500 million light years at 75,000 miles per second. At this rate a galaxy at a distance of 3,720 million light years would be receding from us, or (what comes to the same thing) we should be receding from it, at 186,000 miles per second, which is the velocity of light. Any ray of light emitted by it would begin to chase us with the same velocity as that with which we are running away from it.* It would thus be like a race between Achilles and the tortoise in which the tortoise for once runs as fast as Achilles and therefore could never be overtaken. And if no ray of light emitted by it could ever reach us, it would never be seen by us. Even though space might have an infinite number of galaxies in the infinite recesses of its depth, we could never see more than a few, namely, those within a radius of 3,720 million light years from us, and even these would eventually disappear from our ken.

Since all the nebulae at present visible are receding from us, one day they too would reach the limit of our vision at the critical distance of 3,720 million light years. When this happens they will pass out of our horizon and we shall never see them any more. Apparently, therefore, our universe is doomed to be gradually but systematically impoverished. Nor is the approach of such cimmerian nights, when we shall cease to see any galaxy in the sky, a very long way off after all. The cosmic broomstick that we see at work may sweep the heavens clean of all the galaxies that we now observe in about 10,000 million years, which is only a fifth of the life-span of an average star. After that, our galaxy would have an eternity of solitude.

It is now time to summarise the new Genesis that cosmology is writing in the language of mathematics. Since I have resolved not to use this language here, I shall give you a rather free translation. Here, then, is the new Genesis:

In the beginning there was neither heaven nor earth,
And there was neither space nor time.
And the Earth, the Sun, the Stars, the Galaxies and the whole Universe were confined within a small volume like the bottled genie of the Arabian Nights.
And then God said, 'Go!'

---

* You may find this puzzling in view of the statement previously made that the velocity of light remains constant with respect to any moving observer. This is only true in the special theory of relativity (which refers to a particularly simple system of space and time) but not in Einstein's general theory of relativity. The point is too technical to be elaborated further here.

And straightway the Galaxies rushed out of their prison, scattering in all directions, and they have continued to run away one from another ever since, afraid lest some cosmic Hand should gather them again and put them back in their bottle (which is no bigger than a pin-point).

And they shall continue to scatter thus till they fade from each other's ken —and thus, for each other, cease to exist at all.

You may object that in the foregoing account I have smuggled in God by an underhand trick. Very well. You may substitute 'And then something happened' for the words 'And then God said, "Go!"' The substitution will make no difference to the meaning of the passage though it may deflate a bit its somewhat exalted tone. But I brought in God because some eminent scientists think that the relation of natural science to religion should now be re-examined to provide *lebensraum* for God in its scheme of things. Not so long ago it used to be the boast of a scientist that he had no need of the hypothesis of God. Now he finds that he can no longer do without an Almighty Creator. I do not know for sure whether this is a sign of progress or merely a reflection of the present-day political and economic perplexities of the common man, from which his only hope of deliverance seems to lie in divine intervention.

But to revert to our new Genesis, I should now add that the great danger that threatens its validity is the possibility that the relativity theory on which it is based might not be applicable to the universe as a whole. Realising this danger, Milne broke new ground when he suggested that locally valid principles, like that of general relativity or other equivalent theories of gravitation, could be deduced from still more fundamental world-axioms which could be regarded as 'true' *a priori*.

Now what are these world principles which can be taken to be 'true' *a priori*? Milne's first principle is that the descriptions of the universe as a whole, and consequently the laws of nature, as given by different observers located at the nuclei of the galaxies, are the same. Milne admits that it is very unlikely that the actual universe is such that its contents would be described in the same way from every nebular nucleus taken as an observing-point. But his object is to construct a 'science of laws of nature in an ideal universe in which the various nebular nuclei or *fundamental particles* provide identical descriptions of its contents.' Once such a science has been created it is relatively easy to 'proceed to a more realistic state of affairs, if we want to, by embroidering perturbations or variations on the ideal universe.'

Taking his cue from Hubble's observation of nebular recession referred to above, Milne further postulates that this ideal universe consists of a swarm of particles (nebulae) which at some given time started receding from one another with uniform velocities. Milne calls this idealised system

of mutually separating particles a substratum. He then pushes his analysis
of time reckoning a stage farther, beyond Einstein. As we saw, Einstein
had shown that there was no unique order of temporal events valid for
every observer. It is possible for two events A and B to be simultaneous
according to my reckoning whereas one may precede the other when observed by Micromegas from his native star Sirius. Milne accepts this
result but goes on to add that there is no natural uniform scale of time
measurement either. The idea of a uniform scale of measurement arose
from measuring lengths. It is, however, quite inapplicable to measurement
of time, for the following reason.

If you want to measure a length, you will have first to fix a standard,
e.g. a standard metre or a yard. The act of measurement will involve
superimposing the standard metre or yard alongside the length to be
measured and observing how many times the standard metre or yard
'goes into' the measured length. Thus the process of length measurement
depends on the possibility of superimposing one length over another so
that the two ends of the superimposed lengths coincide, or, what comes to
the same thing, on the possibility of producing equivalent lengths. Now
Milne has pointed out that there is no standard duration of time by which
you can measure the lengths of various time intervals. To say that the
period of time of the earth's diurnal rotation or of a pendulum swing is
uniform, and therefore can serve as standard for measuring durations, is to
beg the question. For obviously you cannot as it were 'freeze' the period
of a pendulum swing that is just finished and put it alongside another that
it is about to execute and see if the two durations are really coincident.
Nevertheless, in spite of the impossibility of placing one duration alongside another in order to establish their equivalence, Milne has shown that
it is possible for different observers in different parts of the universe so
to correlate their time reckonings as to make them in some sense
equivalent.

Although his method of correlating time measurements of different
observers involves some pretty recondite mathematics the basic ideas are
fairly simple. In fact, his theory is merely the arithmetisation of the practice of an ordinary observer recording his own perceptions. Any observer,
'an ego', is aware of something he calls the 'passage of time'. This means
that if he observes two point-events* A, B, then he can always say whether
B occurred after A or before A, or simultaneously with it. Between any two
non-simultaneous point-events we can interpolate an infinity of other
point-events and all of them will be after A but before B if B was later than
A, just as you can interpolate an infinity of other points between two

---

* We imagine that the events take place instantaneously so that they are events of
zero duration—very like the geometrical points which are 'lengths' of zero magnitude.

points *A*, *B* of a line such that any interpolated point is to the right of *A* but left of *B* if *B* is to the right of *A* (see Fig. 59 below).

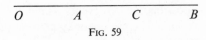

$$O \qquad\qquad A \qquad\qquad C \qquad\qquad B$$

FIG. 59

In other words, the flow of all the events in his consciousness is a linear continuum whose mathematical counterpart is a set of points on a straight line. We could, therefore, represent any event (*C*) as a point *C* on a straight line *AB* (Fig. 59). Choosing any point *O* as our origin we can correlate all points to its right with the positive real numbers and all points to its left with the negative real numbers, subject to the condition that the numbers t_1, t_2 correlated with the events *A*, *B* satisfy the relation $t_2 > t_1$ if *B* is later than *A*. Milne calls any such correlation of events in an observer's consciousness with real numbers a 'clock arbitrarily graduated', and the real number *t* associated with any event *C* the 'epoch' of that event.

So far we have considered a single observer. But there are an infinite number of other observers in the universe. Consider another such observer. He too can correlate events in *his* consciousness with the continuum of real numbers, that is, he too can set up another arbitrarily graduated clock in his own neighbourhood. Under what conditions may these two arbitrarily graduated clocks be said to keep the *same* time? To answer this question we restate the data of the problem in a different form. Fundamentally an observer can observe by his own clock the time of occurrence of any event taking place in his own neighbourhood, the time of occurrence and observation being the same. If the event occurs elsewhere the time at which it actually occurs and the time at which it is observed by the observer will be different. In this case, the observer can only observe the time (by his own clock) at which the news of its occurrence reaches him by a flash of light or a radio signal or otherwise. If, then, he wants to observe another observer's time or clock, all he can do is to strike a light himself and watch for it to illuminate the second observer and his clock. If the second observer co-operates and reflects back to the first observer the light ray immediately after it has illuminated his own clock, the first observer can observe three times:

    (i) the instant of time (by his own clock) at which he flashes the light signal;

    (ii) the *reading* of the second observer's clock at the moment it becomes visible to him; and

    (iii) the instant of time (also by his own clock) at which he perceives the second observer when his light ray has been reflected back to him.

He thus obtains a triplet of time observations and by repeating this sequence of operations, he can obtain any number of such triplets of time readings. According to Milne this sequence of triplet readings is sufficient to afford a measure of the spatio-temporal history of the second observer relative to the first, and also the history of the second observer's clock as it appears to the first. In fact, subject to some fairly general considerations, the two observers can so correlate their clocks that time measurements made by them can be considered in some real sense equivalent. An outline of Milne's method defining a time equivalence is described in Appendix II, and we shall here merely note the basic idea underlying his method.

Milne first lets an observer correlate his own perception of events with an arithmetical aggregate, that is, a set of real numbers. The second observer in turn does the same. Now while one observer's yardstick for measuring time cannot be superimposed on to that of another, the arithmetical aggregates which are the correlates of the two observers' consciousness of the passage of time can be. It is thus that Milne succeeds in defining the notion of equivalence of distant clocks in any kind of relative motion.

Having defined equivalent or 'congruent' clocks, Milne proceeds to show that the idea of equivalent or congruent measuring rods is neither valid nor necessary. Hitherto physicists had relied on two fundamental instruments for probing into the mysteries of the universe—a clock for measuring time and a rigid rod for measuring distance. Milne proposes to dispense with the rigid measuring rod and claims that distance can be measured solely by means of a clock. The method suggested by him is precisely the one now used by meteorologists for measuring distances by means of 'radar-technique', whereby the distance of a reflecting surface is measured by the time taken by a radio wave to return to the emitter. Thus, 'equivalent' clocks give all basic measures of space as well as time. But in attempting to find one way of regulating equivalent clocks or defining an equivalence, Milne actually discovered an infinity.

Of this infinity of modes of describing an equivalence there are two which are physically the most significant on account of their formal simplicity. Milne has given them special names—the t-time and τ-time. If t-scale were adopted by an observer, he would find that the universe was 'created' at a definite instant of time some 4,000 million years ago when all the nebulae in the universe were packed in a pin-point. Rushing out of this point the swarm of nebulae would appear to be receding from one another with uniform relative velocities. An observer located at the centre or nucleus of a nebula can describe the whole system in terms of his private Euclidean space. In this space he would appear to be at the centre of an expanding sphere whose surface moves radially outward with the speed of light. Inside this sphere each nebula has a velocity proportional to its distance from the centre.

Now, as we have seen, Milne imposes the further condition that all observers located at the nuclei of all the nebulae of this swarm are on a par with one another. A consequence of this condition is that the universe is infinite but appears to every observer (in his own private Euclidean space) to occupy a finite volume. Hence the space appears to him infinitely overcrowded near the expanding edge of the universe, the last millimetre containing an infinite number of nebulae, just as to an observer at sea all the oceanic waters appear to be concentrated at the edge of his horizon. But this is only an apparent effect, for to another local observer, at or near the former's horizon, everything seems normal—the concentration of oceanic waters now receding to *his* horizon.

However, a different system of measurement of time would give another picture altogether. For if, instead of kinematical time, t, the observer adopted Milne's dynamical τ-time, the time recorded by, say, the rotating earth, he would find that the universe is not expanding nor was it created at some finite past. Time would appear to stretch backwards and forwards for ever in agreement with the commonsense world-view. But this system of time reckoning forces a modification of our concept of space, which is no longer Euclidean but hyperbolic. In other words, if we accept the common-sense view that space is Euclidean, we have to adopt Milne's kinematic scale of time with all its difficulty of a finite past but infinite future and point-singular 'creation' 4,000 million years ago. On the other hand, if we accept the naïve view that time has neither a beginning nor an end, we are obliged to complicate our notion of space by making it hyperbolic.

*   *   *   *

We have now outlined the main features of two types of cosmological models—one based on Relativity Mechanics and the other on *a priori* reasoning. While they agree in some details regarding the origin and evolution of our universe as a whole, they seem to be in flat contradiction with astrophysical theories. Thus both the relativity and *a priori* cosmological model theories agree that the universe originated from a state of unimaginably dense concentration of matter, in which all the nebulae were packed within a pin point. At a definite epoch of time, about four or five thousand million years ago, the nebulae rushed out of this 'singular' state. During the subsequent evolution of the universe they have continued to recede from each other till a time will come when they will pass out of each other's ken.

On the other hand, the astrophysical theories claim that in the beginning the universe was without form and void—a mass of extremely tenuous gas spread more or less uniformly throughout all empty space. The nebulae were formed from gaseous clouds which condensed out of this gas in the void. The stars condensed out of the nebulae by a similar process of condensation, just as the nebulae condensed out of the primeval gas. The trend of universal evolution was thus in the direction of increasing condensation in some localities from a state of more or less uniform tenuity everywhere. Moreover, this process of condensation of the nebulae and stars could not have been completed in a few thousand million years, which is the maximum age attributed to the universe by the cosmological model theories. The astrophysical and cosmological model theories are therefore in complete discord with one another in almost every respect; and there is apparently no way of reconciling them.

9

CHANCE AND PROBABILITY

As we have seen, some branches of mathematics grew out of man's attempt to satisfy his economic and social needs, but others have sprung from sheer curiosity and even from a desire to invent amusing ways of killing time. For instance, the Greeks despised the practical application of mathematics so much that they considered the very name 'geometry' ridiculous, because it meant 'measurement of the earth'. About two millennia later, Louis XIV of France reduced the political power of his nobles and put an end to their internecine warfare, but left them with wealth and leisure and encouraged them to congregate at court; games of chance, such as cards, dice and roulette, then became very fashionable. This may have contributed to the interest taken by mathematicians at about this time in the formulation of a precise probability calculus for assessing the likelihood of uncertain events. Of course, every gambler has a primitive intuition which enables him to see in certain cases whether one event is more likely than another, and indeed, without this intuition no gambling would be possible. Thus, one could not play poker if it were not intuitively clear that 'full house' is less likely than a 'double pair' and more likely than a 'four'. However, this primitive intuition is sometimes misleading, as the famous French gambler, Chevalier de Méré, found about three centuries ago.

Chevalier de Méré was fond of a dice game which was played in the following way. A die was thrown four times in succession and one of the players bet that a six would appear at least once in four throws while the other bet against. Méré found that there was greater chance in favour of the first player, that is, of getting a six at least once in four throws. Tired of it, he introduced a variation. The game was now played with two dice instead of one and the betting was on the appearance or non-appearance of at least one double-six in twenty-four throws. Méré found that this time the player who bet against the appearance of a double-six won more frequently. This seemed strange, as at first sight the chance of getting at least one six in four throws should be the same as that of at least one double-six in twenty-four. Méré asked the contemporary mathematician, Fermat, to explain this paradox.

Fermat showed that while the odds in favour of a single six in four throws were a little more than even (actually about 51:49), those in favour of a double-six in twenty-four throws were a little less than even, being 49:51. In solving this paradox Fermat virtually created a new science, the Calculus of Probabilities. It was soon discovered that the new calculus could not only handle problems posed by gamblers like Méré, but it could also aid financial speculators engaged in marine insurance.

By the beginning of the nineteenth century, Laplace systematised the theory of probability and showed, to use his own words, that it was nothing but common sense reduced to calculation. The fundamental idea underlying his work may be explained by means of a simple example. Suppose we have an urn containing three similar balls, of which two are black and one white. Suppose we draw one ball at random. Obviously, we may chance to pick any one of the three. There are thus three ways of drawing a ball, of which two give black balls and one white. Hence, argued Laplace, the probability of drawing a black ball is two chances in three, or 2/3. Similarly the probability of drawing a white ball is 1/3. More generally, if there are in all n possible ways in which an event E can happen, of which m are favourable to a certain outcome, then the probability of that outcome is m/n. Thus, in the urn experiment cited above, the event E, drawing a ball, can happen in three ways of which two are favourable to a certain outcome, *viz.* the emergence of a black ball. Accordingly, $n = 3$ and $m = 2$. Hence the probability of the emergence of a black ball in a random draw from the urn is 2/3.

Let us now complicate the problem a little by introducing another urn with four similar balls of which three are black and one white. For the sake of definiteness, we shall call the first urn A and the second urn B. (See Fig. 60.) We first select at random one of the two urns and from the urn thus selected we draw a ball. What is the probability that the ball drawn is black? We might argue that we have two urns with $3 + 4 = 7$

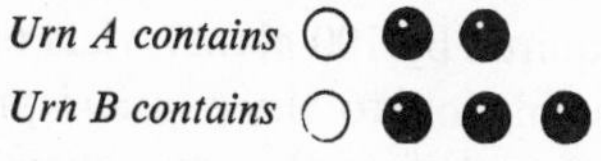

Urn A contains ○ ● ●
Urn B contains ○ ● ● ●

Fig. 60

balls, of which $2 + 3 = 5$ are black. Any one of these seven balls might be selected and, of these seven, five are black. Hence the probability of drawing a black ball in the experiment is 5/7. However, no expert in probability calculus will accept this figure as the correct value of the probability of selecting a black ball. His reason is that while there is no doubt that there are in all seven ways of drawing a ball, nevertheless these seven ways are not equally probable.

There is more chance for any of the three balls in the urn A being selected than any one of the four balls in the urn B. For we first pick up one of the two urns at random and the probability that A or B is selected is $1/2$. If A is selected, we can pick up any of its three balls. This means that $1/2$ probability that falls to the lot of A-urn balls is equally shared by the three of them. Hence the probability that an A-urn ball is selected is $1/2 \times 1/3 = 1/6$. On the other hand, the probability of selecting a ball of the urn B is obtained by partitioning equally the probability $1/2$ of picking it among the four balls contained therein. It is, therefore, $1/2 \times 1/4 = 1/8$. Laplace's m/n rule given above for calculating the probability is valid only if all the n ways in which the event can happen are *equally* probable. As in this case each of the former three out of the seven possible ways are more likely than each of the remaining four, we cannot use the m/n formula for evaluating the probability of drawing a black ball in the experiment.

Now, in this case, we could show by pure reasoning that all the seven possible alternative ways in which the event could happen are not equally likely. There are cases where we may not be able to say *a priori* whether all possible ways of its happening are or are not equally likely. Suppose, for instance, we have a die. To all appearance the probability of casting an ace with it is the same as that of throwing any other figure. All the six possible outcomes are equally probable and the probability of throwing one of them, say, an ace, will, therefore, be $1/6$. However, as a result of actual trials we may find that in 100 throws it turns up ace, say 50 times. We should, then, have serious reason to suspect that the die is not true. How shall we measure the probability of throwing an ace with it?

In such cases, when we have grown wiser after the event, we reject the previous value $1/6$ calculated from *a priori* considerations and work out a new value on the basis of the results of our trials. If we obtain 50 aces in 100 throws, we assume that the probability of throwing an ace with our loaded die is the ratio of the number of times it falls ace to the total number of throws, that is, $50/100 = 1/2$. This probability, $1/2$, worked out after the experience acquired by 100 actual trials is known as *a posteriori* probability in contradistinction to the *a priori* probability $1/6$, derived from *a priori* considerations before the experiment. In the case of a true die both the *a priori* and *a posteriori* probabilities are equal or almost equal whenever the number of trials on which the latter is based is sufficiently large. In fact, this is the test of its trueness.

In most cases we have no means of ascertaining the *a priori* probability of an event, as, for instance, the probability that a man aged thirty will die during his thirty-first year. In such cases we have to remain content with only *a posteriori* probability. In this particular instance, we take all or as many as possible men aged thirty in a country at any time, and observe how many of them die in the following year. If m persons out of n die, the probability that a person aged thirty will die during his thirty-first year is assumed to be the same as that of drawing a black ball from an urn which contains m black balls in a total of n. The latter is clearly m/n. Some people do not like this way of getting probabilities and have given elaborate definitions, which we will examine in the next section.

*   *   *   *

The vocabulary of science consists of two kinds of terms. First, there are those which are borrowed from the language of ordinary speech and are given a more or less exact scientific meaning quite distinct from that of general usage. Second, there are others which are specially coined and have, therefore, no meaning other than their scientific connotation. Thus such words as 'force', 'work' and 'energy' in physics, or 'limit' and 'continuity' in mathematics, belong to the first category; while 'entropy', 'electron', 'meson', 'differential coefficient', *etc.*, belong to the second.

The latter type of term usually arises after the particular branch of science in which it occurs has already progressed so far as to have to take account of situations not ordinarily experienced. New terms have, therefore, to be specially invented to meet these situations, as ordinary speech provides no words to describe them. The first type of term, on the other hand, is defined at the very outset by refining certain vague notions derived from our daily experience. For instance, when we make an exertion, such as lifting a weight or moving a stationary cart, we experience a muscular sensation which gives rise to the idea of force. In physics we refine this vague notion and so define force that we can measure it quantitatively in any given case. However, we are not always so lucky. For sometimes we find that the scientific refinement of the term of ordinary speech can be carried out in several conflicting ways so that we literally have what the French call an *embarras de choix*.

Thus, three different ways of refining our primitive notion of time have been proposed by Einstein, Whitehead and Milne, and we still do not know which of them is the most appropriate for physics. In fact, at times, the struggle for a precise and scientific definition of a term in everyday use is so fierce that the term itself becomes a casualty in the battle of words that ensues over the claims of rival definitions designed to clarify its significance. Such, for instance, is the case with the term 'mind', which behaviourist psychologists find totally meaningless, or with the term 'meaning', which some semioticians (that is, experts in the theory of 'signs' and their 'meaning') consider too vague to be of any use in semiotics.

Just as we have primitive notions of force, time, energy, *etc.*, which physicists have refined for scientific use, so also we have a primitive notion of probability which statisticians and mathematicians have been trying to clarify for the past three centuries without coming to an agreed decision even today. In ordinary speech we often use the word probability somewhat loosely when we wish to express the strength of our expectation or belief. Thus we may speak of the probability of tossing a 'head' when we throw a coin, or we may speak of the probability of Julius Caesar having visited Great Britain. The two cases are obviously different. One difference between them is that while the former event is repetitive the latter is not. For we may throw the coin not only once but over and over again and obtain as large a series of throws as we please. In the latter case, on the contrary, we obviously cannot multiply Julius Caesars *ad lib.*, in order that they may visit Great Britain. This difference between the two types of cases is the basis for distinguishing two distinct meanings of the word *probability*—the technical meaning of physical science, and the non-technical meaning of everyday speech. When we wish to use probability in its technical sense we have first to take care that we apply it only to repetitive events—that is, events whose successive repetitions generate statistics of their outcome—such as throws of coins and dice, selections of cards, revolutions of roulette wheels, accidents of a particular kind, *etc.* Secondly, we have to repeat the event a large number of times and observe the proportion of cases in which it leads to the result whose probability is required. Thus, in the case under consideration, we may throw the coin, say 1,000 times and observe it fall showing 'heads', say 500 times. The proportion of 'heads' in this series of 1,000 throws is, then, $500/1,000 = 1/2$. This proportion is technically known as the frequency ratio and is the measure of the probability that the coin will toss 'heads' in a throw.

However, when we apply the term probability to a non-repetitive event or an isolated case, *e.g.* Julius Caesar's visit to Great Britain, it is impossible, at any rate in any obvious way, to generate a sequence of trials and thus measure the probability of its occurrence by means of a frequency ratio. We have, therefore, to estimate it by a more or less intuitive appraisal of such evidence as we may have. Since in such cases a universally acceptable quantitative estimate of the degree of our confidence or lack of confidence in the statement cannot be given, probability when used in this sense cannot form part of a scientific assertion. To avoid confusion it is, therefore, better to restrict the use of probability in a scientific sense to repetitive events only, and to use the word *credibility* when we wish to speak of our expectation of non-repetitive events.

Not that all authorities accept the validity of this distinction between 'probability' and 'credibility'. Lord Keynes and the Cambridge Geophysicist, Harold Jeffreys, for instance, reject out of hand the legitimacy of this distinction as also that of the frequency definition of probability. They consider that fundamentally there is just one notion of probability corresponding to what we have called 'credibility'. Thus they postulate that if any given evidence H justifies a rational belief of degree p in a statement A, there is a logical relation between A and H which is known as the probability relation. But since nowhere in their treatises on Probability do they give any indication as to how this degree of rational belief ought to be assessed in any given case, their definition has been of as little use to statisticians and mathematicians as Bergson's notion of time to the physicists.

At first sight the Keynes-Jeffreys rejection of the frequency definition of probability in the case of repetitive events might appear rather odd. For what could be more reasonable than measuring probability by means of frequency? But on closer analysis it must be admitted that the frequency definition is by no means free from some very serious objections. For, when we defined probability as a frequency ratio we did not indicate how many trials we should make for calculating it, beyond the rather vague stipulation that it should be large. In the sample cited above we took 1,000 trials and obtained 1/2 as our frequency ratio. Suppose we had taken a still larger number of trials, say, 10,000 and formed a new estimate of the frequency ratio. Would it be the same as the previous one? Not very likely. If not, which of the two frequency ratios should we adopt as the measure of probability?

In general, since the estimate of the frequency ratio and therefore, of the probability of the event in question, depends on the number of trials, we must specify how many trials we should make for estimating the probability. Obviously, whatever number of trials we may choose to specify, our choice will be quite arbitrary. We try to get over this difficulty by saying that the larger the number of trials the 'better' the estimate of the frequency ratio. This may be expressed more precisely by stating that probability is the *limit* of the frequency ratio when the number of trials is infinitely increased. What it means is this: if you take a sufficiently large number of trials, any further increase in their number will give you approximately the same value for the frequency ratio. Thus if you took, say, 10,000 trials you would not appreciably improve your estimate of the frequency ratio by, say, doubling the number of trials. But when probability is so defined we get into worse difficulties. For logically there is no reason whatever why you should get approximately the same estimate of the frequency ratio, whether you based it on ten thousand trials or on twenty. Even if we ignore this objection, we have to contend with another which is even more serious.

Suppose we have an infinite set of trials of an event, *e.g.* tosses of a coin. We can partition this fundamental set into several sub-sets each consisting of an infinite number of trials. For instance, we could gather all the odd trials of the set in one sub-set and even trials in another. Both the sub-sets obviously contain an infinite number of trials. More generally, we could divide the fundamental set of trials into three, four, five or more sub-sets each consisting of an infinite number of trials. Each of these sub-sets of infinite trials will have its own limiting frequency ratio. The question then arises whether the frequency ratios corresponding to each of the infinite sub-sets into which we divide the main set are equal among themselves or not.

If the limiting frequency ratios of the various sub-sets are all equal, *however* we may partition the fundamental set of trials we have only one value of frequency ratio which can reasonably be taken to measure the probability of the event in question. But if, on the other hand, the frequency ratios of the sub-sets are not all equal, we have several different values of the frequency ratio from which to choose one as a measure of probability. Which of them should, then, be chosen raises an insoluble problem, as each of these different values has as much (or as little) right to represent the probability of the event as any other. It follows, therefore, that the consistency of the frequency definition depends on our ability to prove that the limits of frequency ratios of all the infinite sub-sets of trials into which we may partition the fundamental set of trials are equal to one another. But unfortunately all attempts to prove it generally must now be considered to have failed, in spite of some very brilliant attempts by A. H. Copeland, Abraham Wald, Jean Ville and others.

A way out of the difficulties of the 'frequency' and 'credibility' theories of probability has recently been suggested by the advocates of axiomatic theory. They claim that confusion met with in most discussions of probability theory at the present time arises because the 'formal and empirical aspects of probability are not kept carefully separate.' They insist that there are two distinct problems, 'problem I' which aims at setting up a purely formal calculus to deal with probability numbers, and 'problem II' which consists in translating the results of the formal calculus to empirical practice. Problem I is purely mathematical and independent of problem II. Thus, the sole criterion admissible in judging any axiomatic theory (problem I) is that of self-consistency, leaving out of consideration whether theorems derived from the proposed set-up are capable of translation so as to be relevant to empirical practice. But this view disregards an important consideration which is relevant and which will now be dealt with.

In an axiomatic theory of geometry, for example, we construct a formal model in which the entities such as points, lines, etc. are created by abstracting their essential properties from similar entities of our daily experience. We thus retain their fundamental and essential properties and discard what appears to be accidental. The theorems of the formal model can then be applied to empirical points and lines without much difficulty and can give results that are at least approximately valid. In the case of an axiomatic theory of probability, the very question that the formal system of axioms sets up is an adequate replica of the phenomenon of randomness that we find in the real world is left aside.

In geometry we can all agree that the system of axioms implicitly defining points, lines, *etc.* is an idealisation of the essential properties of their grosser counterparts in the real world. But an axiomatic theory of probability has to satisfy us that it adequately reflects the phenomenon of randomness in the real world, if it is not to be a mere exercise in logistics or axiomatics or a 'free creation of mind'. Now what it does is to postulate a set of axioms for combining certain numbers called probabilities of certain undefined things called 'events'. But it provides no way of establishing a connection between the probabilities that are thus defined implicitly by the axioms and the actual frequencies with which the 'events' are observed in the real world. We shall take, for instance, the system of axioms proposed by Kolmogorov.

In presenting these axioms he says that there are other postulational systems of the theory of probability, particularly those in which concept of probability 'is not treated as one of the basic concepts but is itself expressed by means of other concepts. However, in that case, the aim is different, namely, to tie up as closely as possible the mathematical theory with the empirical development of the theory of probability.' In other words, he admits that if the set of axioms is a sufficient basis for the logical and formal development of probability calculus, it has to be supplemented by other means in order to establish contact with the real world. It therefore follows that, in probability theory, problems I and II cannot be artificially divorced and have to be considered conjointly. Consequently the solution of problem I is of limited interest unless it also provides the best available model for solving problem II.

The need for a close tie-up between problems I and II in probability theory is not always sufficiently stressed. It is this need that has led Dr. Good in his recent book to advocate that an axiomatic theory (if it is to be directly applicable) 'should always be supplemented by a set of clearly stated rules' and also 'not so clearly phrased suggestions'. This approach has some advantage in that it pushes the purely axiomatic method as far as it can go without making any new or veiled assumptions—mathematical or otherwise. But these new assumptions come in alright when we supplement the axiomatic theory by 'rules' and 'suggestions'. The problem of 'foundations' then hinges on the 'rules' and 'suggestions' imported later.

This is not to deny the value of the purely axiomatic part of the theory. It is important to ensure that in an axiomatic theory the axioms posed are self-consistent, that they are as few as possible for the development of theory, *etc.* But it is relatively easy to build an axiomatic theory of probability by considering certain things called 'events' and assigning to each such 'event' a number called 'probability'. One could then define 'independent' and 'dependent' 'events', *etc.*, and postulate axioms to express the fundamental rules for operation with such numbers. One could even define these notions and express their definitions and axioms in the language of algebra or logic.

There is no great difficulty with the purely mathematical problem of showing the self-consistency of the set-up and proving that it contains the minimum number of definitions and axioms necessary for developing the formal calculus. But all the practical difficulties re-appear the moment one decides to apply this theoretical set-up to any real phenomenon. In fact, even in geometry, where one sets out by defining abstract geometrical entities like points, lines, planes, *etc.* (which themselves, by the way, are suggested by our daily experience), we are not concerned merely with the self-consistency of the axioms and the abstract set-up but also with the question which of the various possible set-ups (Euclidean or Lobachevskian, for instance), conforms to our actual experience.

It is pertinent to remark that Lobachevskian geometry evoked little interest at the time it was invented, in spite of its logical consistency, and was seriously studied only when it was realised that it might serve as a better model for physical space than Euclidean geometry. If the problem of applicability of the formal or axiomatic theory is of some importance in geometry, it is absolutely fundamental in a discussion of the foundations of probability. Besides, axiomatisation, after all, is only setting up in a symbolic form certain features abstracted from human experience after a careful and profound analysis thereof.

Axiomatisation is possible in a domain which has already been so well thought out that a universal agreement on the elements to be abstracted and expressed in axiomatic form can be obtained among mathematicians and physicists dealing with these concepts, and this is the case, for instance, in geometry. Where this condition no longer obtains it serves merely to mask the circular character of the definitions under a veil of elaborate logical symbolism. Now the one essential feature of our daily experience that every axiomatic theory of probability must abstract and embody in its axioms is the phenomenon of randomness. And it seems to me that this is precisely the feature that an axiomatic theory would fail to embody. To see the reason for this we must begin with an analysis of the nature of chance.

We are apt to divide phenomena into two mutually exclusive categories —necessary and contingent. The older and classical view was that everything in nature is strictly predetermined so that chance does not exist objectively. It is merely an expression of our ignorance as regards the real causes. It followed, therefore, that chance was a purely subjective category which could be eliminated progressively with the advance of knowledge. Unfortunately for the determinists the reverse has actually happened. The role of the contingent, instead of diminishing, has increased with greater knowledge. As a result the pendulum has now swung to the opposite extreme, *viz.* to the view maintained by some logical positivists that there is no necessitation or exclusion except the formal necessitation and exclusion in logic.

If this were true, all 'elementary' propositions or statements would be completely independent of one another and any two such propositions would, as Wittgenstein maintained, give to each other the probability 1/2. In other words, the probability that anything having the simple or 'elementary' property A has also some other simple or 'elementary' property B is 1/2. Actually both these extreme views are wrong. Contingency is as objective a category as necessity. Nor are they mutually exclusive or external to each other. As Engels remarked, 'chance is only one pole of an interrelation, the other pole of which is called necessity'. The significance of this remark is best appreciated when we consider how that which is pure chance for a single throw of a coin becomes contingency for more than one and statistical necessity for a large number of such accidental throws. This is how in reality as well as in the knowledge of reality, chance and necessity interpenetrate into one another at every level.

In view of this continual interplay of chance and necessity it is impossible to define chance in purely formal, logical or axiomatic terms. All that we can do is to take as a starting point for our theory of probability certain concepts based on our intuitive appreciation of human experience, in the same way that arithmetic arose out of the experience of matching or establishing a one-to-one correspondence between similar classes. Now the concepts from which mathematical theory of probability has developed from the time of Pascal and Fermat down to Poincaré and Borel in our own day and which are at the base of all probability theories, are two, which will now be considered.

First, probability judgments are concerned only with repetitive events which have a basic similarity, *e.g.* throws of dice, drawing of balls from urns under similar conditions, *etc.* We cannot definitely predict the outcome of a single event but on account of the aforementioned interplay between contingency and necessity we can make probability statements about groups of such trials. It follows, therefore, that probability of an event always pertains to a group of trials to which the event may be said to belong. Now how are we to ascertain the probability of an event E in a group or series of trials? If nothing is known about the way in which the event E can happen, then equally nothing can be said about its probability. Probability calculus has no magic formula whereby it can transmute blank ignorance into some definite knowledge. The Laplacian principle of non-sufficient reason by which we derived Laplace's m/n rule earlier leads to a number of paradoxes into which we need not go.

But one kind of knowledge that can lead to knowledge of probabilities is the fact that the possible consequences of the event are *known* to be resolvable into a number of 'equipossible' alternatives. How this knowledge may be acquired in any given case is another matter and need not detain us at the moment. Once this set of equipossible alternatives in which the event can take place is known, it is easy to calculate the probability of any given combination of these equipossible alternatives. For instance, suppose we have an urn containing N balls of which n are black and $N-n$ white. We draw a ball at random from it. What is the probability that the ball drawn is black? Here the event—drawing a ball—is resolvable into N different equipossible alternatives as the *known*, or at any rate tacitly assumed, conditions of a draw (*e.g.* that the balls after 'thorough mixing' are drawn by someone blindfolded, that the ball drawn is replaced before the next draw, that the balls are completely identical to the sensation of touch and are distinguishable only by their colour, *etc.*) definitely ensure that all the N alternatives are completely symmetrical and therefore equipossible.

Consider now another event such as throwing a die. What is the probability that it will fall with ace uppermost? Here our knowledge of the conditions of the throw is not enough to tell us whether the six possible alternative ways of a fall are equipossible or not. The older indifference theory treated the alternatives in both these instances as symmetrical and therefore equipossible. In the former case it was right, not because we did not know of any reason favouring one alternative rather than another (as the theory suggested), but rather because we knew that certain definite conditions of draw (mentioned above) ensured that the alternatives in question were equipossible. In the latter case it went astray because it treated our *ignorance* of the structure of the die and the conditions of its fall as a sufficient basis for taking the various alternatives as equipossible. We could treat the alternatives as equipossible only if we *knew* that the die was 'fair'—that is, that it was a perfect cube whose geometrical centre coincided with its centre of gravity and the mode of throw was such that it did not favour the fall of one face rather than another.

This is a far cry from the blank ignorance on the basis of which the indifference theory tried to assign equal probability to all the possible alternatives. If we do not have this knowledge, we have to consider the group of trials produced by a series of its falls. Suppose we have a series of N falls of the die of which n falls result in an ace. Here, then, we know that each of the N consequences is an ultimate event in itself and therefore equipossible. The probability of the die falling ace in this series of N trials is thus n/N. It is true that this definition of probability is relative to a specific finite series of trials and gives us no measure of probability in case of a (future) open series of trials, which is what is usually understood by the term. We shall deal with this issue later.

For the present, our purpose is to explain how the alternative ways in which an event can happen are judged equipossible. It is not because we know of *no* reason why one alternative should be preferred over another. Our ignorance of the *conditions* favouring one alternative to another is no warrant for assuming that there are none such. On the other hand, it is definite knowledge of the way in which the event can happen that can tell us whether the alternatives are equipossible or not.

It is true that in a vast majority of cases our knowledge of the conditions which govern the various alternative outcomes of an event is too meagre to tell us whether the various alternative ways of its happening are equipossible or not. Except for cases of drawing balls from urns or games of chance we do not have adequate knowledge of the detailed mechanism of the events happening to be able to resolve the alternative ways of its occurrence into equipossible alternatives. Yet this is what we must do before we can measure the probability of any alternative. In fact, before we can apply mathematics to any phenomenon whatsoever we must either enumerate or measure. When we enumerate we treat each of the discrete individuals of the group in question as on a par with one another—that is, equal in respect of the attribute under enumeration. When we measure any quantity whatsoever, whether distance, time, mass, utility, *etc.*, we tacitly assume that it is possible to find a portion of it that will fit another.

In other words, we assume that it is possible to establish superposition. In geometry, for example, this is the well-known principle of congruence. Even in the measurement of time, where it is not possible to 'freeze', as it were, an 'hour-bar' and put it alongside another, some equivalent of it has to be assumed before we can measure time, and this is done by *assuming* that periodic cycles through which a clock passes are equal. It is true that in the case of time reckoning this is an assumption. But so also is it the case in geometry where it is now realised that the invariance of a standard gauge or yardstick after transport from one point to another is as much of an assumption as the invariance of the period of a stationary clock. Without some such principle, measurement of continuously varying attributes would be out of the question. That is why some equivalent of it has to be assumed in the measurement of the probability as well.

Two types of cases arise according as the set of alternative ways in which the event can occur forms a discrete or continuous set. If the set of various alternatives in which the event can happen is discrete, we must find a way of enumerating its outcomes in such a manner that each item enumerated is on a par with any other—that is, is equipossible. If the set of various alternatives is continuous, we must discover an 'equipossible gauge' with which to span the entire continuum of its possibilities in much the same way as we use a standard yardstick to measure distances. Now the set of equipossible alternatives may be discrete, continuous, finite or infinite. If it is discrete and finite, it contains within itself the principle of its metrical relations *a priori* as a consequence of the concept of number. Laplace's m/n formula then suffices to give us a measure of the probability of each alternative.

Suppose, for instance, we have an urn model containing n identical balls of which m are white and the rest black. As we saw earlier, the probability that a ball drawn is white is m/n. No one can seriously deny that this is clear from the conditions of the trial described above. Consider next the case of a die. We mentioned that, if we did not know that it was a 'fair' die, the only way of obtaining a set of equiprobable alternatives is to take a set of N trials in which each trial is an ultimate unit and therefore equiprobable. But what is the probability of an ace in a hypothetical set of open series of trials?

Here we encounter an insurmountable difficulty. If we consider the discrete set of infinite trials, each trial in this (discrete) infinite series is an ultimate unit and therefore equiprobable. If the probability of each one of the equiprobable infinite alternatives be taken as zero, then so is the sum of the probabilities of all the alternatives instead of unity. And if it be not zero, though no matter how small, the sum is infinite. The frequentists try to escape this difficulty by claiming that probability of an outcome is the limit of the frequency of that outcome in N trials when N tends to infinity. Now this has meaning only if the series is given by a regular formula. But in that case we can equally define probability in terms of equiprobable cases without appeal to frequency limit.

Take, for instance, the probability of selecting an even integer from the open (infinite) class of integers. Here, although the set of various alternatives in which the number can be selected is discrete and infinite, it is possible to divide the set into two equipossible classes—the class of even and the class of odd integers—which enables us to calculate the required probability in terms of equiprobable cases. But if, on the other hand, the series is random, such as we obtain by an infinite series of throws of a die, it can never be realised in its entirety. All that we know is a finite section of it. In such cases nothing is gained by trying to define probability as the limit of frequency. The only way out is to consider a finite number N of trials and to base probability of an outcome by treating each of these trials as an ultimate equipossible unit as indicated above.

Two objections may be raised against it. First, it makes the measure of probability depend on the number N chosen. This, however, is no serious handicap as in any case we have to make do with approximations not only in this matter but also in the measurement of everything else. Second, it gives us no measure of the probability of the outcome in any similar future series of trials. We usually assume that the value given by a series of N trials would also approximately be the same as that given in a future (large) series. This, however, is the problem of induction, and its justification is a matter that arises however probability is defined. As the problem of induction is beyond the scope of this book, we shall not go into it here.

It is true that this is equivalent to assuming that the probability of the event is the same as that of drawing a white ball from an urn containing N balls of which n are white. In other words, we have recourse to Shewhart's method of defining a random order by means of 'some chosen random operation'. The operational model actually chosen by him is identical with the one postulated above except that Shewhart explicitly requires that the sequence is 'drawn one at a time with replacement and thorough mixing by someone who is blindfolded'.

Churchman has raised two objections against Shewhart's operational model defined above. First—and he concedes this is a trivial one—'replacement' and 'thorough mixing' 'are ill-defined concepts within our language'. Second, and this he claims is a deeper criticism, is the question as to how we can know that a certain sequence of operations 'defines' a concept. In other words, it is contended that 'Shewhart at best gives *one* operation that defines randomness and admits to be the only one. The definition of randomness is therefore not any one of these examples, and its true meaning must be in a non-operational concept. Operations are only means to an end, and no single means is *necessary* for the pursuit of an end.' If this mode of defining randomness is considered too limited, Churchman has suggested no better alternative. While I have much sympathy with his viewpoint that it is futile 'to reduce the generality of definition to the specificity of sensation', it seems to me that in this particular case the operation in question embodies all the essential features of a random event that probability theory has to handle.

We consider next the case when the set of equally probable alternatives is non-enumerably infinite or has the power of a continuum. Now in a continuous manifold the ground-form of metrical relations has to be sought from outside by the imposition of a measure function on it. In this connection it is pertinent to recall a remark Riemann made in his well-known essay on the hypotheses underlying geometry. He said that the measure of every part of a discrete manifold is determined by the number of elements belonging to it, so that a discrete manifold contains within itself the principle of its metrical relations, *a priori*, as a consequence of the concept of number. Hence there is no ambiguity involved in the determination of the probability measure of any alternative when the set of equipossible alternatives happens to be discrete.

The continuous manifolds, on the contrary, do not contain within themselves the principle of the measure relations of their constituent parts. The character of their metric has to come from outside. Once this measure function is devised it gives us the probability of each alternative. If we change the measure function of the continuous set of our equally probable alternatives, we alter the probability measure of that alternative. That is why continuous probabilities are nowadays almost universally treated by mathematicians under the title 'measure theory', a measure function being imposed arbitrarily on the continuous manifold of equally probable alternatives.*

Apparently, there is no way of getting over the arbitrariness of such a measure unless the conflict between the two dialectical polarities, the continuum and the discrete, is somehow resolved and the metric of the continuum discovered from within as in the case of discrete manifolds. The fact that a measure function has to be imposed on a continuous manifold from without enables us to deal with the criticism that classical theory leads to paradoxical results. These paradoxes arise because the manifold of equipossible alternative ways in which the event can happen is continuous and its metric may therefore be chosen in several different ways. In actual practice, the arbitrariness of the measure function imposed on the continuum of equipossible alternatives is no great handicap. The reason is that in many cases additional knowledge of the conditions of the problem suggests the measure function.

*   *   *   *

The great bulk of probability calculus is merely concerned with devising formulae for calculating the probabilities of complex events, given those of the elementary events which combine to make them. All these formulae are based on two fundamental rules—the addition and the product rules— just as the basis of all arithmetic is ordinary addition and multiplication.

We first give the addition rule of probabilities. Suppose a trial or an experiment results in one of the several mutually exclusive events E_1, E_2, ..., E_i. Such, for instance, will be the case if we draw a ball from an urn containing balls of different colours. The result of a single draw can only be a ball of one colour, and if it happens to be, say, red it cannot, at the

---

* We recall here that if a set is measurable, its measure may be defined in an infinite number of arbitrary ways. See page 140.

same time, be another, say, blue. In other words, an event, E_1, like draw-ing a red ball, excludes the possibility of another such as E_2, *viz.* drawing a blue ball. This is the meaning of the phrase 'mutually exclusive events'. If the probabilities of any two such events, say, E_1 and E_2 are p_1 and p_2 respectively, then the probability of the composite event,* either E_1 or E_2, is the sum

$$(p_1 + p_2).$$

For the sake of definiteness, let us keep to the urn experiment. If there are in all N balls of which m_1 are red and m_2 blue, then the probability p_1 of the event E_1, *viz.* drawing a red ball, is obviously m_1/N. Similarly p_2, the probability of the event E_2, *viz.* drawing a blue ball, is m_2/N. Now if we considered the composite event E, *viz.* drawing either a red or a blue ball, it could happen in $(m_1 + m_2)$ ways out of a total of N equally probable ways. The reason is that there are in all $m_1 + m_2$ balls which are either red or blue in a total of N. Consequently the probability of the composite event E is

$$\frac{m_1 + m_2}{N} = \frac{m_1}{N} + \frac{m_2}{N} = p_1 + p_2\dagger.$$

More generally, given any number of mutually exclusive events $E_1, E_2, \ldots,$ E_s with probabilities $p_1, p_2, \ldots, p_s$ respectively, the probability that *any* one of them results is

$$p_1 + p_2, \ldots + p_s.$$

Unlike the addition rule, the product rule of probability is not con-cerned with the disjunction of *mutually* exclusive events but with the probability of a composite event produced by a succession or conjunction of two or more consecutive events. Suppose, for instance, we want to ascertain the probability of a composite event E consisting of a succession of two aces in two consecutive throws of a die. Here the event E_1 is the appearance of an ace in the first throw and E_2 is its appearance in the second throw. In order that E may occur it is necessary that E_1 first occurs and having occurred is followed by E_2.

In this case the occurrence of E_1 has no effect on the probability of E_2 as the two throws are independent. But this need not always be the case. Take, for instance, the case of the urn A containing two black and one white balls cited above. What is the probability of the composite event E that in two successive draws the two balls drawn are both black?

* Such a composite event is also known as a disjunction.
† It is because of the addition rule of probabilities that in an honestly run lottery you are able to double your chance of a win by buying two tickets instead of one.

Let E_1 be the event that the first draw yields a black ball and E_2 the event that the second draw gives a black ball. In order that E may occur it is necessary that E_1 occurs first and is followed by E_2. But if E_1 occurs, we reduce the number of black balls in the urn and thereby the probability of E_2.* Thus the probability of E_2 depends directly on the occurrence or non-occurrence of E_1. The two events are not independent. In either case the probability of the composite event E is the product of the probability p_1 of E_1 and the probability p_2 of E_2, *when E_1 has already occurred*. The last clause italicised is of no importance if E_1 and E_2 are independent. Thus in the former case when E_1, E_2 are independent the probability of a double ace in two successive throws of the die is $p_1p_2 = 1/6 \times 1/6 = 1/36$. In the latter case, when the probability of E_2 depends on whether E_1 has occurred or not, the probability of drawing two black balls in two successive draws is p_1p_2, where p_1 is the probability of E_1 and p_2 is that of E_2 *provided E_1 has occurred already*. p_1 is clearly 2/3. If E_1 has already occurred, we are left with only two balls in the urn, of which one is black and one white, the second black ball of the urn having been already withdrawn as a result of the occurrence of E_1. p_2 is therefore 1/2. Consequently the probability p of the composite event E is $2/3 \times 1/2 = 1/3$.†

These two rules, the addition and product rules, enable us to calculate the probability of any disjunction or conjunction of elementary events when those of the latter are known. The actual solution of many problems in probability calculus may require considerable mathematical skill, but in principle these two rules always suffice. Take, for instance, the experiment described above (page 217) of drawing a ball from one of the two urns A and B selected at random. We first calculate the probability of the event E that the ball selected is a black ball belonging to the urn A. This is a composite event produced first by the selection of the urn A followed by drawing a black ball from it. The probability of the former, *viz.* selecting the urn A, is 1/2. Having selected A, the probability that the ball drawn is black is 2/3, there being two such balls in it. The product rule, therefore, gives the probability of E as $1/2 \times 2/3 = 1/3$.

Similarly, we can work out the probability of the event E' that the

---

* It is assumed that the second draw is made without replacing the ball drawn in the first draw.

† It is because of the product rule of probabilities that dividends of double totes in races are much bigger than those of a single win. For suppose the chance that the horse you decide to bet on in the first race wins is one in ten. Suppose further that the chance of a similar win in the second race is one in nine. Then according to the product rule your chance of a double win is only

$$\tfrac{1}{10} \times \tfrac{1}{9} = \tfrac{1}{90},$$

that is, one in ninety. Since the product rule results in a heavy slump in your chances of a double win, the amount you stand to gain in case you do win is correspondingly higher.

ball selected is a black ball from the urn B. It is $1/2 \times 3/4 = 3/8$. Now the addition rule gives the probability of the disjunction of events E or E', that is, the event that the ball drawn is a black ball belonging either to A or to B. It is $1/3 + 3/8 = 17/24$. This is the correct value of the probability of drawing a black ball in the experiment instead of the erroneous value $5/7$ given earlier.

James Bernoulli, who turned the speculations of Fermat and Pascal on Méré's problem to good account for the benefit of the mercantilists and speculators, proved a remarkable theorem named after him, by means of these two rules. Bernoulli's theorem enables us to calculate the probabilities that an event will occur, 0, 1, 2, ... or n times in n trials given the probability of a single trial. Suppose we draw a ball from the urn A considered above. Since it contains two black and one white balls, the result of a single trial is either a black ball *or* a white ball. As we have seen, the probabilities of the two events are 2/3 and 1/3 respectively. Consider now a set of two trials.* Clearly we can have one and only *one* of the following four *arrangements* where the symbol (B) denotes a black ball and (W) a white ball:

First draw	Second draw
(B)	(B)
(W)	(B)
(B)	(W)
(W)	(W)

What is the probability of each arrangement? Since the probability of drawing a black ball in each draw is 2/3 and that of a white ball 1/3 the probability of each one of the aforementioned arrangements may easily be calculated by the product rule. Thus the probability of the first arrangement, *viz.* that of obtaining a black ball in both the draws, is $(\frac{2}{3})(\frac{2}{3}) = (\frac{2}{3})^2$. Similarly the probability of the second arrangement, *viz.* a white in the first draw and a black in the second, is $(\frac{1}{3})(\frac{2}{3})$ and so on. We now summarise the probabilities of each arrangement.

	Arrangement		
	1st draw	*2nd draw*	*Probability*
Both black	(B)	(B)	$(\frac{2}{3})(\frac{2}{3}) = (\frac{2}{3})^2$
One black and	(W)	(B)	$(\frac{1}{3})(\frac{2}{3}) = (\frac{1}{3})(\frac{2}{3})$
one white	(B)	(W)	$(\frac{2}{3})(\frac{1}{3}) = (\frac{1}{3})(\frac{2}{3})$
Both white	(W)	(W)	$(\frac{1}{3})(\frac{1}{3}) = (\frac{1}{3})^2$

* We assume that the ball drawn in the first trial is replaced in the urn after noting its colour so as to make the conditions of the second draw identical with those of the first. This is an important condition as it ensures that the probability of drawing a white (or black) ball remains the same from one draw to another.

Now if we disregard the order in which the white and black balls actually occur and merely want the probability of obtaining one black and one white ball in a set of two trials, the probability of such a *combination* is given by the addition rule. For it is produced either by the arrangement $(W)(B)$ *or* by $(B)(W)$. Its probability is therefore the sum of the probabilities of these two arrangements, that is,

$$2(\tfrac{1}{3})(\tfrac{2}{3}).$$

The probabilities of all the possible three *combinations* to which the aforementioned arrangements lead, if we disregard the order in which the white and black balls are drawn in the set of two trials, are therefore:

Combination	Probability
Both black	$(\tfrac{2}{3})^2$
One white, one black	$2(\tfrac{1}{3})(\tfrac{2}{3})$
Both white	$(\tfrac{1}{3})^2$

If we add up the three probabilities, the sum is 1 as, of course, is natural because we are certain to have one or other of these three combinations. In fact, we have

$$(\tfrac{2}{3})^2 + 2(\tfrac{1}{3})(\tfrac{2}{3}) + (\tfrac{1}{3})^2 = (\tfrac{2}{3} + \tfrac{1}{3})^2 = 1^2 = 1.$$

It also follows that the probabilities of the three possible combinations are the terms of the binomial expansion $(\tfrac{2}{3} + \tfrac{1}{3})^2$.

Consider now a set of three draws with replacement as before. The number of possible arrangements is now 8 instead of 4 and the result of a draw can be one and only one of them. As before, the probability of each arrangement is given by the product rule. All the eight possible arrangements and their corresponding probabilities are as shown below:

	Arrangement			Probability
	1st draw	*2nd draw*	*3rd draw*	
All black	(B)	(B)	(B)	$(\tfrac{2}{3})(\tfrac{2}{3})(\tfrac{2}{3}) = (\tfrac{2}{3})^3$
Two black and one white	(B)	(B)	(W)	$(\tfrac{2}{3})(\tfrac{2}{3})(\tfrac{1}{3}) = (\tfrac{2}{3})^2(\tfrac{1}{3})$
	(B)	(W)	(B)	$(\tfrac{2}{3})(\tfrac{1}{3})(\tfrac{2}{3}) = (\tfrac{2}{3})^2(\tfrac{1}{3})$
	(W)	(B)	(B)	$(\tfrac{1}{3})(\tfrac{2}{3})(\tfrac{2}{3}) = (\tfrac{2}{3})^2(\tfrac{1}{3})$
One black and two white	(W)	(W)	(B)	$(\tfrac{1}{3})(\tfrac{1}{3})(\tfrac{2}{3}) = (\tfrac{1}{3})^2(\tfrac{2}{3})$
	(W)	(B)	(W)	$(\tfrac{1}{3})(\tfrac{2}{3})(\tfrac{1}{3}) = (\tfrac{1}{3})^2(\tfrac{2}{3})$
	(B)	(W)	(W)	$(\tfrac{2}{3})(\tfrac{1}{3})(\tfrac{1}{3}) = (\tfrac{1}{3})^2(\tfrac{2}{3})$
All white	(W)	(W)	(W)	$(\tfrac{1}{3})(\tfrac{1}{3})(\tfrac{1}{3}) = (\tfrac{1}{3})^3$

Again, if we disregard the order in which the white and black balls appear in the various arrangements and want the probability of a given *combination* of black and white balls, we find that the three arrangements $(B)(B)(W)$,

$(B)(W)(B)$, $(W)(B)(B)$, for example, lead to a single combination, *viz.* two black and one white balls in a set of three trials. The probability of this combination is therefore the sum of the probabilities of its three composite arrangements, *viz.* $3(\frac{2}{3})^2(\frac{1}{3})$. Similarly the three arrangements $(W)(W)(B)$, $(W)(B)(W)$, $(B)(W)(W)$ lead to the combination, two white and one black ball, with probability $3(\frac{2}{3})(\frac{1}{3})^2$. All the four possible combinations in a set of three draws have therefore the probabilities:

Combination	Probability
All three black	$(\frac{2}{3})^3$
Two black, one white	$3(\frac{2}{3})^2(\frac{1}{3})$
One black, two white	$3(\frac{2}{3})(\frac{1}{3})^2$
All three white	$(\frac{1}{3})^3$

Again, we may verify that the sum of the four probabilities is unity, as one or other of these four combinations is certain to occur. In fact,

$$(\tfrac{2}{3})^3 + 3(\tfrac{2}{3})^2(\tfrac{1}{3}) + 3(\tfrac{2}{3})(\tfrac{1}{3})^2 + (\tfrac{1}{3})^3$$
$$= (\tfrac{2}{3} + \tfrac{1}{3})^3 = 1.$$

This also shows that the four probabilities are the terms of the binomial expansion $(\frac{2}{3} + \frac{1}{3})^3$. If we proceed in this manner we can easily show that in a set of n draws there can be only $(n + 1)$ possible *combinations* (though 2^n arrangements) and that the probability of each combination is as shown below:

Combination	Probability
All n black	$(\frac{2}{3})^n$
$(n - 1)$ black, one white	$n(\frac{2}{3})^{n-1}(\frac{1}{3})$
$(n - 2)$ black, two white	$\dfrac{n(n-1)}{2}(\frac{2}{3})^{n-2}(\frac{1}{3})^2$

- -

One black, $(n - 1)$ white	$n(\frac{2}{3})(\frac{1}{3})^{n-1}$
All n white	$(\frac{1}{3})^n$.

As before, the probabilities of the various combinations are the terms of the binomial expansion $(\frac{2}{3} + \frac{1}{3})^n$. Consequently the probability of a combination of r white balls and $(n - r)$ black balls is

$$\frac{n(n-1)(n-2) \ldots (n-r+1)}{1 . 2 . 3 \ldots r}(\tfrac{2}{3})^{n-r}(\tfrac{1}{3})^r,$$

or, $${}^nC_r(\tfrac{2}{3})^{n-r}(\tfrac{1}{3})^r,$$

where ${}^nC_r$, as usual, is the coefficient of $(\frac{1}{3})^r(\frac{2}{3})^{n-r}$ in the binomial expansion $(\frac{1}{3} + \frac{2}{3})^n$.

More generally, if the probability of the black draw is q instead of $\frac{2}{3}$ and therefore of a white draw p instead of $\frac{1}{3}$ (where $p = 1 - q$), the probability of a set of n draws of which r are white and $n - r$ black is

$$\frac{n(n-1)(n-2)\ldots(n-r+1)}{1.2.3.\ldots r}p^r q^{n-r},$$

or, ${}^nC_r p^r q^{n-r}$, for short. If we agree to denote the product of successive integers from unity onwards—that is, the product $1.2.3.\ldots r$—by the symbol $r!$ (read factorial r), the expression ${}^nC_r p^r q^{n-r}$ can also be written as

$$\frac{n(n-1)\ldots(n-r+1)}{1.2.3.\ldots r}p^r q^{n-r}$$

$$= \frac{n(n-1)(n-r+1)}{1.2.3.\ldots r}\frac{(n-r)(n-r-1)\ldots 2.1}{(n-r)(n-r-1)\ldots 2.1}p^r q^{n-r}.$$

$$= \frac{n!}{r!(n-r)!}p^r q^{n-r}.$$

Consequently the probability that in a group of n trials an event E occurs r times and fails to occur in the remaining $n - r$ trials is simply

$$\frac{n!}{r!(n-r)!}p^r q^{n-r}.$$

This result is known as Bernoulli's theorem. Obviously Bernoulli's theorem is equally applicable to any series of independent trials of events whose probability (p) remains constant in each trial. Thus if in a set of n independent trials an event E occurs r times and, therefore, not-E (or E') occurs $(n - r)$ times, the probability of obtaining such a set of n trials is ${}^nC_r p^r q^{n-r}$.

Bernoulli deduced from this theorem what is known as the law of Large Numbers. This law means, in effect, that if in the above experiment of n trials, we find that the event E occurs m times and not-E, $(n - m)$ times, then, given any positive number ε, as small as we please, the probability that the ratio m/n lies between $(p - \varepsilon, p + \varepsilon)$ tends to 1 as n tends to infinity. In other words, it is practically certain that the ratio m/n* will approximate to the probability (p) of the event E, provided the number of trials is sufficiently large. If ever there is safety in mere numbers it is here—in Bernoulli's law of large numbers.

*　　　　*　　　　*　　　　*

* The ratio m/n is also known as the relative frequency of E in a set of n trials.

In order to understand further development of probability theory, as well as its applications to modern statistics, a short preamble on statistics is necessary.

In his *Facts From Figures* (Pelican, 1951), M. J. Moroney has aptly described statistics as the science of deriving facts from figures. This is accomplished in two stages. First, we compile figures about the sort of facts we want to know. Then we try to read the tale the figures tell. There is thus a close tie-up between the two: the sort of enquiry we intend to make determines the type of figures we ought to compile and the sort of figures we compile determines the facts we are able to derive. Although one of the most interesting studies in statistical theory is the design of figures we ought to compile for any given purpose, we shall not dwell on it here. We shall merely assume that the figures have somehow been compiled.

If we have an array of figures about any phenomenon, our first problem is to find a way of summarising the mass of figures presented to us. For no human mind can grasp the significance of a vast array of figures when simply confronted with it. In order not to miss the wood for the trees it is necessary to replace this array by a few selected figures—preferably by a single figure. How shall we proceed so as to discover a single figure that could in some way be taken as a representative of the entire array?

There are several ways of doing this. For example, there is the well-known method of replacing the entire array by its average or arithmetic mean. To calculate it we add *all* the figures in the array and divide it by the number of figures so added—as every schoolboy knows. However , the arithmetic average is only one among several possible ways of replacing a vast array of figures by a single representative figure. In some cases it is clearly misleading to use the arithmetic average as a representative of an entire set. For instance, suppose we had two railway yards. Suppose, further, that in one of these 99% of the wagons suffered a detention of 16 hours and 1% 25 days, while in the second 50% suffered a detention of 18 hours and the other 50% 22 hours.* The average detention in the former case would be about 22 hours while in the latter case 20 hours. Yet on the whole the wagons in the former suffered less detention than in the latter. If you pick a wagon at random from each yard, it is 99 to 1 that the wagon from the second yard has been held up longer. We could say that the wagons in the former have been held up longer only if the detentions were evenly distributed; but that is a big 'if'.

How can we choose a representative figure so as to get over fallacies of

---

* Although artificially constructed it is by no means an entirely fanciful example. In some yards a few wagons suffer prolonged detentions owing to want of wagon labels, need for repairs, *etc.*

this kind? In such cases we replace our array of figures by another kind of average which is not influenced by extreme items of the set. One such average is known as the *median*. It is the detention suffered by the middle wagon if we arrange all wagons in order of their detentions. Suppose we had in our yard 101 wagons with the following detentions:

25 wagons	.	.	.	.	14 hours each
15 wagons	.	.	.	.	15 hours each
12 wagons	.	.	.	.	24 hours each
18 wagons	.	.	.	.	36 hours each
20 wagons	.	.	.	.	40 hours each
11 wagons	.	.	.	.	47 hours each

If we arranged these 101 wagons in ascending order of detentions suffered by them, the detention of the middle wagon, *i.e.* the fifty-first wagon in this ordering, is known as the median. It is easy to see that the fifty-first wagon would be one of the 12 wagons which suffer a detention of 24 hours each. The median value is therefore 24 hours. In the same way we observe that in the case of the two yards cited above the median value for the first yard is 16 hours and that for the second 20 hours. If we compare the median detentions of the two yards instead of their arithmetic averages, we shall avoid the fallacy mentioned above.

Another way of replacing our array of figures of wagon detentions referred to above is to replace it by what is called the *mode*. By mode is meant that value of the detention which pertains to the largest number of wagons. In the example quoted above the largest group of wagons is the group of 25 wagons which suffered a detention of 14 hours each. This is therefore the fashionable or *modal* value; mode in French meaning fashion. The mode—that is, the value of wagon detention common to the largest number of wagons—is thus 14 hours.

From the way in which the median and the mode are obtained it is clear that they are not influenced by the extreme elements of the array. They are, therefore, in some cases better representatives of the entire array of figures than the more usual arithmetic mean. But it will be readily appreciated that even the best representative figure or average can give only a somewhat oversimplified picture of the facts buried beneath the figures. For a deeper penetration into these facts we have to relax the restriction we imposed at the outset, *viz.* to make do with a single figure. Suppose then we were permitted to summarise our array of figures by means of two or more figures instead of only one. How shall we proceed?

In the first place we may take the average as one of the two or three figures permitted us in preparing our summary. How shall we select the other figures? That depends on whether we want to have only one or two additional figures. If we want our summary to consist of two additional figures, we can select the maximum and minimum value from our array of figures in addition to the average. Thus, suppose we were given the following array of figures of detentions in hours to freight trains dealt with in a railway yard like Gaya on a certain day:

$$2, 3, 4, 6, 5, 3, 4, 4, 5, 4.$$

We might replace the entire array by a summarised picture:

Average detention per train	.	4 hours
Maximum detention .	.	. 6 hours
Minimum detention .	.	. 2 hours

But if we wanted our summary to contain only two figures, we could take the difference between the maximum and minimum, *viz.* $6 - 2 = 4$, as a measure of the *range* within which the detentions vary. Thus we could say the average detention per train on that day was 4 hours, with a range of variation of 4 hours. For technical reasons into which we need not go, statisticians have not taken kindly to the idea of using range—that is, the difference between the maximum and minimum—as a measure of the variance or 'spread' of a given array of figures. They prefer to use another measure (called the standard deviation) for this purpose. The idea of standard deviation may most simply be explained as follows.

Taking the figures of the array of detentions to freight trains given above as an illustration, we may say that the average of our figures is 4 hours. If we take the deviation or difference of the individual figures from our average, we have another set of figures equal in number to our original set. These ten deviations are obtained by subtracting the average 4 from each item in succession. They are reproduced below:

$$(2-4), (3-4), (4-4), (6-4), (5-4), (3-4), (4-4), (4-4), (5-4), (4-4),$$
or, $\quad -2, \quad -1, \quad 0, \quad +2, \quad +1, \quad -1, \quad 0, \quad 0, \quad +1, \quad 0.$

We could replace this new set, the set of deviations (or differences from the mean), by a single figure as a representative of the entire set. This latter figure could then be taken as a measure of the magnitude of the variation or 'spread' of the original figures from their arithmetic mean.

How shall we replace our set of deviations or differences from the average by means of a single figure? If we took the average of these differences (each difference with its appropriate sign), we should get zero as the average of these differences as a glance at the figures shows. In fact, this would be so not only in this particular case but in every case on account of the very nature of the arithmetic average. You can try this by taking any set of figures and striking their average. Now take the differences of these figures from the average. Some of these differences will be plus, others minus. If you sum up these differences, having regard to their respective signs, the sum would be zero.

It would therefore seem that it is no use taking the average of these differences or deviations as we shall always get one and the same result, *viz.* zero. To avoid this difficulty we could ignore the minus sign of the differences and strike their arithmetic average *as if* all the differences were positive. We could thus replace the array of differences by means of this average of the differences, all of them being regarded as positive. This would give us a measure of the variance or 'spread' of our original array of figures. For instance, the average value of the deviations quoted above, regardless of their signs, would be 8/10 = 0·8. This figure, *viz.* 0·8 hour, may be taken as a representative of the set of individual deviations from the mean and therefore as a measure of the 'spread' of train detentions round the mean. It is known as the mean deviation.

However, tampering with the sign of deviations brings its own nemesis in the long run. It makes this measure of variance or 'spread' mathematically intractable. Statisticians therefore do not favour this idea. They prefer to solve the difficulty in another way. If some of the differences are positive and others negative, and if the sum of the positive differences always cancels out that of the negative differences, we could make them all positive by squaring the differences. You will recall that the product of two negative numbers is positive so that the square of a negative difference, *e.g.* −2, is +4. We could now deal with the squares of differences which are all positive, and could sum up these squares. Dividing this sum by the number of differences added up we get the average of these squared differences. We now take the square root of this average to come back to the same level of figures as the original. This is known as the *standard deviation* and gives a measure of the 'spread' of the given set of figures round the average.

Using the ten figures of detentions to through trains quoted above as our illustration, we have our array of differences or deviations from the arithmetic average thus:

$$-2, \ -1, \ 0, \ +2, \ +1, \ -1, \ 0, \ 0, \ +1, \ 0.$$

We note that some of them are positive and others negative. We square these differences to make them all positive. The result is

$$4, 1, 0, 4, 1, 1, 0, 0, 1, 0.$$

The sum of these 10 squares of differences is 12. The average is, of course, $\frac{12}{10} = 1·2$. The square root of 1·2 is 1·1, which is thus the value of our standard deviation.

Now you have every right to demand *why* statisticians prefer to use this roundabout method of measuring the 'spread' of the given array of figures to the simple method of calculating the mean deviation. The reason is that the standard deviation enables us to state precisely what proportion of figures would lie within any stated range. In most cases we find that we make very little error if we assume that 50% of the figures of our array lie within the range $a - \frac{2}{3}s$ and $a + \frac{2}{3}s$, where a is the average and s the standard deviation of the array. Similarly 95% of the figures lie within the range $a - 2s$ and $a + 2s$ and 99% within $a - 3s$ and $a + 3s$. This is, no doubt, a rough rule, but it is one which is found to work well enough in practice. In the case of our array of ten figures of detentions to through trains we found $a = 4$ and $s = 1 \cdot 1$, the range $a \pm \frac{2}{3}s$, *i.e.* (3·3, 4·7), contains four items of the array. This works to 40% in place of the 50% postulated by the rule. The error is due to rounding off our figures of train detentions to the nearest hour. If we did not round off these figures, the correspondence would be closer still. Similarly the range $a \pm 2s$, *viz.* (1·8, 6·2), contains all the ten items of the array. This works to 100% instead of 95% postulated by the rule.

In fact, a knowledge of the arithmetic average a and standard deviation s of any given array would enable us to state in most cases, at least approximately, the percentage or proportion of figures lying in any stated range, $a \pm ns$. This proportion by the frequency rule is also a measure of the probability that a figure selected at random from our array lies within the range. If we give to n different values like $n = \frac{2}{3}$, 1, 1·5, 2, *etc.*, we get different ranges. Tables exist which enable us to state the proportion of figures of our array lying in the range $a \pm ns$ for various values of n. This proportion is at the same time the probability that any figure of the array picked up at random lies within the range in question.

*　　　　*　　　　*　　　　*

In an earlier section we outlined a method for calculating the probability of complex events, given the probabilities of the component events. For example, we are given the distribution of black and white balls in an urn and, therefore, the probability of drawing a black or a white ball in a single draw. Knowing this we can calculate the probability of obtaining any preassigned set of results such as drawing two black balls and one white in a set of three trials or drawing one black and six white balls in a set of seven trials, *etc.* But the inverse problem of ascertaining the most probable distribution of the balls in the urn, given the result of a set of trials, is more interesting and much more difficult.

To take an example, suppose we have an urn known to contain a number of similar balls, some of which are white and others black. We draw one ball after another, replacing each ball before the next draw, so as to make the conditions of draw in each case identical. Suppose we made 100 draws and found 55 white and 45 black balls. What is the proportion of white balls in the urn? Deductive logic provides no answer, for theoretically you could obtain such a series of draws from an urn containing almost *any* proportion of white balls. Nevertheless, certain values of the proportion of white balls are more likely to lead to the observed result than others. For example, it can be shown that if p is $1/2$, the probability of drawing 55 white balls in 100 draws is more than 11 times as great as that of obtaining the same result if p were $2/3$. p is, therefore, more likely to be $1/2$ than $2/3$.

The problem, however, is to find the best possible value of p. As we remarked before, as far as the single observation of drawing 55 white balls in 100 draws is concerned, it could have arisen from almost any value of p. We cannot, therefore, expect to find any method of estimation which can be guaranteed to give us always a close estimate of p. All we can reasonably do is to formulate a rule which, if followed all the time, will give us fairly close estimates of p in *most* cases. For instance, if we adopted the rule that p is estimated by the ratio of the number of white balls observed in a series of draws, $55/100$ in the case under consideration, it can be shown that it would lead to a correct or very nearly correct estimate of p in a vast majority of cases.

The celebrated statistician, R. A. Fisher, proposed that the method of estimation chosen should be required to fulfil three conditions. In the first place, the method of estimation should be *consistent*. What it means may be illustrated by an example. Consider a production process turning out ball bearings of a certain diameter d. However perfect the process, it cannot turn out all the balls with *exactly* the same diameter. Their actual diameters would hover round the specification number d within a more or less narrow margin. If we are given a large lot of these ball bearings and we wish to ascertain their average diameter, we may select at random a sample of, say, 50 balls and ascertain the sample average or mean. We may then simply take the *sample* average or mean as equal to the *lot* average (d). This is one way of estimating the lot average.

However, the size of the sample remains arbitrary. We could, for instance, take a sample of 100 or of any other larger or smaller number of balls, instead of a sample of 50, and each such sample would lead in general to a different estimate of d. There is no reason why a sample average of, say, 100 balls should equal that of, say, 50. But if our method of estimating d—i.e. our 'estimator'—is consistent, all these different estimates should approximate closer to the lot average d as the sample size increases. If the given 'lot' is so large that we may treat it as practically infinite, we could, of course, never sample it completely. But the different estimates of its value that we should get would converge in probability to a limiting value d as the number of balls selected in the sample increased indefinitely. In other words, the larger the sample the greater the probability that the estimated value of d differs from its actual value by an arbitrarily small amount. Any method of estimation which fulfils this requirement is called consistent.

Very often we have several consistent methods of estimation. For instance, in the aforementioned example, we took the sample average as an estimate of the lot average. We could instead take the sample *median* as an estimate of the lot average. This, as we explained earlier, means that we first arrange the balls selected in the sample in ascending order of their magnitude—first the smallest ball of the sample, second the next bigger and so on, ending with the biggest ball of the sample. The median then is simply the diameter of the middle ball in this ordering. Thus, if our sample consisted of 51 balls, the median would be the diameter of the twenty-sixth ball when all 51 of them are arranged in order of their diameters. If, on the other hand, the sample size were an even number, *e.g.* 50, there would be two middle items, *viz.* the twenty-fifth and twenty-sixth balls, and the median would simply be the average diameter of these two. If we adopted the median method of estimating the average diameter d of the lot, it can be proved that the sample medians, too, would converge to d as the sample size increased indefinitely. The median method, therefore, is also consistent. Which one of the two methods—the mean method or the median method—then, should we adopt?

To choose between the two we adopt a second criterion of a good estimator. This is the criterion of efficiency. It recommends us to adopt that method of estimation which in practice 'works better'. What do we mean by it? Suppose we decide to adopt the mean method of estimation and work with samples of any fixed size, say 50. Naturally different samples selected from the lot would lead to different values of sample means even though the sample size remained the same. Thus, for instance, five samples of fifty each may have the following means: 41, 38, 39, 42, 40 millimetres. These different values of the sample means have a definite standard deviation. Similarly, if we decide to adopt the median method and work with samples of the same size, *viz.* 50, we shall obtain another series of values of sample *medians* given by the different samples. All these sample *medians* too have a definite standard deviation. Now, as it happens, the fact is that the standard deviation of the sample medians is different from that of the sample means although the sample size in both cases is the same. The more efficient method of estimation is the one that has a *lower* standard deviation. The reason is that it is less subject to sampling fluctuations and therefore leads to more stable estimates. In the present case it can be proved that sample means have a *lower* standard deviation than sample medians. The mean method of estimation is therefore more efficient than the median method.

We can, in fact, compare the relative 'efficiencies' of the two methods. For it can be shown that the standard deviation of sample *medians*, when the sample size is 100, is about the same as that of sample *means* when the sample size is 64. The use of the median method therefore sacrifices about $100 - 64 = 36$ observations in 100. This sacrifice of observations entails, in turn, loss of 'information'* contained in the sample. In general, whatever method of estimation we might choose it would involve some loss of 'information'. Naturally the 'best' method of estimation is the one which does not lose any. Such a method of estimation is said to be sufficient. In practice this means that if our method of estimation is sufficient, it gives all the 'information' that the sample can supply about the lot average and no other method of estimation can add anything to it. It is, therefore, the best estimate one could ever have. But unfortunately sufficient estimators do not always exist, and in fact they are the exception rather than the rule. Since sufficient estimators do not always exist, other methods of approach have been suggested by various writers.

*   *   *   *

In order to explain the vocabulary required for a discussion of an alternative approach to the problem of estimation and inductive inference, we shall now complicate a little the urn problem cited above. Let us imagine that we have a collection of three different types of urns, each of which contains a certain known proportion of white balls (see Fig. 61). We first select an urn at random and draw our series (with replacement) as before. We are given the number of white balls in the series of draws. What type of urn did we select for drawing our series? Here again, our actual series could have, in principle, arisen from almost any type of urn. But we have to select the most reasonable alternative on given evidence. Now, T. Bayes showed that this problem could not be solved without some additional information. Not only must we know the result of our series of draws and the actual proportions of white balls in each type of urn, but we must also know the relative proportions of the various types of urns in our collection of urns.

The proportions of various types of urns may be equal or we may have

---

* This use of the word 'information' is technical but is fairly closely allied to its everyday meaning.

a preponderance of some types of urns over others. The proportion of an urn of any type to the total number of urns in our collection represents the *initial* chance of its being selected for the purpose of drawing our series. Thus, if our collection had 30 urns of three types, 10 of each type, each type of urn has an equal chance of being selected initially. But if, on the other hand, it had 20 of type *A*, 5 of type *B* and 5 of type *C*, the initial

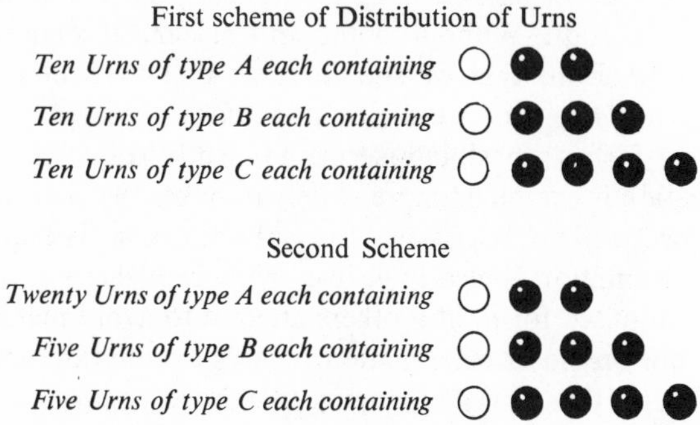

First scheme of Distribution of Urns

Ten Urns of type A each containing ○ ● ●

Ten Urns of type B each containing ○ ● ● ●

Ten Urns of type C each containing ○ ● ● ● ●

Second Scheme

Twenty Urns of type A each containing ○ ● ●

Five Urns of type B each containing ○ ● ● ●

Five Urns of type C each containing ○ ● ● ● ●

Fig. 61

chance of selecting an urn of type *B* would be four times that of type *B* or *C*. This initial chance is technically known as its prior probability. In the examples just quoted the prior probabilities of the three types of urns *A*, *B* and *C* are all equal, being 1/3 in the former case and 2/3, 1/6, 1/6 respectively in the latter case. Bayes showed that, given the prior probabilities of each type of urn in addition to other information already mentioned, we can work out which of the various alternatives is the most reasonable to adopt. But in the absence of any information about prior probabilities the problem cannot be solved.

At this point it may justly be objected that there seems little or no connection between the experiments of drawing balls from urns and the actual problems of inductive inference. But it is remarkable that the practical problems of scientific inference in widely different fields—such as the problems arising in the oxidation of rubber, the genetics of bacteria, fruit fly or human beings, testing the quality of manufactured goods and many more besides, involve the solution of 'urn-problems' of this and similar kinds.* In dealing with more complicated problems we find that when the problem is transcribed in terms of a suitable urn model, we often do not know the prior probabilities of the various types of urns without which Bayes's solution is inapplicable.

* For example, the problem of estimating *p* by observing the proportion of white balls in a series of, say, 100 draws is similar to that of ascertaining the proportion (*p*) of defectives in a production process as a whole by observing the fraction defectives in a sample of, say, 100 articles.

Many scientists try to evade this difficulty by suggesting that if we do not know anything about the initial proportions of the various types of urns, *i.e.* their *prior* probabilities, we may assume them to be all equal to one another, and use Bayes's formula to obtain a solution. Others, recognising the arbitrary nature of this assumption, propose to circumvent Bayes's formula by specially devised postulates. For instance, R. A. Fisher proposed to replace Bayes's formula by what he called the postulate of 'maximum likelihood'. Without going into technical details, Fisher's postulate merely means that we should select that particular alternative which is most likely to produce the result actually observed.

In this form Fisher's postulate seems eminently reasonable, but if its actual application is examined more closely it makes the same assumption about prior probabilities as Bayes. The only difference is that in Bayes's formula the assumption is clearly visible, while in Fisher's formulation it appears in a disguised form. Still others attempt to avoid making Bayes's assumption, but on closer examination Bayes's assumption comes out in the open. It seems that Bayes's theorem, like the legendary phoenix, once burnt reappears from its ashes in another form. What then is the way out of the impasse? Apparently we can adopt only two courses; either we guess the unknown prior probabilities as plausibly as we can and use Bayes's theorem, or try a totally different approach by modifying the problem in the manner recently suggested by Neyman, Wald and others so as to avoid having to use the unknown prior probabilities.

Neyman has suggested that the problem be modified in the same way as Henri Lebesgue modified the concept of ordinary integral. As we saw in Chapter 3, behind the concept of a definite integral there is the intuitive idea of an area bounded by a curve. We can in most cases compute the 'area' enclosed by a curve by means of an ordinary integral. But there are cases where a procedure for computing the 'area' is needed and yet the ordinary integral does not exist. The difficulty was resolved by Lebesgue, who gave a new definition of integral which applies to a wider class of functions to which the earlier concept of an integral could not be applied. But whenever the ordinary integral exists it coincides with the Lebesgue integral. Neyman considers that a similar generalisation of Bayes's theorem is necessary. This generalised theorem should include Bayes's theorem as a particular case when prior probabilities are known, but should be broad enough to work even when the prior probabilities are unknown, without which Bayes's theorem is inapplicable. One such generalisation of Bayes's theorem is Wald's theory of Statistical Decision Functions. To explain his ideas it is necessary to digress a little on statistical theory.

The most important problem in statistical theory is that of inferring the characteristic of a population by observing a random sample selected from it. Suppose, for instance, you were in charge of a manufacturing process producing millions of articles. You would naturally want to know what proportion of the articles produced were defective. One method of ascertaining it would be to test all the articles produced in the past and to compute the fraction defective. But it would be impracticable to do so, not only because the number of articles to be tested is too large, but also because in many cases the test may involve destruction of the article to be tested, as, for example, in testing the tensile strength of steel bars. We can therefore only 'estimate' the characteristic of our 'population' (in this case the proportion of defectives among the entire output of the manufactured products taken as a whole) by observing a specimen sample.

In dealing with this problem Wald has, under pragmatic influence, given up the older point of view, which regarded the question of the value of the proportion defective as something objective and quite independent of the consequences it may entail. Instead, he considers that in ascertaining the proportion of defectives in a lot of manufactured products the question of its 'true' value has no meaning unless its consequences in action are fully investigated. Thus, if you concluded on the basis of certain data that this proportion was quite low, say less than one per cent, you might decide to release the product for sale at regular price. If, on the other hand, you concluded that it was rather large, say ten per cent or more, you might decide to withhold it altogether. For intermediate values between one and ten per cent you might decide to sell it at a reduced price.

Now, since no matter what you do, you can never be absolutely sure that your estimate is 'true', you run a risk of making a wrong decision which will result in some loss. Wald admits that estimation of such a loss when any particular decision happens to go wrong may often run into difficulties; but he cuts the Gordian knot by remarking that it is not a statistical problem but a problem of 'values'. Granting the existence of a 'loss' function corresponding to the entire field of possible decisions, Wald makes the natural assumption that any decision (D) is better than another decision (d), if the expected value of the loss incurred by adopting D is less than that incurred by adopting d. But which among the many possible decisions is the best? Here, there is a fairly wide choice of criteria available and it cannot be stated unequivocally that any one of them is superior to others in all circumstances. However among the various criteria possible Wald suggests that a decision ($\overline{D}$) which minimises the maximum risk is the 'best'. What does he mean by it?

To explain his meaning, let us consider the urn problem cited at the beginning of this section. Here, the datum of the problem is knowledge of the number of white balls in a series of, say, n draws, and we are required to decide from which of the three types of urns, A, B and C, we drew the series. For the sake of simplicity we shall assume that n is unity and that the single ball drawn is white. That is, we propose to decide the *type* of urn selected for making the draw on the basis of a single observation that a ball drawn from it at random is white. Now if we are given the initial chance (or prior probability) of the selection of each type of urn, by means of Bayes's theorem we can calculate the chance that the given series of draws was made from an urn of type A, B or C. On the basis of these chances we can evaluate the 'loss' that we may expect to incur corresponding to each of the three possible decisions that we may adopt. For instance, suppose the initial chance (or prior probability) of the selection of the three types of urns was the same for all—that is, 1/3 for all the three types A, B, C. On the basis of this assumption we can prove by means of Bayes's theorem that the chance that the urn selected was of type A is 20/47, of type B, 15/47 and of type C, 12/47.

In other words, if we happened to decide that the urn selected was of type A, we should expect to be right only about 20 times out of 47 and therefore wrong about 27 times out of 47. Now if we go wrong we incur some 'loss'. Although, as mentioned earlier, the actual estimation of this loss is not a simple matter, yet we may simplify the situation by assuming that every time we make a wrong decision we incur a unit loss, otherwise zero. In other words, the loss that we incur in case we decide in favour of an urn of type A is zero if our decision is right and 1 if our decision is wrong. But the chance of a right decision in this case is 20/47 and that of a wrong one 27/47. It therefore follows that the expected value of the loss in case we decide in favour of an urn of type A is $0 \times (\frac{20}{47}) + 1(\frac{27}{47}) = \frac{27}{47}$. Similarly the expected values of loss in case we decide in favour of types B and C are 32/47 and 35/47 respectively. We note that the minimum of these three values is 27/47, corresponding to a decision in favour of type A. We may now summarise the results of our discussion as follows:

Assumed prior probabilities	Values of expected losses corresponding to the three possible decisions; A, B, C,			Minimum loss	Decision corresponding to minimum loss
	A	B	C		
$\frac{1}{3} : \frac{1}{3} : \frac{1}{3}$	$\frac{27}{47},$	$\frac{32}{47},$	$\frac{35}{47}$	$\frac{27}{47}$	A

But the assumption that prior probabilities are all equal is arbitrary, for the proportion of urns of various types in our collection may as well be quite different. If it is, say, $\frac{1}{9} : \frac{2}{9} : \frac{2}{3}$, a similar calculation by means of Bayes's theorem shows that the chances that the urn selected is of type A, B or C

are 10/61, 15/61 and 36/61 respectively. Consequently the three values of expected losses are 51/61, 46/61, 25/61, of which the minimum is 25/61. In general, to each set of values of the prior probabilities there corresponds a minimum value of the expected loss. Since theoretically the number of sets of values that prior probabilities can have is infinite and, as each set generates its own minimum of expected loss, we have an infinite number of such minimum values of expected loss. We may thus construct a table like the one reproduced below showing the minimum loss corresponding to each possible set of values or prior probabilities.

TABLE OF MINIMUM LOSSES

Assumed prior probabilities	Values of expected losses corresponding to three possible decisions, A, B, C			Minimum loss	Decision corresponding to minimum loss
	A	B	C		
$\frac{1}{3}\cdot\frac{1}{3}\cdot\frac{1}{3}$	$\frac{27}{47}$	$\frac{32}{47}$	$\frac{35}{47}$	$\frac{27}{47}$	A
$\frac{1}{9}\cdot\frac{2}{9}\cdot\frac{2}{3}$	$\frac{51}{61}$	$\frac{46}{61}$	$\frac{25}{61}$	$\frac{25}{61}$	C
$\frac{1}{6}\cdot\frac{2}{3}\cdot\frac{1}{6}$	$\frac{18}{23}$	$\frac{8}{23}$	$\frac{20}{23}$	$\frac{8}{23}$	B
· · · · · ·	· ·	· ·	· ·	· ·	· ·
· · · · · ·	· ·	· ·	· ·	· ·	· ·

Out of this infinity of minimum values of expected loss, each contributed by a possible set of prior probabilities, it is possible that there *may** be a maximum. We pick up that decision as the 'best' which corresponds to this maximum of the minimum losses.† Such a solution is known as a minimax solution of the decision problem and corresponds to a least favourable prior probability of selecting the three types of urns. In other words, if we adopt a minimax solution of a decision problem, we shall minimise the loss that we stand to suffer even under the worst circumstances.

Although we have explained the idea of a minimax solution by means of the notion of prior probability and Bayes's theorem, it must be emphasised that Wald's approach is actually designed to circumvent the use of prior probabilities. He has shown that under some very general conditions a

* We say 'may' because the set of minima in the third column is infinite. If it were finite, there would certainly be a maximum. But an infinite set need have no maximum belonging to itself.

† We may remark in passing that there is a close parallel between Wald's theory of decision functions and Neumann's theory of games. Just as we solve a game problem by selecting a maximum of column-minima of K-matrix, we solve a decision problem by finding the maximum of the column-minima recorded in the third column of our table. As will be shown in the sequel, Wald has developed this parallel still further.

minimax solution of a decision problem does exist and that this solution corresponds to Bayes's solution relative to a *least* favourable prior probability of the unknown characteristic of the population, (*viz.* the type of urn from which the series of draws was made in the case under consideration), even though this least favourable prior probability is actually unknown.

However the main difficulty in applying Wald's theory is three-fold. First, Wald's proof of the existence of a minimax solution of a decision problem is chiefly of heuristic value. It does not enable us to find one in many concrete cases. Second, Wald's principle prescribes in some cases a course of action that would be considered irrational by all reasonable men. Suppose, for instance, a manufacturing process produces articles of which a constant (but unknown) proportion (p) is defective. Suppose further the manufacturer wishes to decide between the following two alternative courses of action:

> *Alternative I*: Sell the lot with a double-your-money-back guarantee for each defective article.
> *Alternative II*: Withhold the lot altogether from the market.

Here it can be shown that the minimax solution of the decision problem is in favour of alternative II so long as the manufacturer considers the possibility $p > \frac{1}{2}$ even remotely possible. Nevertheless, it will be agreed that it would be foolish to dump the lot merely because there is a slight chance of more than half the lot turning out to be defective. Third, a rather more serious difficulty arises when Wald's methodology is generalised to cover cases other than those of industrial application. The reason is that in Wald's theory, the term 'loss' is not restricted to loss of money which an industrialist may incur by making a wrong decision. It is given a very general meaning and includes 'damage' not necessarily measurable by monetary standards.

Thus, if you wanted to estimate the lethal concentration of a certain dosage relative to human beings, and if your estimate was significantly in excess of the 'true' value, your decision would have disastrous consequences. Such a disaster would still be called 'loss' although it could not be measured in money. However, Wald and his followers admit that estimation of 'loss' does run into difficulties in such cases but consider that it is not a statistical problem. Nevertheless, they do assume that this evaluation of 'loss', at least in a simplified and approximate form such as we adopted earlier, is often possible in general even when there is no question of any financial loss.

In spite of these difficulties Wald's theory of statistical decision functions is already a major advance, unifying, as it does, a number of widely different statistical theories such as Fisher's theory of design of experiments and the Neyman-Pearson theory of testing of hypothesis. Fisher's theory is concerned with agricultural experiments designed to compare the relative yields of a number of experimental treatments such as varieties of seeds, manures, *etc.* Suppose, for instance, we have three varieties of wheat seeds. In order that all the varieties be equally represented, we subdivide the entire area of the experiment into 3 × 3 compact plots formed into 3 rows and 3 columns.

If any of the three varieties were sown on any of the nine plots, there would in all be $3^{3^2} = 3^9 = 19,683$ different ways of assigning the three varieties to 3^2 or nine plots. But in all of these 19,683 different ways of assigning varieties to plots all the three varieties are not equally represented (see

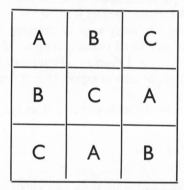 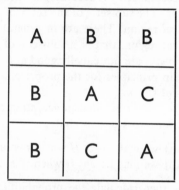

FIG. 62—These two squares represent two ways of assigning three varieties A, B, C to nine plots. The left-hand square is a latin square, as all the three varieties are equally represented in the nine plots. In each column or row each variety appears once and only once. The right-hand square is *not* a latin square, as in the first row variety C does not appear at all whereas variety B appears twice.

Fig. 62). In order to do this we have to limit our choice to only those of these 19,683 ways in which each variety appears only once in each row and column. Such an arrangement of assigning varieties to plots is known as Latin square. Fisher's theory of design and Latin squares here is simply the problem: which of these nineteen thousand odd different ways should We select to obtain optimum results? Wald has shown that Fisher's theory is a particular case of his theory of decision functions.

In the Neyman-Pearson theory of testing a hypothesis the question at issue is the acceptance or rejection of a statistical hypothesis H. Suppose, for instance, a lot consisting of N units of a manufactured product is submitted with the claim that the proportion (p) of defectives in the lot is not more than, say 5%. The problem is to devise a procedure for testing the hypothesis (H) that p is less than or at most equal to 0.05. The procedure prescribed is to select at random a number (n) of articles out of the lot and decide *either* to accept *or* reject H on the basis of observations made on this sample of size n. An elementary theorem of permutations and combinations shows that there are in all

$$\frac{N!}{n!(N-n)!}$$

different ways of selecting a sample of size n out of the lot containing in all N articles. There are thus $S = N!/n!(N-n)!$ different samples that could be picked. A test procedure is merely a method of dividing this totality (S) of all possible samples of size n into two complementary sub-sets, say sets s and s^* together with the statement of a rule prescribing the rejection or acceptance of H according as the sample selected belongs to the sub-set s or s^*. The sub-set s of the set S of all possible samples is known as the *critical* region. There are in general infinitely many ways of choosing a critical region, that is, a sub-set s of the set S of all possible samples. All such ways are not equally good and Neyman and Pearson have suggested certain principles for the proper choice of the critical region, that is, the sub-set s.

They point out that in accepting or rejecting an hypothesis H we may commit errors of two kinds:

(i) we may *reject* H when it is, in fact, *true*;
or, (ii) we may *accept* H when it is, in fact, *false*.

They then calculate the probabilities of committing these two kinds of errors. The probability α of the former is really the probability determined on the assumption that H is true, that the observed sample belongs to the sub-set s or the critical region. The probability β of the latter is really the probability determined on the assumption that H is false, that the observed sample belongs to the sub-set s^*—that is, falls outside the critical region. The probabilities α and β therefore simply mean this: Suppose we draw a large number, say M, of samples of size n. If we follow the rule of rejecting H when any one of the M samples belongs to s and accepting it when the sample belongs to s^*, we shall make M statements of rejection or acceptance of H. Some of these statements will be wrong. If H is true and M large, it is practically certain that the proportion of wrong statements will be approximately α. Likewise, if H is false and M large, it is practically certain that the proportion of wrong statements will be approximately β. Naturally the choice of the critical region or sub-set s should be such as to minimise both α, β, that is, the proportion of wrong statements whether H is true or false. Now it is quite easy to select a critical region for which *either* α or β is arbitrarily small. But it is impossible to make both α and β arbitrarily small at the same time for a fixed value of sample size n. Consequently Neyman and Pearson consider critical regions for which α has a fixed value and out of the totality of all such regions they select the one for which β is a minimum. The quantity α is called the *size* of the critical region and the quantity $(1 - \beta)$, the *power* of the critical region. Neyman and Pearson propose to select out of critical regions of a fixed size α that one which is the most powerful.

Wald has shown that this theory of critical regions is also a special case of the decision function theory. In fact, Wald's theory is so general that it is even possible to interpret a statistical decision problem as a zero-sum two-person game. As we saw in Chapter 7, von Neumann and Morgenstern have developed an interesting theory of games with a view to interpreting economic behaviour. In such a game the strategy of one player is designed to maximise the payment or outcome function $K(j, i)$ while the other endeavours to minimise it. Now Wald suggests that a statistical decision problem can be interpreted as a zero-sum two-person game by setting up the following correspondence:

Two-person Game	*Decision Problem*
Player 1	Nature
Player 2	Experimenter
Outcome $K(j, i)$	Loss function

Wald shows that the analogy between the decision problem and a zero-sum game is complete except for one feature, *viz.* the absence of sharp clash of interests between the two players in a game. While it is true that the experimenter wishes to minimise the loss function, we could hardly say that Nature opposes him by endeavouring to maximise it. However, Wald counters this objection with the suggestion that it is not unreasonable for the experimenter to act *as if* Nature wanted to maximise the loss function. This is, no doubt, playing safe, but the trouble is that it is apt to be *too* safe in many cases.

For instance, we saw earlier how the minimax solution of the decision problem facing a manufacturer wishing either to sell his product with a double-your-money-back guarantee in case of a defective article or to dump the lot was irrationally overcautious. This is because, interpreted as a game, the problem views Nature as a hostile opponent endeavouring to make the player lose as much as possible. But actually in many situations, far from actively opposing the experimenter, Nature may even help him. Some post-Waldian work on compound Decision Functions seems to show that there exist situations in which one would *profit* by discarding the Wald view of expecting either definite hostility or even belligerent neutrality from

Nature and actually planning the experiment on anticipations of positive help.

*   *   *   *

One consequence of the probability calculus is that no successful gambling system can be devised. But as many were slow to realise it, attempts at creating such a system continued to be made for a long time. Indeed, they have not ceased altogether, even now, and sometimes such 'infallible' gambling systems are described in popular magazines or offered for sale at casino meets and gambling dens. Here is one specimen: Double your stake each time you lose till you score a win. A simple calculation would show that, if you followed this rule consistently, you are certain to make a net gain equal to the amount of your initial wager. Suppose, for instance, you bet one rupee initially and lose four times in succession but win on the fifth occasion. You would thus lose fifteen rupees in the first four trials but gain sixteen in the fifth trial. By repeating this process indefinitely you could no doubt become a multi-millionaire in no time.

However, the reason why I do not fly to Monte Carlo and try this system is that there are two very serious snags in applying it. In the first place, it assumes that I have literally an infinite fortune to enable me to double the stakes each time I lose. Secondly, it assumes that the 'bank' against whom I play would be willing to take on my bets each time I raised the stakes. If I repeat this process indefinitely, sooner or later I am bound to have a run of bad luck, which may result in, say, ten successive failures. I should then have to raise my stake to a sum of Rs. 1,024— if I am to continue the game. If I lost 20 times in succession, my next bet would have to be over the million mark. On the other hand, a succession of wins would bring in no corresponding gains as my bets would remain as low as the initial bet.

You could, of course, make the game less exciting and much less catastrophic by raising only slightly the amount of bet after each loss and slightly lowering it after each win, every increase being a little more than the previous decrease. If then you win and lose alternately with complete regularity, your capital would oscillate up and down but would increase on the whole as each increase is slightly larger than the previous decrease. In this way, too, you could win a huge fortune for yourself, but here again the rub is that such regularity does not usually occur and the chance of its occurrence is as small as that of an equal number of straight wins. In all this, of course, it has been assumed that the bank takes no cut. But as a matter of fact, since every roulette wheel has one or more zeros, this raises the odds against the player, so that no matter what system the latter may adopt, his money gradually must leak from his pocket to that of the bank as surely as the heat of a boiler leaks into the surrounding air.

I have used the above analogy purposely as there is a basic similarity between the two phenomena. Both are subject to the laws of probability and if you could beat the laws of chance in a casino you could do many more wonderful things. You could, for instance, create perpetual motion machines and thus run ocean-going liners without fuel or devise self-operating factories. For, if you could reverse the flow of money from the pocket of the casino bank to that of the gambler, you could equally well reverse the flow of heat from the surrounding air into the boiler and thus operate your factory or liner without fuel. The reason for this is that heat processes themselves are very similar in their nature to games of dice and roulette, and to expect that heat will flow from the surrounding air to the boiler is as vain as to hope that money will flow from the pocket of a casino bank into that of a player.

Heat in a body is nothing but the rapid and irregular motion of its constituent atoms and molecules. The more violent this molecular motion is, the warmer is the body. As this motion is quite irregular, it is subject to the laws of chance and the probability of the body being in any given thermal state can be worked out in the same manner as the probability of getting a given combination of, say, ones and sixes when a large number of dice are thrown together. Just as we can work out what is the most probable combination likely to result from throwing a large number of dice, we can also work out the most probable state of motion in a system consisting of a large number of particles moving about irregularly. This state actually corresponds to a more or less uniform distribution among all of them of the total available energy. Thus, if one part of a body is heated so that the particles in this region begin to move faster, one would expect this energy to be distributed evenly among all the remaining particles. However, as the motion of the particles is quite irregular, there is also the possibility that, merely by chance, a certain group of faster moving particles congregate together and thus collect a larger part of the available energy in one region at the expense of the other. This spontaneous concentration of heat in a particular part of the body would correspond to the flow of heat against the temperature gradient, that is, to the flow of heat from the cooler to the warmer portion of the body, like the flow of money from the pocket of the bank to that of the gambler. Neither of these phenomena is excluded in principle; only the probabilities of their occurrence are negligibly small.

Obviously the flow of heat from the warmer to the cooler portion of a body is accompanied by a greater degree of disorder of molecular motion. For while in the previous state the faster-moving molecules flock together, in the latter the faster and the slower are distributed randomly all over. The degree of disorder of molecular motion in any given system of bodies or particles is called *entropy*. Hence the flow of heat from the hotter part of a body to the cooler part tends to increase the entropy of the system or the degree of disorder in its molecular motion. Now if you considered the entire universe as one statistical ensemble of irregularly moving particles, each state is likely to be followed by a more disorderly one, as absolute disorder is the most probable state. It follows, therefore, that the universe as a whole tends towards a state of maximum entropy or maximum disorder. If, as is currently believed, it is this increasing entropy that gives the flow of time its irreversible trend, time itself would cease to flow the moment the entropy of the universe as a whole attained its absolute maximum, towards which it is hastening. It is, indeed, a far cry from Méré's paradox to speculations about the end of Time, but the calculus designed to resolve the one foretells the other.

*　　　　*　　　　*　　　　*

Laplace, the celebrated mathematician of the early nineteenth century, once remarked that, given the initial position and velocity of every particle in the universe at any particular instant, and given all the forces at work in Nature, a super-intelligence could calculate with precision the entire past and future history of the cosmos. Nothing would be uncertain for him; the future, as also the past, would lie before him, unfolded in a vast cosmic panorama. The reason why he could thus apprehend at one glance the history of the cosmos from beginning to end was the fact that its affairs, or at any rate those studied by physics, were governed by a system of rigid causal laws, which ensured that the present state of the cosmos exactly determines its future, as it is itself already determined by the past. It followed, therefore, that a knowledge of the forces animating the universe, as well as its initial state at any instant of time, is all that is required to foresee its future history and comprehend its past evolution.

But the rub was that the cosmic calculation posed by Laplace was for ever beyond the capacity of human mathematicians, even if they could somehow divine the system of causal laws and the initial state of the cosmos at a particular time. For our human—all too human—mathematicians found even the comparatively simple problem of the 'three bodies' (that is, the problem of determining the motion of only three bodies moving under their own mutual interactions) practically insoluble. One could, therefore, hardly hope for a solution of Laplace's problem wherein virtually an infinite number of bodies move under their own actions and reactions.

Finding itself helpless before such a welter of individual motions, mathematical physics, towards the middle of the nineteenth century, struck out in a new direction. Instead of attempting to keep track of each individual particle, molecule or event, it began to focus attention on the statistical properties of large crowds of them. For instance, in the kinetic theory of gases, a gas enclosed in a cylinder was envisaged as a swarm of an immense number of randomly moving molecules. Kinetic theory did not concern itself with the motion of any one of these molecules individually, but rather tried to grasp the motion of the entire assembly of molecules at one blow by a statistical study of the entire group as, for instance, by calculating the *average* energy of the group of molecules as a whole and identifying it with the measure of gaseous temperature.

In this way it was discovered that, in addition to causal laws which produce determinate effects, there are statistical laws derived from a statistical study of crowds of molecules or events. Such, for example, is the second law of thermodynamics, which states that heat flows from a hot to a cold body when the two are placed in contact. As Boltzmann showed, this law could easily be deduced from statistical considerations.

To understand Boltzmann's argument, consider first the following analogy. Suppose we have two committees of representatives of teachers and students. The teachers predominate in one and the students in the other. Suppose, further, that a number of persons picked at random from the first committee are exchanged with an equal number also picked at random from the second. Since the teachers predominate in the first and the students in the second, the exchange is likely to reduce the proportion of teachers in the first and increase it in the second. Consequently the average age of the first group would decrease, while that of the second would increase. This is what is most likely to happen. But it is not altogether impossible that the exchange may result in the influx of more teachers from the second committee in return for a corresponding number of students from the first. If this happens, the average age of the first committee would increase while that of the second would decrease.

To revert to Boltzmann's argument, consider now two chambers containing a gas at two different temperatures. Since the temperature of a gas according to the kinetic theory is merely a manifestation of the average kinetic energy of its constituent molecules, it follows that the molecules of the gas in the hot chamber have on the average greater energy than those of the cold. In other words, there are more faster moving molecules in the former than in the latter. In terms of our analogy the former has more 'teachers' than 'students'. Now when the chambers are connected together, approximately equal numbers of molecules will move from one to the other chamber. The exchange results in the influx into the hot chamber of a larger number of slow-moving molecules (students) from the cold chamber in lieu of the fast-moving molecules (teachers) of the hot chamber. The reverse is the case with the cold chamber. The *average* energy of the molecules of the cold chamber, and consequently its temperature, increases at the expense of the molecules in the hot chamber whose average energy correspondingly decreases. Heat thus flows from the hot to the cold chamber.

However, the assumption that when the two chambers are connected together, a greater proportion of the fast-moving molecules leaves the hot chamber and is exchanged with the slow-moving molecules from the cold chamber, is only 'reasonable' on probability considerations. No doubt it is very likely to be true, but it need not necessarily be always true. For instance, once in a while, the warm chamber may exchange all or most of its slow-moving molecules (students) with only the fast-moving molecules (teachers) of the cold chamber. If this happens the hot chamber will become hotter, and the cold chamber cooler still. The second law of thermodynamics would be violated as we should have a case where heat flowed against the temperature gradient, *viz*. from the cold to the warm chamber. Such a contingency is not impossible, though extremely improbable, as the probability of its occurrence is infinitesimally small. Thus, while the causal law rigidly lays down how any event should happen in the future, the statistical law visualises a manifold of infinite possibilities for crowds of events out of which it picks up those that are mostly likely to occur.

Now, till very recently, the statistical laws, in spite of their obvious utility, were never considered 'really' fundamental. They were tolerated as *ersatz* laws and used in the hope that some day they would be reduced to genuine laws of the causal type. But about forty years ago the Austrian scientist, Franz Exner, put forward the opposite view that the 'assumption that every individual occurrence is strictly causally determined has no longer any justification based on experience.'

Although little attention was paid to it at the time, yet only a few years later a large number of eminent scientists, one after another, came to a similar conclusion. Thus Heinsenberg, Schrödinger, Born, Eddington, Jeans, Weyl, Dirac, *etc.*, expressed the view that modern developments in quantum physics definitely show that statistical laws are the fundamental laws of nature and that all the so-called causal laws are merely the result of statistical regularities of crowds of microscopic events derived by the application of probability calculus.

However, it seems to me that the introduction of the probability notion into fundamental physical theory does not indicate that the microscopic event in question is not causally determined. It merely shows that we do not know whether the necessary and sufficient conditions for it to materialise are fulfilled or not. For instance, take the question of assessing the expectation of human life. There are two distinct methods of ascertaining it. One is the method of medical diagnosis of a doctor who may be able to tell me that I have just a year to live. The second is the method of a life insurance actuary, who calculates the average expectation of life of a whole class of individuals from a statistical study of mortality tables. The reason why the second method works at all is the fact that there are determinate laws which govern the expectation of life of every individual. Only they have still to be unravelled; and, where we are interested only in the average properties of crowds of events, it is futile to attempt the unravelling. Nevertheless, if the causal laws of life and death change, as they do with the more efficient organisation of health services or (in the reverse direction) with the outbreak of war, they force a corresponding change in the actuarial calculations.

In fact, the probability calculus is based on the empirical assumption that the relative frequencies of events of a particular kind in a 'collective' of a large number of such events remain stable. Thus, if the event in question is the survival of an individual, say aged 30, during his thirty-first year, we consider a collective of a large number of individuals aged 30 in a society. We observe the number of those who survive after their thirty-first year and calculate the frequency or the ratio of the survivors to the total. This frequency, which remains the same for some time, represents the probability of the survival of the group. It is true that sometimes we are able to state the probability of an event from *a priori* considerations, as when we say that the probability of tossing 'heads' with a coin is 1/2. But even here, the statement finds its ultimate justification in the fact that the frequency ratio of 'heads' in a large number of throws remains stable at 1/2.

This stability of the frequency ratio would be impossible if every individual event were completely indeterminate and if a sort of Ariadne's thread did not run through the causal nexus tying up the collective as a whole. When this thread snaps, the probability calculus breaks down or, at least, has to reconstruct the calculation *de nouveau*, basing itself on a new probability value. All this is well understood in actuarial practice, where the average expectation of life is continually revised as determined by the conditions of life and health in a changing society. In wave mechanics, however, this aspect is taken care of by making each separate measurement or observation the basis of a new estimate of quantum mechanical probabilities in a highly artificial manner. This artificial manner of their derivation is, then, justified by the experimental success of the underlying *ad hoc* assumptions.

Their success is not denied, even though the present state of quantum theory is considered far from satisfactory by the advocates of indeterminism themselves. But this very success, which implies that the quantum probabilities, mere frequency ratios, remain stable, is an indication that some sort of causal laws governing the quantum phenomena are at work 'behind' the statistical laws. The success of quantum mechanics, such as it is, therefore, does not involve the breakdown of the causality concept any more than the success of actuarial calculations means the breakdown of the medical (causal) laws of health and disease in a community.

The new developments in quantum mechanics, therefore, far from involving the total breakdown of the old determinist principle of causality, are merely a pointer that it can no longer be expressed in purely mechanist terms of classical physics and is in need of a reformulation in terms of categories of a wider and profounder logic as indicated in the sequel.

Before examining some suggested reformulations we may, at the outset, meet the criticism that such a formulation is for ever impossible. For instance, J. von Neumann is said to have 'proved' mathematically that the theory of quantum mechanics can never be so extended and perfected as to yield perfectly determinate predictions. It is thus claimed that the determinate prediction of atomic phenomena is 'impossible not only in present-day practice but in eternal principle'. But as this 'proof' is based on the tacit assumption that the laws of quantum mechanics are valid for all kinds of statistical assemblages, it begs the very question it sets out to 'prove'. It is like trying to show the impossibility of the methods of medical diagnosis by the assumptions of actuarial science. Most indeterminists, therefore, are less dogmatic and accept that no *logical* reasons exist for excluding the possibility that 'quantum mechanics might one day be considered as a statistical part of science imbedded in a universal science of causal character', even though they actually deny any possibility of such a formulation.

This denial is justified by an appeal to Heisenberg's principle of indeterminacy, which, broadly speaking, recognises the fact that every act of observation disturbs the very object it tries to observe. In a way, this principle is not entirely new and was recognised even in classical physics. For instance, it was realised that observing the temperature of a liquid by immersing a thermometer therein does to some extent modify its actual temperature. But this disturbance being of a determinate character, classical theory could calculate it and account for it. What is new in Heisenberg's principle is that the act of observation disturbs the atom, electron, or proton, *etc.*, in an *unpredictable* way. It is this feature that is new to physics and is, indeed, the *kernel* of the indeterminacy principle. However, if Heisenberg's principle were consistently followed to its logical conclusion, there could be no prediction of any sort whatever. Accordingly, the indeterminists themselves do not consistently stick to it and proceed to make scientific predictions by assuming a veiled form of causality under the guise of probability laws, which scientists always found handy, when it was either too much trouble to apply the causal laws to individual phenomena or when the laws themselves happened to be unknown.

There is, however, one sense in which the quantum critique of the old Laplacian conception of causality is right. Laplace, as we mentioned before, envisaged a universe of independent particles whose future course of motion could be written down from a knowledge of their initial positions and velocities. But position at a point in space and velocity at an instant of time depend on two abstractions, namely, that of a 'point' as a dimensionless length and that of an 'instant' as an extensionless duration. The underlying assumption by which these abstractions are evolved is the hypothesis that any length or any duration, no matter how short, can, in principle, be measured. Now the new developments in modern physics seem to show that this assumption is no longer valid in the sub-atomic regions. The only way whereby we can measure the length between two adjacent points is by means of light. But two points, whose distance apart is less than a certain function of the wave length of light by which they are observed, form only one image and are, therefore, seen as coincident. Consequently, the physical means of measurement impose a natural limit to the precision of measurement. Whatever we may do we cannot surpass this limit.

Likewise, two events close together can be distinguished as distinct only if they are separated by a certain minimum duration. If they are closer than this minimum limit, they are observed as one, or are not recognised as distinct events at all, lacking the time for their formation, which, though small, is not zero. Hence the basic assumption of the mathematical theory of physics, *viz.* that time and length measurements can be made with an indefinitely increasing degree of precision, has now come into conflict with the present-day practice of physicists, whose means no longer permit them to cross certain narrow limits. It is, therefore, likely that while the classical abstractions, point-position and instantaneous-velocity have been extremely useful and, in fact, indispensable for the creation of rational mechanics, they are now preventing a fuller apprehension of physical processes on the sub-atomic scale. Science recognises the need for abstractions, but it also stresses the importance of going beyond them and creating new ones in order to have a better grasp of natural processes when new developments show up the limitations of the old. It seems to me, therefore, that the old abstractions, point-position and instantaneous-velocity, along with their inevitable concomitants like electrons, positrons, photons, collisions, attractions, spins, waves, *etc.* (with the possible exception of the quantum of action), have to give way to newer abstractions, which somehow resolve the conflict between the apparent continuity of the macrocosm (or the world of everyday observations), and the absolute discontinuity of the microcosm (or the sub-atomic world). Naturally, such abstractions cannot be cast in the image of anything that we can intuitively derive from our everyday daily experiences of the macroscopic world.

Some eminent scientists seem to think that quantum mechanics has sounded the knell of causality for good and all; and, like the fallen Humpty-Dumpty, not all the king's horses can put it together again. Eddington, for example, says, 'indeterminism (of atomic phenomena) cannot be got rid of by any possible change of the present-day concepts of position and velocity to some other concepts'. Why? Because 'it is not possible to transform the current system of physics, which by its equations links probabilities in the future with probabilities in the present, without altering its observable content.' But this is so only because it continues to stick to the Newtonian abstractions of point-position and instantaneous-velocity, which are apparently no longer suited to the new situations encountered in present-day physical research. As a result, physical theory is forced to resort to probability calculus, as observation enables it to determine only one-half the co-ordinates of the physical configurations exactly, while leaving the other half entirely undetermined. However, if physics could find a way out of the abstractions of point-position and instantaneous-velocity to a more rational synthesis of sub-atomic notions in terms of non-intuitive parameters, it need not have to appeal to probabilities at all, except for developing a theory of errors, as in classical physics.

But to attain to this fuller synthesis of motion, it would be first necessary to create a new mathematics which is able to unite in itself the positive sides of analysis as well as of the theory of discrete manifolds. At present, both analysis based on the notion of continuity and its opposite—the theory of discrete manifolds—are being developed in isolation one from the other. The problem is to create a new mathematics which embodies the reciprocal unity of both. When such a mathematical apparatus is at last forged and the continuum and the discrete synthesised, it might be possible to overcome the antagonism between the wave-aspect and the particle-aspect of both matter and light and construct a more rational scheme of things than the mathematical abracadabra that is present-day quantum mechanics.

Unfortunately, instead of developing the quantum theory on these lines, the quantum physicists are content to adopt a positivist outlook. The positivists claim that science reveals a number of laws connecting our sensations and it is only a 'more or less refined metaphysics' to try to go deeper and explain these laws. As a result, they are led by the logical development of their own thoughts to a peculiar realm of what Reichenbach calls 'interphenomena' and Jeans calls the 'substratum', and tend to a mystical interpretation of the sub-atomic particles. No wonder there are almost as many different schools of wave mechanics as there are of psychology, and, as in psychology, scientific observation is quite unable to decide between them, not only here and now but in 'eternal principle'. All this may not matter to the positivistically minded quantum physicists, who seem to have reconciled themselves to the view that the fundamental processes of nature are inexplicable and that it is futile to try to fit them into the causal scheme of science as a whole. But it is an unsatisfactory state of affairs for those who seek a deeper understanding of the natural processes and firmly believe that nothing in nature is isolated or inexplicable.

*   *   *   *

In an earlier section we gave an account of the 'frequency', 'credibility', 'axiomatic' and Laplace's 'classical' theories of probability. We saw that although all of them were in one way or another unsatisfactory, the last-mentioned theory, amended in the manner indicated, was the most suitable. Now, if we examine closely the chance phenomena to which mathematical theory of probability is actually applied, such as games of chance, genetics, statistical mechanics and statistics, we shall find that the probability definition in terms of equipossible cases is not only adequate but is *the one that is actually used.*

Games of Chance

As we have seen already, probability definition in terms of equipossible cases is not only adequate for all problems arising here but is in fact the only one available. The frequencies actually observed are used only as a test for the verification of initial hypotheses, from which probability is derived (on what may be described as *a priori* grounds) by resolving the outcome into a number of equipossible alternatives.

Genetics

Jean Jacques Rousseau began his inimitable *Confessions* by complimenting himself with the remark that nature broke the mould in which he was cast. Had the findings of modern genetics been known in his day, he would have realised that this indeed was no great distinction. For each one of us—genius or dunce, prodigy or imbecile, beauty or beast, blonde or brunette—is cast in a genetical mould the like of which has never been in all history, nor shall ever be again in all eternity. The reason is that the process of casting genetical moulds is in some ways very similar to dealing card hands and there is actually more likelihood of a given bridge hand being repeated than that of duplicating the genetical mould of any given individual. First, this given individual is the result of mating by just two specific parents out of all the myriads of the past and present, and he could have been produced by no others. But this is not all. These two specific parents could have theoretically produced some 300,000,000,000,000 different genetical moulds, each different from any other in one or more ways.

During conception, when a male sperm fertilises a female ovum, the nuclei of both the sperm and the ovum split into a number of parts to produce a new cell, which is, in fact, the genetical mould* of that particular individual. To understand the mechanics of this mould-making process, imagine a pack of cards with the four kings removed. You will then have 48 cards grouped in 24 pairs, twelve black pairs beginning with a pair of black aces and ending with a pair of black queens, and twelve red pairs beginning with a pair of red aces and ending with a pair of red queens. Now pick up one card out of each such pair, that is, either the ace of spades *or* clubs out of the pair of black aces and so on for all pairs of each denomination and colour. How many different sets of 24 cards could you pick up in this way?

To simplify the problem let us first take a pack containing only four aces —two pairs in all, a black pair and a red (see Fig. 63). Out of the pair of black aces we could pick the ace of either spades or clubs. This gives two ways of picking a black card. Similarly, there are two ways of picking a card of the second (red) pair, and since each of the former two ways can be combined independently with the latter there are in all $2 \times 2 = 2^2 = 4$ ways of picking two cards out of four, as a glance at Figure 63 will show. If you had three pairs of cards, two pairs of black and red aces and a third pair of, say, black twos, a similar calculation would show that you could pick up three cards out of the six in $2 \times 2 \times 2 = 2^3 = 8$ different ways (see Fig. 64).

It is not difficult to see that for every increase in the number of pairs in our pack the number of different ways increases by a power of two. So, out of the full pack of 48 cards of 24 pairs with which we originally started, we can pick up 24 cards in the manner described in $2^{24} = 16,777,216$ different ways. Now if you substitute for the cards what geneticists call 'chromosomes', you have a pretty close approximation to what actually happens when a sperm or ovum is formed from germ cells by a process of cell division. For a germ cell consists of 24 pairs of different chromosomes, very much as our pack of cards consists of 24 pairs of cards of the same

---

* This mould is admittedly very plastic and is modified a good deal by the baking it subsequently receives in the environmental furnace, but it carries within it all the legacy that the parents have to give.

A PAIR OF BLACK ACES   A PAIR OF RED ACES

Gives rise to **2** ways of   Gives rise to **2** ways of
choosing a black ace.   choosing a red ace.

Each of these two ways combine to form four different
ways of choosing a black and a red ace from the two
given pairs.

Fig. 63

A PAIR OF BLACK TWOS

Gives rise to 2 ways of choosing a black two

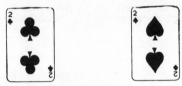

Each of these two ways can combine with each of the four
ways of choosing a black and a red ace from the previously
given pair of black and red pairs in $4 \times 2 = 8$ ways:

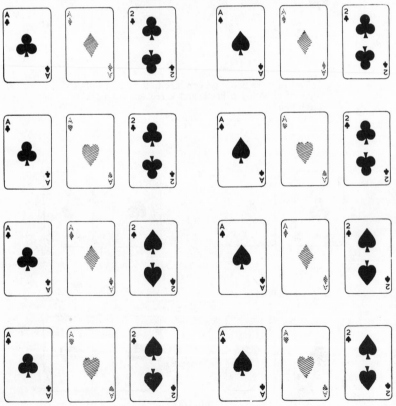

Fig. 64

denomination and colour. When it divides, one chromosome out of each pair separates from its associate in the pair and the result is the production of two sperms (or ova) each containing only 24 single chromosomes.

The process of sperm (or ovum) formation out of the germ cell is thus closely analogous to that of picking 24 cards out of 48 in the original pack. That is why a sperm (or ovum) may be any one of the $2^{24} = 16,777,216$ different combinations of 24 chromosomes out of the 48 in the germ cell. But any one of these 16 million odd different sperms may fertilise any one of the same number of different kinds of possible ova at the time of conception to form a genetical mould of 24 pairs of chromosomes. Consequently, the total number of different genetical moulds that any given pair of parents could theoretically cast is $2^{24} \times 2^{24} = 2^{48} = 281,474,976,710,656$ or, in round figures, $300,000,000,000,000$. In other words, the chance of even a given pair of parents ever duplicating any one of these genetical moulds is so remote that it could happen only once in about $(300,000,000,000,000)^2 = 9 \times 10^{28}$ times. This is many *million* times *rarer* than duplicating a given bridge hand in a bridge game and about 200 times rarer than dealing *all the four players a complete suit of cards*.

This shows how the probability calculus, and in particular the definition of probability in terms of equipossible cases, helps us solve one problem of human genetics, the problem of what Amram Scheinfeld has called 'The Miracle of "you".' But the same calculus and the same concept of probability can handle much more complicated genetical problems. Naturally the manifold of permutations and combinations possible with a set of 24 pairs of chromosomes in the human germ cells, out of which human sperms and eggs are formed by a process of bifurcating one chromosome out of each pair, is truly prodigious. But to add to the complication the chromosomes themselves are by no means the ultimate factors of heredity. They are really packets of many gelatinous beads closely strung together. It is these beads called *genes* which are the ultimate carriers of whatever genetic legacy a parent has to pass on to its offspring. The technics of mathematics required to deal with this additional complication are therefore rather stiff; but the probability concept in terms of equipossible notion underlying the technics remains the same as in the case of ordinary card and dice games. That is why the probability calculus designed to deal with card or dice games is at the same time adequate for handling problems of modern genetics.

Statistical Mechanics

In statistical mechanics we study the statistical properties of ensembles of particles. As we saw before, mathematical physics, unable to keep track of the individual motions of the molecules of a gas enclosed in a chamber, was obliged to study the statistical properties of the entire swarm as a whole. Suppose, for instance, we have a gas chamber containing N gas molecules in all. According to the kinetic theory of gases, each of these molecules is in a state of incessant motion. At any instant each of them has a position specified by three position co-ordinates (x, y, z) and a velocity specified by three components (u, v, w) along the three co-ordinate axes. The state of each molecule is thus specified by 6 numbers. Since there are N molecules in all, the state of all the gas molecules in the chamber is specified by 6 N numbers.

Now, if you imagine an imaginary space of $6N$ dimensions, that state of the gas as a whole could be represented by a single 'point' in this super-space of $6N$ dimensions. Each point in this space is specified by $6N$ 'co-ordinates', the set of $6N$ specification numbers required to fix the state of the gas as a whole at any particular instant. Such a superspace of $6N$ dimensions is called phase space. Let us divide this superspace into two regions of equal 'volume' and let r molecules be in one and the remaining $N-r$ be in the other. In how many different ways could we divide N molecules into two classes so as to have r in one region and $N-r$ in the other?

An elementary theorem of combinations in algebra shows that this number is:

$$\frac{N!}{r!(N-r)!}.$$

This means that to *each* macrophysical state of the gas in which the two regions of equal volume of the phase space contain r and $(N-r)$ molecules there correspond

$$\frac{N!}{r!(N-r)!}$$

different microphysical states, each of which is considered equiprobable *a priori*.* Thus, here again the notion at the base is that of equiprobability. The position has not changed with the discovery of discontinuities in physical processes as a result of quantum theory. In the classical theory the subdivision of the phase space into equal volume elements was arbitrary but in the new mechanics certain regions are prohibited to the system by quantum conditions. Nevertheless, the statistics of the classical quantum theory are based on the same notion as the earlier statistical mechanics, *viz.* that all arrangements obtained by assigning similar particles to regions of phase space of equal volume are equally probable *a priori*.

---

* For the sake of simplicity we have divided phase space into two parts of equal volume, but the same argument holds even if we divide it into any number of parts of equal volume.

The earlier statistical mechanics assumed that one can distinguish between similar elements or particles. The present form of quantum theory or wave mechanics abandons the principle of distinguishability. But this does not introduce any new concept of probability. It has merely the effect that instead of treating each arrangement of molecules as *a priori* equiprobable we now treat each *distribution* as equiprobable since we no longer can distinguish one arrangement from another. In other words, all the $\dfrac{N!}{r!(N-r)!}$ different arrangements which give rise to the distribution r molecules in one region and $N - r$ in the other are treated as indistinguishable and all possible distributions are treated as *a priori* equiprobable instead of all arrangements.

Thus, in the Bose-Einstein statistics assumed by Bose to derive Planck's law for heat radiation by a statistical theory of quanta, it is assumed that every distribution (instead of arrangement) is *a priori* equiprobable. The further assumption made was that a region in phase space can have any number of particles. Fermi extended Pauli's exclusion principle to molecules of an ideal gas and assumed that no two molecules can occupy the same region. But none of these innovations affects the notion of probability—that is, the basis of the new mechanics.

Statistics

It now remains to comment on the probability notion actually employed in the theory of mathematical statistics. It is generally considered that the mathematical theory of statistics actually uses the frequency notion. This is due to the fact that in a vast majority of cases our knowledge of the conditions governing the event does not permit us to enumerate the equiprobable ways in which the event can occur. In such cases the only way we can form estimates of its probability is to gather empirical evidence of its actual happening in a number of trials. But this is a problem of evaluation and not of definition. In tackling this problem, *viz.* that of evaluation, we consider a discrete finite set of N trials and treat each trial in this group as equiprobable so that relative frequency of a particular outcome is also the ratio of the equipossible alternatives favourable to that outcome to the total number (N) of such alternatives.

It is, therefore, clear that even in cases where no *a priori* estimate of probability is possible, the *a posteriori* estimate thereof based on frequency acquires meaning only in terms of the equipossible notion. In fact, if frequency theory were adopted consistently, *a priori* estimates, even in cases where knowledge of conditions of symmetry such as in games of chance makes such estimates possible, would be meaningless. All that we should have would be mere empirically ascertained frequencies and the bottom would be knocked out of the foundations of the theory of testing hypotheses. If, for example, we wish to design an experiment to test the claim of the lady, cited by Fisher, that she could discriminate whether the milk or the tea was added first to the cup, we have to compare the empirically ascertained frequency of right answers with their *a priori* probability.

If the frequency of the right answers in a long series of trials was significantly higher than their *a priori* probability on the basis of a denial of her claim, this would be a strong reason in its favour. But as there is no place for any *a priori* probability in the frequency theory, reasoning of this kind has no validity in frequency theory. Whenever we test a statistical hypothesis, all that we do is to deduce the probability that a given function of the observations will occur on the assumption of that hypothesis in exactly the same way as we deduce the probability of getting a given distribution of white and black balls in a series of draws from an urn with known composition content. If this probability is very low, we reject the hypothesis; otherwise we proceed as if it were true. The hypothesis in question may involve a number of presuppositions, some explicit others implicit, and it is on the basis of these considerations that probability is deduced by resolving the various outcomes into a number of equipossible alternatives.

It is true that we do not have a conclusive argument in any case, for any value of this probability (except 0 and 1) is compatible with the hypothesis. But the main point at issue is that unless probability is defined in terms of the equipossible alternatives by means of what may be called the *a priori* presuppositions of the hypothesis in question, there is no logical basis for tests of hypotheses in statistical theory. In other words, a consistent adherence to frequency definition leaves no room for a theory of testing statistical hypotheses. For we can then no longer explain empirically ascertained frequencies and have to content ourselves with merely recording the difference of frequencies.

We therefore observe that the classical definition of probability, based on the notion of equipossible cases but amended in the manner indicated above, is adequate for all present-day applications of probability theory.

10

LOGIC AND MATHEMATICS

THE non-mathematician is apt to believe that the mathematician is in possession of a set of infallible principles of reasoning which enables him to reach truth without ever going astray. For a long time this belief was universally held. But when, with the invention of the calculus, mathematicians began to work with queer sorts of notions like that of the infinitesimal—a paradoxical entity that was sometimes zero and sometimes not zero at the convenience of the calculator—faith in the infallibility of mathematical reasoning began to wane. Fortunately, however, the economy of Western Europe at the time was in its ascendant progressive phase. It was then in the midst of transforming itself from a static feudal society of serfs and seigneurs into a dynamic society of the bourgeoisie and the merchant adventurers.

Technological problems presented by such a society could not be solved by disputatious men wrangling on the nature of the infinite and the infinitesimal in the hidden retreats of their monasteries. They could be tackled only by practical men who observed things for themselves and experimented in their laboratories. These men did not set out to create a static hierarchy of 'eternal' truths but adopted such methods as seemed likely to lead to the solution of current problems before them. As a result, following D'Alembert's maxim, 'push on, and faith will follow', they fashioned the Infinitesimal Calculus into a magnificent instrument for handling problems of flux and change. Its methods, no doubt, sometimes led to absurd results, but on the whole they were adequate to do the job for which they were devised. It was only during the nineteenth century, after the calculus had already advanced quite far, that the work of purging it of contradictions and looseness of reasoning was undertaken by Cauchy, Weierstrass, Riemann and others.

As we have already seen, their work introduced new canons of rigour in mathematical reasoning. In fact, so great was the success of this new mode of reasoning that, towards the close of the nineteenth century, it appeared as if the calculus had been finally rid of all contradictions. But alas! the appearance was an illusion. For, the discovery soon after of new paradoxes of the infinite showed that mathematical logic was not infallible after all.

Unfortunately, the efforts of contemporary mathematicians to resolve these paradoxes have not been as successful as those of their forbears of the early and mid-nineteenth century. The main cause of this failure appears to be the quasi-theological desire of some modern mathematicians to set up an absolute logic or super-theory, whose principles are absolutely sacrosanct and from which all 'law and order' should emanate. In their search for such an unchangeable pattern of 'eternal' truths, which may be extended but not controverted, these mathematicians have developed two distinct schools—the logicalist and the formalist schools. The former endeavours to achieve this result by reducing mathematics to an absolute logic, a sort of infallible, supreme code against which there can be no appeal. The latter, on the other hand, considers that this reduction, even if successful, would still not suffice to free mathematics of paradox. It therefore endeavours to set up another Absolute—a Super-theory ('meta-mathematics')—above both logic and mathematics. We shall examine the formalist viewpoint in greater detail later. Here, we shall show how the logicalist school proposes to restore 'absolute' certainty to mathematics by attempting to reduce it to logic.

Now, as we all know, logic* is that branch of philosophy which deals with the nature and criteria of reasoning. In other words, it tells us the correct way of deducing valid conclusions from given premises. The ancients set much store by it, for (so ran the argument), our senses might deceive us but our reasoning never. From the fact that our senses sometimes mislead us—as, for example, when we see a mirage or experience an hallucination—they concluded that human senses could give only a misleading conception of 'reality' and should therefore not be trusted. This was, for instance, Plato's view, when he compared human beings to chained cave-dwellers condemned to watch shadows on a wall without ever being able to see the 'real' objects that cast them. How, then, was this 'reality' to be apprehended? Plato considered that the ultimate nature of the universe could be comprehended only by 'pure' thought, contemplation and reasoning. Logic, the science of deductive inference, was therefore the sole instrument available to man for understanding the nature of 'reality'; for the laws of logic had a kind of superior authority to which 'reality' must itself conform. It is a relic of this ancient tradition that 'illogical' and 'contradictory' are even today taken as synonymous with the unreal and non-existing.

In spite of Plato's insistence on the primacy of reason over the senses, most of the ancient Greek philosophers were restrained in their tendency to deduce scientific laws from *a priori* conceptions by their delight in sensuous observation. The Middle Ages inherited this tendency without

---

* There is also another sense in which the word logic is used by modern logicians. We shall refer to it later (see page 279).

the Greek restraint, and the result was a disaster. For the medieval philosophers—the 'scholastics'—concluded that all the fundamental problems
of human knowledge could quite literally be solved by a sort of verbal
'dialectics' based on purely syllogistic reasoning. So they posed certain
questions and disputed endlessly about them according to certain rules.

No doubt many of them were very able men whose debates and discussions conditioned men's minds to the notion of a rational universe, an
essential prerequisite of modern science. But as neither the questions nor
the rules of disputation were concerned with the actual needs of men and
of life at the time, scholasticism largely lost itself in barren oddities which
led nowhere. For instance, they would ask what is the 'thing' or the
'essence', and giant *summas* would be written in futile wrangles over the
question. In fact, some of them were concerned merely with 'disputation'
itself and made no attempt to arrive at any conclusions. Thus, in his
masterpiece, *Sic et Non*, 'Yes and No', the scholastic Abelard argued at
length a great variety of 'theses', but gave no indication on which side of
the fence he himself sat.

Such was medieval scholasticism. It flourished unchallenged for centuries because its social and economic roots lay in the stagnant feudal
society of the Middle Ages. In such a society everything seemed to be
eternally fixed or pre-ordained from the position of the Vicar of Christ and
the seigneur down to the meanest serf. Naturally, in such a static society,
there could not arise the idea that man could mould his environment
according to his conscious purposes. Nor, consequently, the methodology
appropriate to such a world view. However, by the close of the thirteenth
century the gradual but steady growth of international trade had already
paved the way for the break-up of feudal society. By this time international
trade in Western Europe, which a few centuries earlier was a mere trickle,
had become a veritable flood, and led to prosperity never known before.
As a result, a new way of life gradually began to emerge all over Europe
and with it the intellect of Europe burst out of its scholastic shell.

The emergence of this new way of life gave rise to a number of technological problems of trade, industry and navigation which could not be
solved by 'pure' thought and meditation. You may be able to speculate
in the quiet solitude of your study about the nature of God, Angels, Void
or the Heavens without bringing disaster to yourself or anyone else, but
it is otherwise if you have to determine the longitude of your ship while at
sea. If you rely on 'pure' thought to determine your longitude, sooner or
later you will take your ship on the rocks. Thus it was that observation
came into its own because the social problems that arose could be solved
in no other way. So great and rapid was the success of the new empirical
method in physics, chemistry and other sciences that scientists, one and

all, gave up the Platonic method of pure speculation in favour of empirical observation.

But the philosopher was still not quite satisfied. On the one hand, he knew that there was nothing mandatory in the laws of logic; that is to say, there was no reason why 'reality' should necessarily conform to the laws of thought. On the other, he saw that although human senses were not always to be trusted, they alone gave us whatever knowledge we had of the outside world. Whence, then, could 'absolutely certain' knowledge come— from mental experience or from observation? And the old dilemma concerning the priority of one over the other arose all over again.

Modern philosophy proposes the answer that logic and mathematics alone can give 'absolutely' certain knowledge. But as absolutely certain knowledge in an ever-changing world is not easy to acquire, the knowledge which logicians obtain from logic is of a rather trivial kind. It is 'absolutely' certain precisely because it tells us nothing at all. For logic is only concerned with deducing valid conclusions from given premises. What is inferred as the conclusion is at the very outset contained in the premises from which the deduction proceeds. Thus, if all A's are X and *if B is an A,* then it *follows* that B is X. The deduction merely reiterates what lies already hidden in the premises. It thus establishes a purely formal relation between the symbols B and X, given the relation between the symbols A and X, and B and A. A logical inference like, 'if all A's are X and if B is an A, then B is X', is a 'tautology'; that is, a formula 'whose truth is independent of the elementary propositions contained in it'. This means that it is true even if in the real world the objects corresponding to the symbol A do not have the character or quality corresponding to the symbol X.

Thus, if in our example, A stood for 'all men', X for 'mortal' and B for 'Socrates', 'Socrates is mortal' would still be true even if, in actual fact, all men (like Swift's *Struldbrugs*) lived for ever. A tautology, therefore, is that combination of elementary propositions which is always 'true' no matter whether its constituent propositions are true or false. Since logic studies the relations between propositions independently of what each proposition is about, its first problem is to identify those elementary propositions which are always true irrespective of the truth or falsity of their constituent propositions. Obviously, the truth of a tautology depends on the form of its elementary propositions rather than their truth or falsity and in the last analysis it is only a roundabout way of saying A is A.

Now the logistic thesis is that 'pure' mathematics is nothing but logic. As in logic, mathematical propositions do not say anything about the material objects about which its propositions apparently talk. They merely state in different terms what is already contained in its axioms. In other words, they are tautologies. But the view that all the theorems of pure mathematics which fill so many volumes are only a roundabout way of saying that A is A seems, to say the least, very surprising. Nevertheless, some mathematicians of the logistic school as, for example, Bertrand Russell, undertook to prove it in detail.

To prove the identity of logic and mathematics it is necessary (a) to show that all concepts of mathematics can be derived from those of logic, and (b) to construct a logic which yields all of mathematics but none of its paradoxes or contradictions, such as that of the 'set of all sets' in Cantor's theory, and those of Zeno, Burali-Forti and others. As a first step towards proving (a) Russell tried to define number in logical terms, since the basis of mathematics is the number system. Starting from natural numbers, that is, the positive integers, we can successively generate all other types of numbers such as negative, fractional, irrational and complex numbers. According to Russell, the natural numbers, however, cannot be further analysed in mathematical terms, but in terms of the logical concept of a 'class', that is, an ensemble which consists of all individuals having a certain property.

To define natural numbers, he defines the notion of 'similar' classes. Two given classes are considered similar if the elements of one can be put in 'one-to-one correspondence' with those of the other in exactly the same way as the blind Cyclops matched his flock of sheep with a heap of pebbles. Number is then defined as the 'class of all classes that are similar to it'. This looks pretty stiff. But what it really means is that if you considered, say, all classes of pairs, *e.g.* pairs of shoes, pairs of boys, pairs of socks, pairs of virtues, pairs of angels, *etc.*, all these classes of pairs have one common quality, *viz.* that of being a 'pair' and this common characteristic of the class, of all these classes of pairs, serves to *define* the number two.

This definition seems to suggest that we have somehow reached the ultimate '2' as the class of all pairs. But whether any given class is a pair or not can be known only if we can match its elements 'monogamously' with those of another class already known to be a pair. In other words, a definite matching process is necessary as a test before we can pronounce any class to be a pair, which is just what Russell seems to wish to avoid. There are many other difficulties which Russell's definition raises but we shall not go into them here.

Now, you may object with Poincaré that integers are indefinable and it is futile to get behind them. But Russell and his followers refuse to accept this position, as being too 'vague' and even 'metaphysical'. They affirm that natural numbers could and should be *precisely* defined. Unfortunately, however, the idea underlying Russell's definition is that entities remain permanent, which is false. For, as Engels said, 'The world is not to be comprehended as a complex of ready-made things, but as a complex of processes in which apparently stable (*i.e.*, *permanent*) things, and also their mind-images in our heads, go through an uninterrupted change of coming into being and passing away'.*

Thus, for example, this man you see now is, strictly speaking, not the 'same' as he was a moment ago. He is no longer the same complex of cells and their functioning as he was. Russell and his followers might reply that he remains the 'same' for all practical purposes, and if they did we should be satisfied. But they do not seem inclined to compromise the precision of their theory by making any concession to what we may call 'practical purposes'. Moreover, they do not wish to make part of deductive logic any 'empirical' proposition, that is, any experience gathered in human practice. They thus cut themselves off from the only source of knowledge that we have, *viz.* human observation. So the endless and futile search for a 'pure' and 'precise' definition of natural numbers continues in a truly scholastic fashion. The Polish logician, Chwistek, himself a follower of Russell's logistic school, somewhat ruefully summed up the latest position as follows:

'All other attempts to develop arithmetic (*that is, to define natural numbers*) are either fragmentary and therefore not entirely clear, or are based upon certain metaphysical suppositions which contradict the principles of sound reason. The first objection applies to the axiomatic method which was employed by Peano and Hilbert, the second to such systems as that of Whitehead and Russell and even to my theory of constructive types' (italics inserted).

Thus ends Russell's quest for a 'precise' definition of number which everyone takes for granted. Of course, clarity of thought and precision of expression are as precious to a thinker as gold to the kings. But a Midas touch in precision, the reward, which Russell and his followers ask of their deity, Logic, can be as ruinous as the gift of Bacchus to King Midas.

If the attempt to derive the concept of number from the logical concept of class did not quite succeed, the effort to construct a paradox-free or consistent logic capable of yielding all of mathematics did not fare any better. To construct such a logic Whitehead and Russell made extensive use of a mathematical symbolism devised to mathematise logic. We explained in Chapter 2 how logic could be mathematised by means of an appropriate symbolism and how the calculus of reasoning that results is symbolically identical with that of numerical computation in the binary scale. In Chapter 6 we showed how the logical laws of contradiction and

---

* This idea did not, of course, originate with Engels. It is one of the oldest ideas in the history of philosophy.

excluded middle could be expressed in algebraic form—the algebra of sets. This process of symbolising logic confers on it great power and has therefore been followed up energetically by the logicalists, who have perhaps tended to push it to unwarrantable extremes. They have thus used their symbolic logic to further the logical analysis of mathematics itself—that is, to provide a logistic 'foundation' for mathematics. This fusion of mathematical logic with the logical analysis of mathematics has, no doubt, illumined the process of mathematical deduction, but it has not been able to fulfil the main object for which it was designed. We shall indicate here very briefly the main developments in this field.

First, a rich symbolic vocabulary is created to enable the use of symbolic methods of operation in logic similar to those used in algebra and other branches of mathematics. In Chapter 2 we used the symbols (v) and $(.)$ to denote the logical sum and product of propositions and the symbols I and O to denote the truth and falsity of a proposition. In their *Principia Mathematica*, Whitehead and Russell use many more symbols besides these four, but the chief among them is the symbol $\supset$ used to denote the logical idea of *implication*. When we say that a proposition p *implies* another proposition q we mean, in ordinary parlance, that *if p is true, then q is true*. But what if p is *not* true? In that case we, the lay folks, do not care whether q is true or false. But the logician, who is not concerned with the truth or falsity of elementary propositions so much as with their mutual relations, cannot afford to ignore the case when p is false. Consequently he subsumes these two cases—the case when p is false and q is *true and* the case when p is false and q is *false*—in the same notion of implication. Accordingly, when he says that p implies q he *excludes* only the last mentioned of the following four possible cases, *viz.* the conjunction of the truth of p and the falsity of q:

$$p \text{ implies } q \equiv p \supset q \equiv \begin{cases} \text{Case I:} & p \text{ true, } q \text{ true,} \\ \text{Case II:} & p \text{ false, } q \text{ true,} \\ \text{Case III:} & p \text{ false, } q \text{ false,} \\ \text{Case IV:} & p \text{ true, } q \text{ false.} \end{cases}$$

A corollary of this extension of the notion of implication is the curious result that a false proposition *implies* any proposition. With the help of symbols like v, $.$, $\supset$, *etc.*, it is possible to express in symbolic language certain basic or fundamental propositions called primitive propositions,* on which all subsequent deduction is to rest. Whitehead and Russell actually assumed five such propositions but Bernays reduced their number to four by showing that one of them could be deduced from the remain-

---

* These primitive propositions are tautologies.

ing four. We shall cite here only two of these four as a specimen just to show what they are like. They are:

$$\text{Primitive I: } p \vee p \supset p.$$
$$\text{Primitive II: } p \supset p \vee q.$$

Primitive I is the principle of tautology which when translated into ordinary language would read 'if either p is true or p is true, then p is true.'

Primitive II is called the principle of addition and translates into 'if p is true, then either p or q is true'.

At first sight they look pretty trivial but quite complicated results can be deduced from them by logical deduction; but here we face a difficulty. Since the ultimate object is to develop a theory of inference in order to provide a foundation for mathematical deduction, we have to decide what kind of deduction is to be permitted to deduce a theory of mathematico-logical deduction. Whitehead and Russell consider that two rules of procedure—the rule of substitution and the principle of syllogism—are valid for deducing logical formulae from the set of primitive propositions. The rule of substitution is simply this. If we have a tautology and replace therein any given symbol p everywhere it occurs by any other symbol or combination of symbols, the result is another tautology. For example, if we substitute 'pvq' for p in the first primitive or tautology, *viz. $pvp \supset p$,* we obtain a new tautology, *viz. $(pvq) \vee (pvq) \supset (pvq)$.*

The syllogistic principle stipulates that anything implied by a true proposition is true. In other words, if $p \supset q$ *and if p* is true, then q is true. This is required because the logical notion of implication is wider than its counterpart of everyday speech in that it also subsumes the two cases (i) p false, q true and (ii) p false, q false besides the 'normal' case of p true and q true. But the symbolisation of logic, setting up of primitive tautologies from which all deduction is to emanate, and the two rules of procedure for deducing theorems from the primitive tautologies do not suffice to rid mathematical reasoning of contradiction and paradox. To do so, Whitehead and Russell have to introduce another theory, the theory of types. The essence of this theory is that paradoxes originate from a vicious circle due to tacitly assuming that a function which defines a class or set can itself satisfy the function. For instance, in the paradox of the village barber referred to in Chapter 5, the function defining the class of shavers is the class of individuals not shaved by the barber. To make the barber, by which the class of shavers is defined, a member of this class itself, is to argue in a vicious circle.

To take another example, consider the paradox in Cantor's theory of transfinite numbers mentioned in Chapter 5. It arises because it permits us to speak of the 'set of all sets' which it then puts on the same level as an ordinary set. As we saw, Cantor defines a set as a collection of some given individuals. The sets that we may form by combining one or more individuals is a set of the first level. From sets of level 1 we may form sets of sets, that is, sets of level* 2 and from sets of sets, 'sets of sets of sets', that is, sets of level 3 and so on *ad infinitum*. The theory of types requires that if a set *s* belongs to a set *S*, then *s* must be a set (or individual) of a level lower than *S*. If we adhere to this rule, the set of all sets becomes a phantom entity inadmissible by the rules of our inference.

The Russell-Whitehead theory of types does the job it was meant to do, *viz.* eliminate paradox and contradiction from mathematical reasoning. But the trouble is that in trying to avoid paradox it runs into other difficulties. For instance, it would seem to indicate that irrational numbers are of higher type than rational numbers for the former are defined as classes of rational numbers. To obviate these difficulties Whitehead and Russell introduce another axiom, the Axiom of Reducibility, which has been criticised as arbitrary and extra-logical because it is not a tautology as other primitive propositions of the *Principia* system are. It is true that the theory of types can be eliminated by using a sufficiently complex system of symbols as has been shown by the investigations of Chwistek, Church and Quine. But these investigations in turn have given rise to their own crop of tangles. The necessity for clearing them up has weakened the thesis of the identity of mathematics and logic. But this is not all. There are other axioms, such as the axiom of infinity and axiom of choice or multiplicative axiom, which too, like the axiom of reducibility, are non-tautological though necessary for the foundation of mathematics. We shall not examine them here in detail as the only point we wish to emphasise is that the introduction of all these non-tautological extra-logical axioms seems to show that it is not possible to reduce all of mathematics to logic.

*   *   *   *

In our examination of the Russell-Whitehead attempt to reduce mathematics to logic, we considered logic as that branch of philosophy which deals with the nature and criteria of reasoning. According to this view, there are certain *a priori* principles of reasoning which it is the business of logic to study and formulate. There is, however, another sense in which the word logic is now used by modern logicians. The germ of this view of logic is to be found in the work of Russell's celebrated pupil, Ludwig Wittgenstein, although he himself remained an adherent of the former view.

Unlike his master, Wittgenstein has expounded all his ideas in a single

---

* Thus if a pair of individuals is a set of level 1, the set of such sets of pairs is a set of level 2.

short book entitled *Tractatus Logico-Philosophicus*; but as it has been written in the form of cryptic aphorisms, it is not always very intelligible. Nevertheless, brief as the *Tractatus* is, it seeks to clarify many aspects of a number of different subjects by starting from the principles of symbolism and logic and then proceeding to apply them to Epistemology, Physics, Ethics, and finally, to what he calls the Mystical. But, after a very learned discussion, which at times is barely intelligible, he comes to the curious conclusion that nothing correct can be 'said' in philosophy so that philosophical discussion is a futile and senseless pastime. However, it is well worth while going a little way with Wittgenstein, as his work is the first draft of the prescription which logical positivists have been elaborating ever since for curing what Sartre has called the sickness of language. This sickness descended on our language almost with a bang during the two world wars, when the world and language began to appear as hopelessly divided from one another. In the words of Antoine Roquentin, the hero of Sartre's *La Nausée*, the word remained on the speaker's lips and refused to go and rest upon the thing, making language an absurd medley of sounds and symbols beyond which flowed the world—an undiscriminated and incommunicable chaos.

Wittgenstein considers that the main cause of this sickness of language is inadequate understanding of the conditions which a vigorous and healthy language, that is, a logically perfect language, needs must fulfil. But what is a language? Language, of course, is the means whereby men communicate to one another their ideas. In the extended sense used by Wittgenstein and his followers, it includes any system of signs or symbols used for the purpose of intercommunication. These signs may be sounds (spoken words), marks on paper (printed words), or any other signalling device, *e.g.* flashlight, *etc.* Now it is well known that everyday language, such as we all use in our daily affairs, is by no means perfect. Excellent though it be for the purpose for which it was evolved by the combined social effort of the community, it does not always enable us to say exactly what we intend to convey.

This is very often the cause of misunderstandings, especially when we interpret the messages of prophets, thinkers, leaders and important personages who have something particularly significant to say to the world. Indeed, a single line in their work may sometimes require a Talmud of commentary to elucidate what it 'really' means. And yet it may still remain meaningless to many. Wittgenstein claims that all such confusions arise from an inadequate understanding of the nature of a logically perfect language. Not that he has constructed such an ideal language for universal use. Far from it. But he considers that an understanding of the conditions of a logically perfect language will dissipate all confusion.

Now what are the conditions of a logically perfect language? In the first place, the rules of grammar of such a language must explicitly exclude nonsensical combinations of signs (or words). What it means is that every language has to have a grammar or, more particularly, a syntax, which lays down rules for combining different signs (or words). But these grammatical rules are not as precise as an ideally perfect language would require. For instance, the laws of English syntax explicitly exclude such a word-combination as 'heavy is spectrum this' but *not* 'this spectrum is heavy', although both are nonsensical word-combinations. The reason why the syntax of English language often permits such nonsensical word-combinations as the latter is that its syntax is not as complete as that of a logically perfect language.

The rules of syntax, or syntactical rules, of a logically perfect language, must exclude *all* nonsensical word-combinations. In such a language its grammar or syntactical rules would not allow us to form *any* meaningless word-combination such as the one cited above. Secondly, a symbol or a sign must not be ambiguous; it must have a definite and unique meaning. Ordinary language, such as English, of course, abounds in words with multiple meanings. Such, for instance, is the case with the words 'rubber' and 'graft'. The former may mean in one context as eraser for rubbing off pencil marks, and in another winning two games in a round of two or three games of bridge. The latter likewise may mean a shoot inserted in a slit in another stock, a piece of transplanted living tissue, or illicit spoil in politics.

These are, no doubt, ordinary ambiguities which are often clarified by the context in which they actually occur. But most languages also abound in words having 'emotive meanings', that is, expressions without any clearly discernible referential quality to any class, entity or category belonging to a common objective world. Even if we ignore altogether such verbal ghosts as some men like to conjure when, carried away by a sort of enchantment, they begin to tread on the margin of meaning and 'hover over the brink of unideal vacancy', we shall still be left with enough words and phrases having no clearly recognisable meaning to provide adequate vocabulary for all sorts of 'doubletalk', 'doublethink' and 'blackwhite' languages, in some ways similar to George Orwell's 'Newspeak'.*

For example, 'democracy', and 'socialism' are no longer identifiable forms of government, nor 'good' and 'true' universally recognised values. That is why Ministries of Truth disseminate lies and Bureaus of Enlightenment propagate obscurantism. Then again, there is a literary style of

---

* Considering this phenomenon, one might be tempted to believe in the linguist E. H. Sturtevant's paradoxical theory that language was invented for the purpose of lying and deceiving.

writing which uses words for their obscure harmonics, giving them vague and sometimes private meanings contrary to their true meanings of everyday speech.* Thus, on both counts, that is, whether we consider its syntax or vocabulary, every language of daily use, such as English, falls far short of the ideal and is in need of a theory of symbolism for carrying out a comprehensive programme of what Morris calls 'debabelisation'.

Wittgenstein himself has not gone very far in this direction but he does seem to give the appearance of having prepared the ground for such a reconstruction. This is done by making a very subtle and scholastic distinction between what he calls 'saying' and 'showing'. What he means thereby is this. If I make a statement asserting a certain fact, there must be something in common between the structure of the signs in the statement and the structure of the fact. Or, as he himself says, 'in order to be a picture (of a fact) a fact must have something in common with what it pictures.' 'What the picture must have in common with reality in order to be able to represent it after its manner—rightly or falsely—is its form of representation'. That which has to be in common between the sentence and the fact cannot (so he contends) be itself in turn *said* in language, though it can be *shown*.

To take an example, suppose I said, 'The sun crosses the meridian at midday'. On the one hand, I have the 'picture' of the fact, that is, a certain combination of words, and on the other the 'fact', *viz.* the crossing of the meridian by a luminous heavenly body. The two have something in common whereby the 'picture' or the statement can serve as a substitute for the 'fact'. And this common element is inexpressible in language, though it 'shows' itself. This scholastic distinction between 'saying' and 'showing' becomes of paramount importance in Wittgenstein's theory because he overlooks its actual origin. Its origin is simply the social practice whereby men have agreed to represent the luminous heavenly body by the word-sign, *sun*, and so on for other objects, actions, processes, thoughts, *etc.* Thus, the 'picture' represents a 'fact', not because there is some mysterious unutterable element common between the two, but simply because the community of human beings agree to take the 'picture' as a substitute for the 'fact'. Moreover, the agreement is social, otherwise there can be no intercommunication.

---

* These attempts are all of a piece with that 'long, immense and reasoned deranging of all the senses' which, according to Rimbaud, is the poet's main task and which has inspired most surrealist painting and sculpture, such as Duchamp's false pieces of sugar actually cut in marble and designed to show the unwary visitor (who suddenly discovers their unexpected weight) the self-destruction of the objective essence of sugar in a blaze of instantaneous, almost Buddha-wise illumination. The ancient Zeuxis-myth was content to make the birds peck at his grapes: our surrealist contemporaries are more ambitious!

Wittgenstein, however, ignores this obvious explanation of the phenomenon, whereby we are able to represent a fact by a socially agreed convention. On the contrary, he goes on to combine this distinction between 'saying' and 'showing', the most fundamental thesis in his theory, with the Principle of Verification to produce patently invalid conclusions. Superficially his principle of verification appears innocent enough and indeed very scientific. For it merely requires that 'in order to discover whether a picture of a fact is true or false (*i.e.* to verify it) we must compare it with reality'. Hence the process of verification involves a comparison of the 'picture' with its corresponding 'fact', or 'of a configuration of signs with a configuration of the objects signified'. Now such a comparison or verification may either be made by myself alone in all its details or by the collective efforts of a socially organised group. Thus, if I had to verify the statement referred to above, *viz.* 'the sun crosses the meridian at noon', I might go to an observatory and personally see the sun crossing the 'spider's web' in the eyepiece of a telescope, or I might accept the experiment performed by a set of qualified astronomers.

Although, nowadays, every scientific verification is actually a social process—it could hardly be otherwise—Wittgenstein insists that I cannot verify a proposition except by reference to facts presented in *my* experience. It is true that he does not say it in so many words. Nevertheless, to use his own phraseology, he 'shows' it clearly enough even though he may not 'say' it. For verification, according to Wittgenstein, can mean only 'verification in my own experience'. And since it can mean nothing else, it need not be said, for why say something that goes without saying? In fact, it would, for this very reason, be meaningless to say so, for 'entities must not be multiplied without necessity' in accordance with the principle of economy (Occam's razor) that he advocates. So he is led head-on to solipsism, the doctrine that nothing is real except *my* 'sense data' and that the universe around me is nothing but *my* dream. Thus he says, 'In fact what solipsism *means* is quite correct, only it cannot be said, but it shows itself. The world is *my* world, shows itself in the fact that the limits of the language (the language which I understand) mean the limits of my world'.

In this way Wittgenstein fulfils his philosophic destiny of 'drawing a limit to thinking or rather not to thinking, but to the expression of thoughts'. And what is the result? One must say nothing except what can be said, *i.e.* 'the propositions of natural science, that is, something that has nothing to do with philosophy and then always when someone wished to say something metaphysical, to demonstrate to him that he had given no meaning to certain signs in his propositions.' But his advice to confine oneself to propositions of the science only or the 'totality of true propositions', as he calls them, is in the end self-defeating. For he excludes thereby all those hypotheses from 'the sphere of natural science' which are not true. And since we can never tell a true hypothesis from a false with absolute certainty, we can never know whether or not it belongs to the sphere of natural science. Obviously, therefore, we can no longer speak of even scientific 'laws' any more than of philosophic propositions, as both, after his own manner of speaking, can only be 'shown' but not 'said'.

We saw earlier how a search after scholastic 'precision' by Russell led to metaphysical mystification in logic and mathematics. We now see how Wittgenstein's attack on metaphysics is self-destructive as it leads only to a new Metaphysic of Silence whose last commandment is 'whereof one cannot speak thereof one must be silent.' Is it an accident that Wittgenstein's reasoned analysis of language leads him to the same sort of mystico-metaphysical silence as Rimbaud's 'reasoned derangement of all the senses'?

*   *   *   *

Although Wittgenstein was the first to enunciate a programme of the logical analysis of language, he himself did not go very far in carrying it out. As we have seen, the line of thought he followed led him into the blind alley of solipsism. Carnap proposes to remedy this defect by a more thoroughgoing analysis of language. Suppose we wish to study any given language O. The language O is called *object-language*. But before we can say anything about O, we must presuppose some language M in which we propose to talk about it. The language M, in which the results of our investigation are to be formulated, is known as *metalanguage*. Thus, in Siepmann's French Grammar, the object-language is French and metalanguage is English. In Nesfield's Grammar, on the other hand, both the object-language and metalanguage are the same, *viz.* English. Now, any object-language O presupposes the existence of three factors. In the first place, there is the speaker (or writer) and his listener (or reader). Secondly, there is the word, the spoken sound (or written mark) which serves as a conventional sign for denoting some object, property or action. Finally, there is the object, property or action itself, denoted by the word-sign.

Of course, a valid analysis of the functioning of any object-language ought to consider all the three factors together in all their mutual actions and interactions. But, to start with, we may simplify our problem by ignoring some of these factors. If we study our object-language in abstraction from the speaker (which includes his listener as well) we are left only with words and their meanings. A study of the relations between the sentences of an object-language and their meanings without any regard to the persons using them is known as Semantics.

Semantics divides the words (or signs) of the object-language into two categories—descriptive signs and logical signs. Those words (or signs) which designate things, properties of things, relations among things, actions, *etc.*, are called descriptive signs. The other signs are taken as logical signs. They are used only for connecting descriptive signs to form sentences and do not by themselves designate things, properties of things, *etc.* Such, for example, are the words 'if', 'any', 'all', 'some', 'not', 'is', 'are', and so on. But semantics by itself is not enough to clear confusion from philosophy and give its propositions 'absolute' certainty. While it does away with one cause of confusion (the speakers and listeners of the language), it has to put up with the nuisance of the factual meaning of words. Carnap therefore proposes to 'abstract from the second factor also and thus proceed from semantics to syntax'. If we take into consideration only the expressions (strings of word-signs) or sentences, leaving aside the objects, properties, states of affairs, or whatever may be designated by the word-signs occurring in the expressions, we get what Carnap calls the syntax of the object-language under examination. Carnap, of course, realises that as 'the meaning of words is the basis of the whole semantical system, it might seem as if nothing would be left if we eliminated the second factor as well'.

*   *   *   *

To understand Carnap's answer to this objection, a short digression on a paradoxical remark of Bertrand Russell's concerning the nature of mathematics might help. Mathematics, said Russell, is the only science where one never knows what one is talking about nor whether what is said is true. At first sight Russell's aphorism might appear to degrade mathematics to the triviality of a meaningless gibberish, but what he means to say is that in 'pure' mathematics we deal with various entities without knowing or even caring to know their 'meaning'. Take, for instance, the branch of mathematics usually called geometry, where we talk about entities such as points, lines, planes, circles, *etc.* Now, if we attempt to define what we mean by these terms, we often encounter pretty serious difficulties.

Thus, the text-book definitions of geometrical terms, *e.g.* of a point as 'that which has position but no magnitude', or of a straight line as the 'shortest distance between two points', are not quite satisfactory even though they may be good enough for a beginner. Since the attempt to define the meanings of mathematical terms leads sometimes to confusion or even contradiction, Russell and his followers propose that 'pure' mathematics should have nothing to do with the meanings of these terms. All that is necessary for the development of mathematics is to start from certain arbitrary definitions and axioms (or 'formal' rules) according to which these terms are to be used in the sequel, and to stick to them consistently throughout.

So, in 'pure' geometry, the geometer should not concern himself whether a 'point' means an infinitely small dot and a 'line' an infinitely thin scratch on a piece of paper, but should merely specify the axioms or 'formal' rules according to which he proposes to use them in developing his subject. For instance, he may proceed by postulating that there are certain things called 'points', 'lines', 'planes', *etc.*, such that two 'points' determine a 'line', two 'lines' determine a 'plane', and so on. From these and other similar axioms explicitly set up at the outset, he should derive all the conclusions that follow logically. In this way, he would never need to know what he means by these terms, or whether what is said of them is true. He would thereby avoid having to answer certain awkward questions regarding the nature of mathematical entities or concepts, although his theorems would be denuded of all 'content' or 'meaning' as a result and would possess a purely formal structure—a sort of internal relatedness among themselves.

Carnap has recourse to Russell's method of formalised discourse in order to overcome the objection mentioned above, namely that nothing would be left if we attempted to proceed from semantics to syntax by leaving aside the meanings of words as well from our consideration. He says, in effect, that just as the attempt to define the 'meaning' of mathematical terms leads to confusion and contradiction, so also the attempt to read 'meaning' into the propositions of logic and philosophy gives rise to philosophic difficulties and 'pseudo-propositions'.

He therefore proposes that, in logic and philosophy, we must deal only with the relations between different propositions but never with the relation between a proposition and the fact it represents. In other words, we may never inquire into the meaning of philosophic propositions. But how, one may ask, are we to deal with propositions which are not allowed to have any meaning or whose meanings we are forbidden to investigate? In order to meet this, Carnap distinguishes two modes of speech. In the one that he calls the 'material' mode of speech, statements concerning 'objects', 'facts', or 'states of affairs' have 'sense', 'content', or 'meaning'. In the second, that he labels the 'formal' mode of speech, propositions have no 'meaning' or 'content'. They say nothing at all about the world at large and are mere conventions governing the usage of certain words or terms in subsequent discussion.*

Take the proposition which in the 'material' mode says that 'time has

* Here, one may even recognise a parallel with Mallarmé, who sought to turn language into something of a pure non-referential structure on its own, a sort of pure *being*— that is, an incantation wherefrom the ordinary senses of the words have been systematically drained away. But with Mellarmé, at any rate, the object is not communication *per se*, whatever else it may be.

neither a beginning nor an end'. According to Carnap, a discussion of this thesis in the 'material' mode, that is, as a proposition referring to the flow of time in the real objective world, would give rise to 'insoluble difficulties and contradictions'. This is, indeed, true, as Kant showed in what he called the First Antinomy of Pure Reason when he demonstrated that it could be proved equally conclusively that time has a beginning and that it has none. Carnap therefore suggests that we should translate this thesis into the 'formal' mode: 'Every positive and negative real number expression can be used as a time co-ordinate'.

In this form—that is, in the 'formal' mode of speech—we do not assert anything at all about the real objective time of the physical universe being endless in both directions or not. We merely set up a 'convention' or a 'formal rule', which we propose to use in our discourse on something we choose to call 'time'. As long as we consistently stick to the convention we have set up at the outset and refrain from giving it a 'meaning' or 'content', no contradiction can arise. Carnap therefore claims that the so-called 'pseudo-propositions' so common in philosophy and philosophical analysis arise because of our usual habit of speaking in the 'material' mode. If such pseudo-theses are to be avoided, we must 'avoid the use of the material mode entirely', that is, we must cease to speak meaningfully.

Now if we adopt Carnap's suggestion and define our terms in the formal metalanguage, by means of expressions of the object-language without any reference to the meanings of those expressions, all we are left with is a system of syntactical rules of formation and transformation according to which we may choose to string together the signs (or words) of the object-language. Obviously these rules could be chosen arbitrarily if no regard is to be paid to any interpretation of the words of the object-language. It follows that we are at liberty to set up almost any system of syntactical or logical rules, any kind of logic or logical calculus. 'To be sure', Carnap hastens to add, 'the choice is not irrelevant; it depends on whether the logical calculus so invented yields on interpretation by the addition of semantical rules, a rich language or only a poor one. But, in principle, there is before us the boundless ocean of unlimited possibilities.'

There is, of course, no objection in principle to introducing abstractions in scientific and mathematical analysis. We are thereby able to concentrate on essentials, to the exclusion of complicating or irrelevant details. But the abstractions introduced by Carnap in his analysis of language are self-destructive. For, if we analyse language in abstraction from the two factors, namely, the speakers and the meanings of the words used by them, it ceases to be a language altogether. For language is a 'whole' created by the meaningful speeches of its speakers. If we abstract our language from the speakers and their meanings we do not have an analysis of language but of the private vocabulary of a grammarian who chooses to set up his own syntax (or logic). This point may be made clearer by means of an analogy.

Suppose we wish to develop a theory of triangles. If we construct our theory by ignoring one of the sides of a triangle, we do not have a theory of triangles but one of angles. If we ignore the second side as well, we have a theory of finite segments. Such abstractions are quite futile, as in making them we destroy the whole entity (triangle in this case) we set out to study. The abstractions proposed by Carnap in his analysis of language are of the same type. They destroy the very object under investigation. If Carnap, nevertheless, does manage to give the appearance of having constructed a semantical and syntactical theory of language, it is because he tacitly substitutes himself as the interpreter of the words of the object-language he proposes to analyse.

His theory of semantics and syntax is, therefore, merely a solipsistic analysis of language, that is, language as understood and practised in one's own private comprehension, just as the solipsist interprets the objective world as a reflection of his own sensory experience. As a result, Carnap's semantics or syntax is not a theory of language as we know it—that is, as a vehicle of social intercommunication. It is rather a hotchpotch of syntactical conventions that one sets up artificially for one's own private use or comprehension, ignoring other people as well as the objects of the material world.

Unfortunately, even when Carnap and his followers thus retire into the inmost recesses of their own private understanding, the 'eternal' truth they set out to trap eludes their grasp. For every syntactical (or logical) system which the wit of man has been able to devise is, on close examination, found inevitably to be tinged with the imperfection of its human creators. Consequently, the logicians keep on inventing increasingly heavier and stiffer brands of logic, which others keep on tearing to pieces. The history of mathematics and logic during the past fifty years is indeed a brave record of the births and deaths of numerous such monolithic systems which were believed to be infallible at the time they were created but were subsequently found wanting.

For instance, about forty-five years ago, Whitehead and Russell seemed to have all but succeeded in reducing mathematics to an absolute logic; but within a decade of the publication of their *Principia Mathematica*, Ramsey and Chwistek exposed a number of contradictions in the *Principia* logic. Unfortunately, neither Ramsey's application of Wittgenstein's ideas, nor Chistwek's theory of constructive types, both of which were designed to save the *Principia* system from shipwreck, had any better luck. So mathematicians began to devise newer and still more ponderous logics. Such were the logics of Curry and Church which, in their turn, were proved inconsistent by Kleene and Rosser. Of the five different systems of logics enumerated by Lewis and Langford, in their *Symbolic Logic*, not one was found by the authors sufficiently 'precise' to embody 'acceptable principles of deduction.'

Paradoxical as it may seem, all these attempts to set up an absolute and infallible logic have come home to roost in the work of the Austrian logician, Kurt Gödel, who seems to have proved that if any system that includes arithmetic contained a proof of its consistency, it would also contain one of its own inconsistency. It would thus appear that mathematical logicians have not succeeded in making logic invulnerable any more than Thetis did in making Achilles immortal. Every one of their dips in the philosophical Styx has conferred immortality but with its own equivalent of Achilles's heel.

* * * *

If the logicians' attempts to set up an Absolute Logic have not entirely succeeded, the efforts of the formalists to create a static hierarchy headed by another Absolute—in this case metamathematics—have also run into difficulties. The formalists endeavour to reduce mathematics to a game of manipulating symbols, mere marks on paper like $+$, $-$, $=$, $\supset$, *etc.*, in accordance with certain formal rules. The logic of handling these symbols, *viz.* metamathematics, thus becomes an Absolute—a sort of super-theory designed to 'justify' mathematical reasoning by showing that it does not lead to any inconsistency or contradiction. Now, consistency in a system of statements means a kind of coherence displayed by the set so that these statements and their consequences do not clash. For instance, if I make the three statements,

(i) No prophet was ever born in Patna;
(ii) Guru Gobind Singh was born in Patna;
(iii) Guru Gobind was a prophet;

the set is obviously inconsistent. For, from the second and the third, I can 'deduce' a conclusion contrary to the first statement. The formalists contend that 'pure' mathematics is concerned with a set of initial statements, called 'axioms', from which we derive a number of consequences.

The formalist problem, then, is to show that the initial statements and *any* possible consequences such as we may derive from them do not clash as the three statements in the foregoing illustration do. But derivation of consequences from the initial statements presupposes a method of drawing conclusions or a theory of inference. Such a theory has, therefore, to distinguish between a right and a wrong way of drawing inferences. If it is arbitrarily decided that a certain way of drawing inferences is the right way, it will lead only to a meaningless play with the symbols isolated from the practical problem of scientific methodology. If, on the other hand, we decide that our actual way of handling statements and arguing from given premises is the right way, the meta-theory which the formalists try to set up is built on the very foundation which the formalist super-theory is intended to examine.

It therefore seems that a thorough-going formalist foundation of mathematics is impossible, though formalist theory is very valuable on account of the light it throws on the interrelations between the distant parts of mathematics, and in particular on the relation between logic and mathematics.

This difficulty of the formalist theory may also be stated in somewhat different terms. To 'justify' mathematics, the formalist endeavours to set up a meta-theory—that is, metamathematics. But, to justify the latter, he would need a still higher super-theory, a sort of meta-meta-mathematics, and so on. In this way he involves himself in an endless regress from which there is no escape. In other words, if a metamathematical theory of proof is to be formalised as a set of symbols manipulated according to specified rules, then there must be an infinite regress of such rules, assuming that the metamathematical theory is to be completely described. This is the heart of Gödel's theorem referred to earlier.

It appears, therefore, that the endeavour of the logicalists and the formalists to establish mathematics as an Absolute or a monolithic system of Eternal Truths is not likely to succeed completely. With increasing awareness of this, there has also occurred a certain weakening of the formalist thesis that mathematics is a game played with *arbitrary* symbols according to *arbitrary* rules subject only to the condition of consistency. It is now claimed that this *play* is *like* a game—which is, of course, not quite the same thing as to *be* a game. Nevertheless, since the symbols remain *arbitrary*, subject to *arbitrary* rules of combination and at the same time free of all meaning *except* such as they may acquire by being placed in juxtaposition with one another, formalism does run into difficulties in explaining the applications of mathematics.

However much it may claim to be a pure structure, independent of any application, reference, or content, its value (which is undoubtedly great), depends precisely on the fact that it does not ignore application altogether. Even Hilbert, the leader of the formalist movement, considers that metamathematics deals with mathematics in the same way as the 'physicist investigates his apparatus'. But since mathematical proofs, in so far as they prove anything significant, deal with certain aspects abstracted from some sphere of reality, formalist metamathematics cannot completely cut the umbilical cord that ties mathematics to reality.

For even if these aspects are embodied in an axiomatic form and treated as pure structures of meaning-free symbols independent of any outside reference, the meaning comes into the symbols all the same as a result of the mutual connections of the various symbols as expressed in the initial axioms. What actually happens is that formalist analysis is able to purify meanings of symbols in order to study their properties, very much as a chemist purifies a substance in order the better to study its behaviour. The axiomatic or formalist approach, insofar as it is significant, is therefore really a sophisticated analysis of mathematical concepts to show what is involved in our knowledge of them even though it cannot explain their origin. Consequently even a formalist, try as he may, cannot escape meaning, content (and therefore application of a sort), any more than Francis Thompson could flee the Hound of Heaven:

> I fled Him, down the nights and down the days;
> I fled Him, down the arches of the years;
> I fled Him, down the labyrinthine ways
> . . .   .
> But with unhurrying chase,
> And unperturb'd pace,
> Deliberate speed, majestic instancy,
> They beat—and a Voice beat
> More instant than the Feet—
> 'All things betray thee, who betrayest Me.'

*   *   *   *

Since the logicalists' and formalists' analysis of number and mathematical proof raises such a storm of difficulties, we are justified in seeking some other way out. It seems to me that the intuitionists are right in asserting outright that the whole numbers are given us immediately in intuition and that it is vain to try to get behind them. At most, we may consider them as generated by successive additions of unity to other numbers already formed. Thus, starting with unity, we generate the number 2 by the addition of unity to itself, the number 3 by that of unity to 2, and so on indefinitely. The most important element in this construction of whole numbers is, therefore, the concept of unity, a concept given us by our immediate intuition.

At the root of this concept of unity is the fact that most things in the universe around us do manage to retain a measure of stability or permanence—at least to our human scale of observation—even though they are really in a state of ceaseless flux. Once the series of whole numbers is constructed by successive acts of addition of unity in the manner described above, it is quite possible to base mathematics entirely on this notion of integers by a sound analysis of the constructions and processes of mathematics, described in ordinary language and ordinary mathematical symbols, plus a few logical symbols, as has been done by Paul Dienes. But the intuitionist solution of the foundation problem raises the question of the validity of a fundamental law of classical logic which Euclid employs frequently in his proofs of geometric theorems.

When Euclid is hard put to demonstrating a theorem, he usually adopts the following strategem. He says, in effect, that if you are not willing to concede the truth of his theorem, suppose it is *not* true. Starting from the assumption that the theorem in question is false, he deduces by a sequence of logical steps an absurd conclusion, whence it is evident that the theorem must be true, *Q.E.D.* This procedure is a useful logical device and without it a good deal of geometry and other parts of mathematics would be as good as lost. Unfortunately, the intuitionist analysis of the foundations of mathematics has called in question the validity of this device. The reason is that it is based on the tacit assumption that every statement is either true or false, and if it cannot be false it must be true.

At first sight it might seem astounding that anyone should challenge this assumption, which is technically known as the law of excluded middle. Nevertheless, a deeper examination shows that an unqualified acceptance of the law does give rise to difficulties. In the first place, it is not true that *every* statement must either be true or false. For instance, the statement 'virtue is red', is neither true nor false; it is only a meaningless string of words. However, this objection is of no importance, for we may legislate that the law is true of all meaningful statements, particularly as in logic and science we usually deal with only meaningful statements. Consequently, a rule that is true of all meaningful statements, though not of *all* statements, is none the less valuable for being thus limited.

A more serious difficulty is the fact that it may not sometimes be true of even meaningful statements. For example, if I state that X's hair is black, it cannot be said that there are only two alternatives, namely that either the statement is true (that is, X's hair is black) or it is not true (that is, his hair is not black), for it may be that X is bald. Nevertheless, we may still save the law of excluded middle by adding the clause, '*if X's hair exist*' it is either black or not black. But then, we have to explain what we mean by the verb 'exist' and what is involved in the notion of 'existence'.

Unfortunately the idea of 'existence' has been a source of endless confusion in philosophy for the past two millennia, beginning with Plato's *Theaetetus*, and it has not cleared up even today. The root cause of trouble lies in the multiplicity of different meanings usually associated with the notion of 'existence'. There is one sense of the word 'exist' when I say, 'Tigers exist', and quite another when I assert the 'existence' of virtue, a bank overdraft, an electron or the ether. A detailed analysis of the various meanings of the word 'existence' would lead us too far away from our present theme. We shall, therefore, confine ourselves merely to explaining the notion of 'existence' as used in mathematics.

Now, in mathematics, we consider an entity as 'existing' if it fulfils at least one of the following two conditions. First, we may be able to point to it by means of a definite process or construction. Second, we may be able to show that the denial of its 'existence' leads to an absurdity. For instance, if I want to assert the existence of the square root of, say, 25, I specify the process of square root extraction and point to the number 5, which, multiplied by itself, gives the product 25. On the other hand, as an example of the second kind, suppose I wish to assert the existence of an integer N such that any integer n greater than N can be represented as the sum of at most four primes (that is, of numbers not divisible by any number other than unity). I can do this only in an indirect fashion by showing that the denial of the existence of a number with such a property leads to an absurdity, as the Russian mathematician, Vinogradoff, has recently done.

It has not been possible, at any rate for the present, to point to any such number N in the same direct manner as in the case of the square root of 25. In the former case, we have a tangible entity whose property we have asserted; in the latter we only show that if the entity possessing the postulated property does not exist, we can deduce a contradiction. Some distinguished mathematicians have recently put forward the view that the second, indirect and non-constructive method of proving the existence of mathematical entities is inadmissible, or at least of doubtful validity in many cases where it is employed.

The reason for their objection may be made clear by a deeper analysis of the aforementioned statement, *viz.* X's hair is either black or not black. This statement involves an implicit reference to two collections into which we choose to divide the possible colour attributes of all kinds of hair. In one of these collections, we include those hairs which have the colour black and in the second those of all other colours excepting black. Without such a frame of reference, the full content of the negative statement cannot be determined. Now, in this case both the collections of the attributes black and not black are static and have fixed meanings, so that they do not alter during the course of the argument. In such a situation, the law of excluded middle holds that if X has hair, it belongs to one or other of these two mutually exclusive attribute collections. It cannot belong to both or neither of these two collections at the same time. However, there are collections which, even when 'well defined' by means of an exact phraseology, do not remain static but are (so to speak) in a perpetual state of formation. Such, for instance, is Richard's collection of all numbers definable with less than one hundred words.

There are only a finite number of words listed in any dictionary. Out of these we can form various combinations of one hundred words or less. Many of these combinations will be meaningless strings of words; others will be meaningful but will not define any number. Only some of these combinations will define numbers, *e.g.* the word combination 'five multiplied by six, reduced by seven'. Let us retain only such combinations of words as define some real number. We will thus have a finite number of numbers definable with one hundred words or less. Some one of these finite numbers must be larger than all the rest. Now, 'increase this largest of these finite numbers by one'. This is a number defined with less than one hundred words, as may be verified by counting the words within inverted commas in the foregoing sentence. It therefore belongs to Richard's collections. But it is by construction not a member of this collection.

The paradox arises because, on the one hand, the collection is conceived of as a completed totality of numbers definable with less than one hundred words, while on the other, these definitions constitute a collection in a continual state of formation which generate newer definitions from those already formed. In other words, the terms used to define numbers have not sufficient fixity of meaning. Unfortunately, classical logic is applicable only when its concepts have fixity of significance to serve as the terms of its syllogism. It fails when it has to handle statements whose implicit frame of reference involves concepts or collections in a perpetual state of dynamical evolution. To deal with them successfully, we require a profounder logic, the logic based on unitary principle.

Now, in mathematics there are collections which cannot be regarded as static, mere aggregates of individuals formed by a definite rule. One such collection is the continuum of *all* real numbers lying between two given numbers. The idea of the continuum is easy to explain, though difficult to define in a mathematically rigorous fashion. Suppose, for instance, we have a metal rod. When it is heated, it expands from its initial length, say 15 inches, to another length, say 15·25 inches. We assume that it could not have done so without passing through *all* possible lengths between 15 and 15·25 inches. This is the practical intuitive idea of the continuum. Mathematically, its characteristic is that it consists of *all* real numbers between 15 and 15·25.

We consider *any* real number between 15 and 15·25 as a definite object (of thought) if we are given a definite rule for determining all the consecutive digits of its 'complete' decimal representation. For instance, a real number like $15\frac{1}{10}$ or 15·1 is a definite object because its complete decimal form is 15·1000000 . . ., an endless chain of zeros following unity in the first decimal place. Similarly we consider a real number like $15\frac{1}{7}$ as definite because we can represent it by means of the recurring decimal 15·142857,142857,142857, ...

These rules, however, do not create *all* the infinite decimal forms or real numbers. For, in addition to the real numbers *devolved* by such rules, we may also conceive of real numbers in another way, *viz.* by determining the successive digits of its decimal form by arbitrary acts of free choice instead of definite rules. Thus, for instance, after we have determined any digit, say, in the fifth place, we are free to pick the next from any of the ten digits, 0, 1, 2, 3, 4, 5, 6, 7, 8 and 9. Such a decimal form which is in a perpetual state of dynamical evolution may be called an *evolving* decimal form, after Paul Dienes. The continuum, then, is the complete class of both *devolving* and *evolving* decimal forms, that is, forms determined by definite rules as well as by arbitrary acts of free choice. Consequently, its constituent numbers obey no regular law of formation, and to that extent the concept of the continuum, like Richard's collection, is a dynamically evolving collection.

When, therefore, the intuitionists deny the validity of the application of the law of excluded middle to statements whose implicit frame of reference is the continuum, they are enunciating merely the fundamental law of a new and profounder logic, that the principle of contradiction breaks down when applied to collectives in a ceaseless state of becoming. Unfortunately, the intuitionist reasoning, whereby Brouwer and his disciples support this conclusion, is obscure and only partially correct. This seems to give an air of paradox to an otherwise correct point of view.

Thus, as an instance of the breakdown of the law of excluded middle, Brouwer cited a statement concerning the decimal representation of the number π. It is known that the ratio of the circumference of a circle to its diameter is the same for all circles, whatever the radius. This number is denoted by the Greek letter π. Its value is 3·14159 . . . and we have definite rules whereby we can calculate it correct to any number of decimal places. In the Science Museum of Paris, they have written its value up to several hundred decimal places.* Nowhere in this finite representation of π as a decimal form do we find the group of digits 0, 1, 2, 3, 4, 5, 6, 7, 8 and 9 in their natural order. Now, if we imagined the value of π to be worked out to an infinite number of decimal places, would the following proposition be true or false?

'That somewhere in the infinite decimal representation of π there occurs the group of digits 0, 1, 2, 3, 4, 5, 6, 7, 8, 9 in their natural order.'

Brouwer holds that this statement cannot be considered either true or false as it cannot be verified in a finite number of steps. But as π is a

* Its value has now been worked to 2,035 digits by means of the ENIAC computer.

definite real number whose digits in the decimal form can be worked out to any number of places we like, Brouwer's query is a determinate problem admitting a 'yes or no' answer, although actually we are unable to decide the issue. This practical difficulty does not invalidate the application of the law of excluded middle in this particular instance, though Brouwer is right in denying its universal validity and querying the formalist assumption that the answer to every mathematical question is a simple 'yes or no' even if it is 'well put'. It seems to me that the validity of Brouwer's objection lies more in the application of this principle to statements whose implicit frame of reference consists of collections which, like the continuum, are always in a continual state of formation and remain for ever unfinished, rather than in the distinction he makes between known and unknown results.

*   *   *   *

To recapitulate, one of the fundamental problems of the foundations of mathematics is the reform of mathematical reasoning so as to avoid paradox or contradiction. Mathematical reasoning has been haunted by the fear of paradox since the days of Pythagoras and Zeno. As we saw, Pythagoras found to his dismay that the diagonal of a unit square cannot be expressed as a ratio of two integers, and Zeno astounded Athens by apparently proving that Achilles could never overtake the tortoise. Paradoxes in many ways similar to these have been discovered from time to time, and as recently as the close of the nineteenth century a whole series of them were uncovered by Burali-Forti, Russell, Richard, König, Berry and others. The logical positivists claim that, if mathematical reasoning is to avoid paradox and contradiction not only here and now but 'for ever', then it must be reduced to logic which is nothing but language—or rather the syntax of a language—made perfect. But the conditions of a logically perfect language are two. First, the rules of its grammar must explicitly exclude nonsensical combinations of signs (or words). Second, every symbol or sign in it must have a definite, unambiguous and unique meaning.

The first to lay down these conditions of a logically perfect language was Wittgenstein. But, as he was led head-on to solipsism, the doctrine that nothing is real except the thinker's sense-data, and that everybody and everything else around are nothing but his dream, most logical positivists now consider this consequence of Wittgenstein's reasoning to be a great defect. Carnap undertook to remedy it by making a distinction between what he called the 'material' and 'formal' modes of speech. When we use the material mode of speech, our statements and propositions have meaning and content. But when we use the 'formal' mode of speech, we do not assert anything at all about the real things or events about which our statements or propositions appear to talk. We merely set up 'conventions' or 'formal rules' which we propose to use in our discourse on the 'things' or 'events' in question. As long as we consistently stick to the conventions we have set up at the outset and refrain from giving our statements a 'meaning' or 'content', no contradiction can arise. Carnap therefore claims that the only way to eliminate the so-called 'pseudo-propositions' of philosophy is to 'avoid the use of the material mode entirely'.

Carnap is right in insisting that all questions of philosophy have a linguistic aspect, because we have to use language to express them. For instance, the question regarding the nature of space has a linguistic aspect because we could express its properties by using the language of Euclidean geometry, provided we correlated the measurements of time by different observers in the manner suggested by Milne.* Or, alternatively, we could use the language of Riemannian geometry, provided we reckoned time according to Einstein. To this extent it could be asserted that philosophic theses with regard to the nature of space are questions of linguistic forms. But Carnap goes too far when he claims that the philosophic question regarding the real nature of space ends here and can have no other aspect than that of the linguistic form just discussed.

Obviously, objective space, the theatre of perceptible phenomena occurring in the universe around us, has 'real' properties quite apart from the question of linguistic conventions that we may choose to adopt for their description. To dismiss such questions as mere 'metaphysics' or as 'pseudo-propositions' devoid of all sense and having no relevance to the philosophy of science, is to emasculate both philosophy and science and to degrade them to a mere hotchpotch of conventions. It is true that language is the only instrument we have for expressing and communicating scientific and philosophic propositions. But if the reality of the external universe happens to be at times too complex to be trapped in the neat linguistic expressions of our making, we have no right to reject this 'reality' itself as a metaphysical phantom and confine our thinking activity to a barren formalisation of language devoid of all 'content' or 'meaning'. Thus, if the thesis that 'time is endless' (to use the 'material' mode of speech) leads to verbal contradiction, we cannot get over this difficulty merely by denying the reality of time itself and setting up formal rules according to which we propose to use the word 'time', and thus turning it into a 'meaningless' formal construct of an artificial language. For, we do not thereby know whether time as the temporal link between objective world events has a beginning or not, we only have a convention governing the use of the word 'time' in our discourse.

* See Chapter 8.

This is not to deny the importance of the linguistic and even conventional aspects of scientific and philosophic theses. Their study is one of the most important and pressing tasks of philosophy. For not only is language our only instrument of expression and communication, it is also the very warp and woof of our thought. In fact, it dominates and determines our thought so completely that the limits of our language are in many cases the limits of our world, a truth which the authors of Orwell's 'Newspeak' tried to exploit in their efforts to control human thought. Moreover, the defects of our language, the imprecision of its words, the tyranny of its phrases, the rigidity of its structure, and the cumbersomeness of its style may cost us very dear. They may even make us lose battles as they did the Japanese during the last war.*

For these reasons it must be admitted that logical positivism has been of great value in that it has fostered this awareness of the powers and limits of language with a vigour and clarity never attempted before. But for all that, language must remain a means to an end, the end being the understanding of the world around us with a view to changing it for the better, and to the communication of that understanding to others. To make its study an end in itself, as the logical positivists seem inclined to do, is to indulge in a sort of Tarasconnade—that is, a chase for chase's sake—complete with all the paraphernalia of fowling-pieces, game-bags, whips, whistles, hounds and hunting-horns, but in a country where there is no game to be had, not even the lone and legendary rabbit. Let us consider the effect on mathematics of this inversion of means and ends.

In their endeavour to rid mathematical reasoning permanently of paradox and contradiction, at one stroke, the logical positivists are obliged to deny that there is any relation between a mathematical structure and a sphere of reality. As a result, they are led to the view that mathematics is a branch of logic which is only a language all of whose statements are tautologies, or disguised and roundabout ways of saying that A is A. In so far as the positivists consider mathematics as a language, a mere construct of symbols arbitrarily selected, they lose sight of the intimate connection between our understanding of the mathematical truths and their expression in symbolic language for the purpose of communication.

If there is any arbitrariness, it is not in our understanding of them, but rather in the particular symbols used to write them. Whether we write 'two and two make four' or 'deux et deux font quatre', or '$2 + 2 = 4$', may be arbitrary, but not what these expressions mean. Likewise, whether we

---

* In *The Hinge of Fate*, Sir Winston Churchill writes: 'The rigidity of the Japanese planning and the tendency to abandon the object when their plans did not go according to schedule is thought to have been largely due to the cumbersome and imprecise nature of their language which rendered it extremely difficult to improvise by means of signalled communication.'

denote fifty by 50 or L may be arbitrary, but not what the fundamental signs stand for. But even here the immense superiority of the former notation shows that this arbitrariness has its limits. Much of the argument for the arbitrary character of mathematics implicit in the Game theory is really due to the confusion between the means of expressing mathematical truths, which may within limits be selected arbitrarily, and the mathematics itself.

But, this apart, the positivists face a dilemma on their own basis. For if mathematics is an arbitrary construct of symbols, how could it ever be applied and used to predict events? It is no answer, as Kattsoff has remarked, to say that experience *might* not have verified the prediction, for the point is that it did. The only way they can escape this dilemma is by introducing a distinction between pure and applied mathematics. Any such distinction must be false, for pure mathematics has actually arisen as an abstraction from empirical practice. For example, pure geometry stems from the practice of land measurement and pure arithmetic (Boolean algebra) from that of counting. In turn, it is revitalised by some concrete application like pure geometry to physics and statistics, or pure arithmetic to the theory of calculating machines.

To isolate pure mathematics from the applied is like isolating the giant Antaeus from his mother Terra, the earth, contact with which gave him renewed strength from every fall in all his contests. To lift pure mathematics to the high heaven of passive contemplation, away from the mundane applications of the work-a-day world, is to stifle it, very much as Hercules stifled Antaeus in mid-air. Therefore, instead of isolating the pure and applied aspects of mathematics we must endeavour to understand their interrelation and thereby the relation of mathematics to reality. This means that we must not treat mathematics as a game of manipulating arbitrary symbols or paper marks according to arbitrary rules, or even as resembling such a game, but as a product of human societies whose members co-operate socially to advance civilisation and culture. This means that mathematics has sociological, psychological, pragmatic and empirical aspects, besides the syntactical aspect with which alone the positivists seem to be concerned. For we must remember that, while mathematics has to resort to abstractions to secure a foothold for a first peep into reality, it must continually transcend them and go beyond them by taking into account other aspects, previously ignored, to get a fuller view of reality. This is the only way out of the dilemma into which positivism pushes the mathematician.

APPENDIX I

A. Einstein and A. N. Whitehead

Dr. Johnson is said to have pooh-poohed *Gulliver's Travels* with the remark that once you have thought of little men about ten times smaller and big men about ten times bigger than human beings, the rest followed automatically. Johnson, of course, missed the point of Swift's classic and his criticism was unjust. However, a more scientific criticism of Swift's satire might be that he made the untenable assumption that the world of phenomena continues to present in the main the same aspect to infinitely small or infinitely big beings as it does to those built on our own scale of dimensions. For instance, the table on which I write appears to me as an enduring, stable object but to the Lilliputians, or at any rate to intelligent beings made on a sufficiently small scale, it might appear as a swarm of chaotically moving particles with no permanence of structure such as it displays to us humans. On the other hand, to the Brobdingnagians, or better still, to beings as big as Voltaire's Micromegas, who was so huge that all the waters of the Mediterranean did not suffice to wet his heels, the universe might appear very much as it does to an intelligent fish without a fixed habitat and perpetually tossed about by the waves of the ocean.

Now, while it has been a favourite device of some writers to present their own world-view as seen by beings of various sizes, or even by animals of various sorts, no one has attempted to present the world from the eyes of a roving fish.* The main difference between the outlook of such a fish and that of man arises from the fact that the former's environment is in a state of ceaseless flux while that of the latter does show relative stability. It is true, as Engels remarked, that even for man, the world cannot be comprehended as a 'complex of ready-made things, but as a complex of processes'. Nevertheless, in spite of the 'uninterrupted change of coming into being and passing through', which all things suffer, they do appear to remain stable—that is, to retain their identities. It is to this ability to recognise 'sameness' beneath the ceaseless flux of things that we owe the creation of science itself. For a being without even an apparently stable *milieu*, whose environment changed as ceaselessly as the waves surrounding our hypothetical fish, and in whose experience nothing seemed to remain permanent, could not possibly evolve the concept of number, which, as Tobias Dantzig has remarked, is the language of science. The reason is

* Except Rupert Brooke, in his charming but hardly apposite verses entitled *Heaven*.

that the idea of number originated from the empirical practice of counting, which in turn depends on our ability to recognise objects which remain the 'same' with the passage of time, and which are, moreover, discrete.

Likewise, measurement depends not only on our capacity to recognise discrete objects which remain the 'same' but on our ability to perform a series of operations which are also recognised as the 'same'. Thus, if I am able to measure lengths it is because of two reasons. First, I can recognise that the yard or the metre, which I use as my standard of measurement, remains—or at least, appears to remain—the 'same'. If the yardstick ceased to have even this appearance of permanence, I should have no standard whereby to measure lengths. Second, I can recognise that the operation of measuring any given length by successively laying the measuring rod along it and counting how many times it goes between its extremities remains the 'same', whether the length thus measured is here or elsewhere. In other words, transporting the measuring rod from one place to another makes no difference to its length. It would, therefore, seem that a background of 'sameness' or uniformity behind the changing pattern of the complex of processes is essential to the emergence of the concepts of number and measurement on which all science, at any rate physical science, rests.

Now the essence of Whitehead's objection to Einstein's theory of Relativity is that it involves the demolition of this background of uniformity on which science is built. As we saw in the text, Einstein claims that you cannot have space without things, or things without space, any more than you can have a grin without a Cheshire cat. Further, the character of space-time matrix of material events everywhere is determined by the density, momentum and energy of matter and radiation at that point. But the distribution of matter and radiation in the universe is far from uniform. For example, there are places in the universe, such as the interior of the stars known as 'white dwarfs', where matter is almost incomparably denser than in the interstellar voids—or, for that matter, in the interior of giant stars. It follows, therefore, that the character of the space-time framework which is determined by this non-uniform distribution of matter and radiation, is itself non-uniform. In other words, one chunk of space-time is not the 'same' as another.

For this reason Whitehead doubts the possibility of conducting any measurement in a space which is heterogeneous or non-uniform as to its properties in different parts. For such a framework precludes the possibility of any fixed conditions for obtaining a basis for measurement. Whitehead's point might, perhaps, be made clearer by means of an illustration. Suppose I want to measure the length of the arc of the earth's meridian from the North Pole to a point on the equator. If I start measuring this distance by means of a rod, taking the rod first to the North Pole and then gradually laying it along the meridian till I arrive at the equator, I should make a serious error in my measurement if I ignored the fact that the rod, in travelling through the Polar regions to the equator, changes its length due to expansion or contraction caused by temperature differences.

In this particular case, I can correct the error because I can express the length of the rod at any temperature in terms of its length at a suitably selected standard temperature, of, say, zero degree Centigrade. I can, then, say that the length of the meridian arc is about 10 million times the length of the standard metre rod at the standard temperature. In other words, I can state that the meridian arc will accommodate about 10 million metre rods of this kind between its two extremities *assuming that each rod remains at one uniform temperature*, even though we know that the temperature is not the same all along the meridian arc. If there is no possibility of reducing the conditions of measurement to such a uniform basis, no measurement can be made. Einstein's non-uniform heterogeneous space makes it impossible to state the uniform conditions which are the pre-requisites of every measurement. Accordingly, concludes Whitehead, the practice of measurement in heterogeneous physical space, such as Einstein postulates, is devoid of any 'real' meaning.

There is considerable force in this objection, and that is perhaps why some men of science try to explain Einstein's theory by positing a uniform space of *five* dimensions in which the universe is set. But, as Whitehead himself has observed, such a fictitious space, which never enters into experience, cannot get over the difficulty.

A similar point has been made by Milne, many of whose conclusions are similar to those of Whitehead, in the alternative formulation of his *Kinematic Relativity*. Einstein himself has not bothered to present his views as a consistent logical system. He claims that the system of postulates set up by him leads to equations that have proved their superiority empirically by eliminating the chief discrepancies remaining in Newtonian theory, as well as by predicting two experimental results that had never been thought of before. Nevertheless, this experimental confirmation of Einstein's law of gravity does not necessarily guarantee the validity of his postulates, as Whitehead, too, deduced from different assumptions results which are identical with Einstein's within the limits of present-day experimental errors. Moreover, experimental confirmation of a theory at a particular stage of scientific development need not always be a proof of its 'eternal' truth, particularly when the abstractions introduced by it to get to grips with nature show up their own limitations.

Accordingly, Whitehead's criticism of Einstein's theory cannot be rejected as invalid offhand. All the more so, because Einstein's theory is

based on the philosophically anarchic idea that the universe is a con-
glomeration of units moving along isolated 'world-lines' or courses un-
influenced by time as an active 'factor in causation', and that its events are
mere coincidences of these 'world lines'. Now, however useful the abstraction of 'world lines' and their 'coincidences' may be in geometricising
physics, it is not clear, as Whitehead has remarked, how the perception of
light, on which almost every physical experiment ultimately depends, can
be reduced to a perception of coincidences. The same objection cannot be
raised against Whitehead, for he seeks to make the idea of temporal process the basis of all intellectual and scientific thought. To this extent, at
least, it has to be admitted that Whitehead's theory is more organic and
renders a better account of the unity of the cosmos than do Einstein's
abstractions, world-lines and their coincidences.

APPENDIX II

We have shown in the text how an observer can make a triplet of time observations. Consider first only two observers, A and B, and a triplet of time observations of each. If (t_1, t_2, t_3) is one such triplet observed by A, then t_3 is the time (given by A's clock) at the instant he sees B's clock indicating an epoch t_2. Now to each reading t_3 of A's clock there corresponds a *reading* t_2 of B's clock. Let A graph t_3 against t_2, obtaining a relation

$$t_3 = f(t_2) \qquad . \qquad . \qquad . \qquad . \qquad . \quad (1)$$

Now consider any triplet of time readings (T_1, T_2, T_3) observed by B. Here T_3 is the time given by B's clock at the instant B observes A's clock, indicating an epoch T_2. Let B plot likewise T_3 against T_2, obtaining a relation

$$T_3 = \varphi(T_2) \qquad . \qquad . \qquad . \qquad . \qquad . \quad (2)$$

It is obvious that the simplest case in which we can hope to synchronise the clocks of A and B is when A and B move relatively to one another in a straight line and the relation of A to B is symmetrical—that is, that things would remain the same if the roles of A and B were reversed. Mathematically, this will be the case only if the function f is identical with the function φ; or, if $f = \varphi$. If this relation does not hold, the two clocks are not synchronous or congruent. Under what conditions then will the two clocks be congruent?

To answer this question we note that some T's, the readings of B's clock, will coincide with some t's, the readings of A's clock, if we imagine the following experiment: Suppose A strikes a match at time t (by his own clock), and B reflects the ray by means of a mirror. A then again re-reflects the reflected ray back to B and so on endlessly. The result may be shown diagramatically as in Fig. 65.

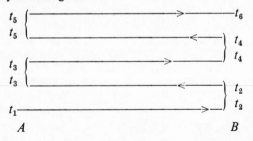

$$A \qquad\qquad\qquad\qquad\qquad\qquad\qquad\qquad B$$

Fig. 65

A's triplets are (t_1, t_2, t_3), (t_3, t_4, t_5), (t_5, t_6, t_7), ... B's triplets would then be (t_2, t_3, t_4), (t_4, t_5, t_6), ... Since the second and third numbers of each of A's triplets are related by (1), we have a series of equations like

$$t_3 = f(t_2),\ t_5 = f(t_4),\ t_7 = f(t_6) \text{ and so on.}$$

In general, for any *odd i*

$$t_i = f(t_{i-1}) \qquad . \qquad . \qquad . \qquad . \qquad . \quad (3)$$

Now consider B's triplets. Since the last two of each triplet are related by (2), we have another series of equations

$$t_4 = \varphi(t_3),\ t_6 = \varphi(t_5),\ t_8 = \varphi(t_7), \text{ etc.}$$

In general, for any *even i*

$$t_i = \varphi(t_{i-1}) \qquad . \qquad . \qquad . \qquad . \qquad . \quad (4)$$

To make the two clocks congruent we must equate

$$f = \varphi.$$

Hence, the two relations (3) and (4) can be combined into one and stated as under:

For any i (odd or even), $t_i = f(t_{i-1})$

In other words, if t_i is any reading of one observer's clock at the *instant* he *sees* the *reading* t_{i-1} of another's clock, a function f must exist such that

$$t_i = f(t_{i-1}).$$

Milne calls such a function f the signal function of the two observers. Now if there are more than two observers, *e.g.* A, B, C, D, ... all moving in one straight line, to each pair will correspond a signal function. Let f_{AB} be the signal function of A and B, f_{CD} that of C and D, and so on. In order that clocks of all these observers be congruent or equivalent, Milne shows that they must satisfy the further condition:

$$f_{AB}f_{CD} \equiv f_{CD}f_{AB}$$

In other words,

$$ff_0 \equiv f_0 f,$$

where f_0 is a given function.

It can be shown that solutions of this equation, in which the unknown is not a magnitude but a functional form, is a group and that $f(t)$ must be of the form

$$f_{AB}(t) = \psi a_{AB}\psi^{-1}(t),$$

where a_{AB} is a positive real number characteristic of the observers A, B corresponding to f_{AB}, and ψ is an arbitrary function and ψ^{-1} its inverse. In other words, to every function ψ there corresponds a way whereby all the observers A, B, C, D, . . . can correlate their time measurements or clocks and make them equivalent or congruent. Such a function ψ is then said to generate an equivalence; that is, a way of synchronising the clocks of different observers so as to make them equivalent. Now Milne's main theorem is that if we took another function φ, that is to say, adopted another way of synchronising the different clocks, then this φ-way of clock synchronisation becomes identical with the ψ-way, if we regraduate the clocks of the observers of the φ-equivalence in an appropriate way. Essentially, therefore, there is only *one* equivalence, and the different equivalences generated by different functions ψ are merely different descriptions of the same kinematic entity.

Of the numerous possible descriptions of this kinematic entity—the equivalence—the most important are two. In the one $\psi(t)$ is simply equated with t and in the other with $t_0 \log t/t_0$ where t_0 is an arbitrary constant. The former defines a way of clock graduation so that an equivalence is described 'as consisting of particles in uniform relative motion separating from a common point of coincidence'. The latter defines another way of clock graduation whereby the *same* equivalence appears to consist of relatively stationary particles. We have thus isolated two measures of time t and τ. It is τ-time that is identified with the Newtonian time of classical physics. But it is the t-time, Milne's time, that is more fundamental. The reason is that τ-time depends on a constant t_0 which appears in the generating function ψ of the τ-equivalence, namely $\psi(t) = t_0 \log t/t_0$. On the other hand, the t-form of the equivalence is simply defined by $\psi(t) = t$ in which no t_0 appears. This is because the t-equivalence has a natural origin of time $t = 0$, the epoch of coincidence of all the observers. Here t is simply the 'age' of the system at that event.

NAME INDEX

SUBJECT INDEX

A CATALOGUE OF
SELECTED DOVER BOOKS
IN ALL FIELDS OF INTEREST

A CATALOGUE OF SELECTED DOVER
BOOKS IN ALL FIELDS OF INTEREST

CELESTIAL OBJECTS FOR COMMON TELESCOPES, T. W. Webb. The most used book in amateur astronomy: inestimable aid for locating and identifying nearly 4,000 celestial objects. Edited, updated by Margaret W. Mayall. 77 illustrations. Total of 645pp. 5⅜ x 8½.
20917-2, 20918-0 Pa., Two-vol. set $10.00

HISTORICAL STUDIES IN THE LANGUAGE OF CHEMISTRY, M. P. Crosland. The important part language has played in the development of chemistry from the symbolism of alchemy to the adoption of systematic nomenclature in 1892. ". . . wholeheartedly recommended,"—Science. 15 illustrations. 416pp. of text. 5⅝ x 8¼. 63702-6 Pa. $7.50

BURNHAM'S CELESTIAL HANDBOOK, Robert Burnham, Jr. Thorough, readable guide to the stars beyond our solar system. Exhaustive treatment, fully illustrated. Breakdown is alphabetical by constellation: Andromeda to Cetus in Vol. 1; Chamaeleon to Orion in Vol. 2; and Pavo to Vulpecula in Vol. 3. Hundreds of illustrations. Total of about 2000pp. 6⅛ x 9¼.
23567-X, 23568-8, 23673-0 Pa., Three-vol. set $32.85

THEORY OF WING SECTIONS: INCLUDING A SUMMARY OF AIR-FOIL DATA, Ira H. Abbott and A. E. von Doenhoff. Concise compilation of subatomic aerodynamic characteristics of modern NASA wing sections, plus description of theory. 350pp. of tables. 693pp. 5⅜ x 8½.
60586-8 Pa. $9.95

DE RE METALLICA, Georgius Agricola. Translated by Herbert C. Hoover and Lou H. Hoover. The famous Hoover translation of greatest treatise on technological chemistry, engineering, geology, mining of early modern times (1556). All 289 original woodcuts. 638pp. 6¾ x 11.
60006-8 Clothbd. $19.95

THE ORIGIN OF CONTINENTS AND OCEANS, Alfred Wegener. One of the most influential, most controversial books in science, the classic statement for continental drift. Full 1966 translation of Wegener's final (1929) version. 64 illustrations. 246pp. 5⅜ x 8½.(EBE)61708-4 Pa. $5.00

THE PRINCIPLES OF PSYCHOLOGY, William James. Famous long course complete, unabridged. Stream of thought, time perception, memory, experimental methods; great work decades ahead of its time. Still valid, useful; read in many classes. 94 figures. Total of 1391pp. 5⅜ x 8½.
20381-6, 20382-4 Pa., Two-vol. set $17.90

YUCATAN BEFORE AND AFTER THE CONQUEST, Diego de Landa. First English translation of basic book in Maya studies, the only significant account of Yucatan written in the early post-Conquest era. Translated by distinguished Maya scholar William Gates. Appendices, introduction, 4 maps and over 120 illustrations added by translator. 162pp. 5⅜ x 8½.
23622-6 Pa. $3.00

THE MALAY ARCHIPELAGO, Alfred R. Wallace. Spirited travel account by one of founders of modern biology. Touches on zoology, botany, ethnography, geography, and geology. 62 illustrations, maps. 515pp. 5⅜ x 8½.
20187-2 Pa. $6.95

THE DISCOVERY OF THE TOMB OF TUTANKHAMEN, Howard Carter, A. C. Mace. Accompany Carter in the thrill of discovery, as ruined passage suddenly reveals unique, untouched, fabulously rich tomb. Fascinating account, with 106 illustrations. New introduction by J. M. White. Total of 382pp. 5⅜ x 8½. (Available in U.S. only)   23500-9 Pa. $5.50

THE WORLD'S GREATEST SPEECHES, edited by Lewis Copeland and Lawrence W. Lamm. Vast collection of 278 speeches from Greeks up to present. Powerful and effective models; unique look at history. Revised to 1970. Indices. 842pp. 5⅜ x 8½.   20468-5 Pa. $9.95

THE 100 GREATEST ADVERTISEMENTS, Julian Watkins. The priceless ingredient; His master's voice; 99 44/100% pure; over 100 others. How they were written, their impact, etc. Remarkable record. 130 illustrations. 233pp. 7⅞ x 10 3/5.   20540-1 Pa. $6.95

CRUICKSHANK PRINTS FOR HAND COLORING, George Cruickshank. 18 illustrations, one side of a page, on fine-quality paper suitable for watercolors. Caricatures of people in society (c. 1820) full of trenchant wit. Very large format. 32pp. 11 x 16.   23684-6 Pa. $6.00

THIRTY-TWO COLOR POSTCARDS OF TWENTIETH-CENTURY AMERICAN ART, Whitney Museum of American Art. Reproduced in full color in postcard form are 31 art works and one shot of the museum. Calder, Hopper, Rauschenberg, others. Detachable. 16pp. 8¼ x 11.
23629-3 Pa. $3.50

MUSIC OF THE SPHERES: THE MATERIAL UNIVERSE FROM ATOM TO QUASAR SIMPLY EXPLAINED, Guy Murchie. Planets, stars, geology, atoms, radiation, relativity, quantum theory, light, antimatter, similar topics. 319 figures. 664pp. 5⅜ x 8½.
21809-0, 21810-4 Pa., Two-vol. set $11.00

EINSTEIN'S THEORY OF RELATIVITY, Max Born. Finest semi-technical account; covers Einstein, Lorentz, Minkowski, and others, with much detail, much explanation of ideas and math not readily available elsewhere on this level. For student, non-specialist. 376pp. 5⅜ x 8½.
60769-0 Pa. $5.00

THE SENSE OF BEAUTY, George Santayana. Masterfully written discussion of nature of beauty, materials of beauty, form, expression; art, literature, social sciences all involved. 168pp. 5⅜ x 8½.   20238-0 Pa. $3.50

ON THE IMPROVEMENT OF THE UNDERSTANDING, Benedict Spinoza. Also contains *Ethics, Correspondence,* all in excellent R. Elwes translation. Basic works on entry to philosophy, pantheism, exchange of ideas with great contemporaries. 402pp. 5⅜ x 8½.   20250-X Pa. $5.95

THE TRAGIC SENSE OF LIFE, Miguel de Unamuno. Acknowledged masterpiece of existential literature, one of most important books of 20th century. Introduction by Madariaga. 367pp. 5⅜ x 8½.
20257-7 Pa. $6.00

THE GUIDE FOR THE PERPLEXED, Moses Maimonides. Great classic of medieval Judaism attempts to reconcile revealed religion (Pentateuch, commentaries) with Aristotelian philosophy. Important historically, still relevant in problems. Unabridged Friedlander translation. Total of 473pp. 5⅜ x 8½.   20351-4 Pa. $6.95

THE I CHING (THE BOOK OF CHANGES), translated by James Legge. Complete translation of basic text plus appendices by Confucius, and Chinese commentary of most penetrating divination manual ever prepared. Indispensable to study of early Oriental civilizations, to modern inquiring reader. 448pp. 5⅜ x 8½.   21062-6 Pa. $6.00

THE EGYPTIAN BOOK OF THE DEAD, E. A. Wallis Budge. Complete reproduction of Ani's papyrus, finest ever found. Full hieroglyphic text, interlinear transliteration, word for word translation, smooth translation. Basic work, for Egyptology, for modern study of psychic matters. Total of 533pp. 6½ x 9¼. (USCO)   21866-X Pa. $8.50

THE GODS OF THE EGYPTIANS, E. A. Wallis Budge. Never excelled for richness, fullness: all gods, goddesses, demons, mythical figures of Ancient Egypt; their legends, rites, incarnations, variations, powers, etc. Many hieroglyphic texts cited. Over 225 illustrations, plus 6 color plates. Total of 988pp. 6⅛ x 9¼. (EBE)
22055-9, 22056-7 Pa., Two-vol. set $20.00

THE STANDARD BOOK OF QUILT MAKING AND COLLECTING, Marguerite Ickis. Full information, full-sized patterns for making 46 traditional quilts, also 150 other patterns. Quilted cloths, lame, satin quilts, etc. 483 illustrations. 273pp. 6⅞ x 9⅝.   20582-7 Pa. $5.95

CORAL GARDENS AND THEIR MAGIC, Bronsilaw Malinowski. Classic study of the methods of tilling the soil and of agricultural rites in the Trobriand Islands of Melanesia. Author is one of the most important figures in the field of modern social anthropology. 143 illustrations. Indexes. Total of 911pp. of text. 5⅝ x 8¼. (Available in U.S. only)
23597-1 Pa. $12.95

THE PHILOSOPHY OF HISTORY, Georg W. Hegel. Great classic of Western thought develops concept that history is not chance but a rational process, the evolution of freedom. 457pp. 5⅜ x 8½. 20112-0 Pa. $6.00

LANGUAGE, TRUTH AND LOGIC, Alfred J. Ayer. Famous, clear introduction to Vienna, Cambridge schools of Logical Positivism. Role of philosophy, elimination of metaphysics, nature of analysis, etc. 160pp. 5⅜ x 8½. (USCO) 20010-8 Pa. $2.50

A PREFACE TO LOGIC, Morris R. Cohen. Great City College teacher in renowned, easily followed exposition of formal logic, probability, values, logic and world order and similar topics; no previous background needed. 209pp. 5⅜ x 8½. 23517-3 Pa. $4.95

REASON AND NATURE, Morris R. Cohen. Brilliant analysis of reason and its multitudinous ramifications by charismatic teacher. Interdisciplinary, synthesizing work widely praised when it first appeared in 1931. Second (1953) edition. Indexes. 496pp. 5⅜ x 8½. 23633-1 Pa. $7.50

AN ESSAY CONCERNING HUMAN UNDERSTANDING, John Locke. The only complete edition of enormously important classic, with authoritative editorial material by A. C. Fraser. Total of 1176pp. 5⅜ x 8½. 20530-4, 20531-2 Pa., Two-vol. set $16.00

HANDBOOK OF MATHEMATICAL FUNCTIONS WITH FORMULAS, GRAPHS, AND MATHEMATICAL TABLES, edited by Milton Abramowitz and Irene A. Stegun. Vast compendium: 29 sets of tables, some to as high as 20 places. 1,046pp. 8 x 10½. 61272-4 Pa. $17.95

MATHEMATICS FOR THE PHYSICAL SCIENCES, Herbert S. Wilf. Highly acclaimed work offers clear presentations of vector spaces and matrices, orthogonal functions, roots of polynomial equations, conformal mapping, calculus of variations, etc. Knowledge of theory of functions of real and complex variables is assumed. Exercises and solutions. Index. 284pp. 5⅝ x 8¼. 63635-6 Pa. $5.00

THE PRINCIPLE OF RELATIVITY, Albert Einstein et al. Eleven most important original papers on special and general theories. Seven by Einstein, two by Lorentz, one each by Minkowski and Weyl. All translated, unabridged. 216pp. 5⅜ x 8½. 60081-5 Pa. $3.50

THERMODYNAMICS, Enrico Fermi. A classic of modern science. Clear, organized treatment of systems, first and second laws, entropy, thermodynamic potentials, gaseous reactions, dilute solutions, entropy constant. No math beyond calculus required. Problems. 160pp. 5⅜ x 8½. 60361-X Pa. $4.00

ELEMENTARY MECHANICS OF FLUIDS, Hunter Rouse. Classic undergraduate text widely considered to be far better than many later books. Ranges from fluid velocity and acceleration to role of compressibility in fluid motion. Numerous examples, questions, problems. 224 illustrations. 376pp. 5⅝ x 8¼. 63699-2 Pa. $7.00

THE AMERICAN SENATOR, Anthony Trollope. Little known, long unavailable Trollope novel on a grand scale. Here are humorous comment on American vs. English culture, and stunning portrayal of a heroine/villainess. Superb evocation of Victorian village life. 561pp. 5⅜ x 8½.
23801-6 Pa. **$7.95**

WAS IT MURDER? James Hilton. The author of *Lost Horizon* and *Goodbye, Mr. Chips* wrote one detective novel (under a pen-name) which was quickly forgotten and virtually lost, even at the height of Hilton's fame. This edition brings it back—a finely crafted public school puzzle resplendent with Hilton's stylish atmosphere. A thoroughly English thriller by the creator of Shangri-la. 252pp. 5⅜ x 8. (Available in U.S. only)
23774-5 Pa. **$3.00**

CENTRAL PARK: A PHOTOGRAPHIC GUIDE, Victor Laredo and Henry Hope Reed. 121 superb photographs show dramatic views of Central Park: Bethesda Fountain, Cleopatra's Needle, Sheep Meadow, the Blockhouse, plus people engaged in many park activities: ice skating, bike riding, etc. Captions by former Curator of Central Park, Henry Hope Reed, provide historical view, changes, etc. Also photos of N.Y. landmarks on park's periphery. 96pp. 8½ x 11. 23750-8 Pa. **$4.50**

NANTUCKET IN THE NINETEENTH CENTURY, Clay Lancaster. 180 rare photographs, stereographs, maps, drawings and floor plans recreate unique American island society. Authentic scenes of shipwreck, lighthouses, streets, homes are arranged in geographic sequence to provide walking-tour guide to old Nantucket existing today. Introduction, captions. 160pp. 8⅞ x 11¾. 23747-8 Pa. **$7.95**

STONE AND MAN: A PHOTOGRAPHIC EXPLORATION, Andreas Feininger. 106 photographs by *Life* photographer Feininger portray man's deep passion for stone through the ages. Stonehenge-like megaliths, fortified towns, sculpted marble and crumbling tenements show textures, beauties, fascination. 128pp. 9¼ x 10¾. 23756-7 Pa. **$5.95**

CIRCLES, A MATHEMATICAL VIEW, D. Pedoe. Fundamental aspects of college geometry, non-Euclidean geometry, and other branches of mathematics: representing circle by point. Poincare model, isoperimetric property, etc. Stimulating recreational reading. 66 figures. 96pp. 5⅝ x 8¼.
63698-4 Pa. **$3.50**

THE DISCOVERY OF NEPTUNE, Morton Grosser. Dramatic scientific history of the investigations leading up to the actual discovery of the eighth planet of our solar system. Lucid, well-researched book by well-known historian of science. 172pp. 5⅜ x 8½. 23726-5 Pa. **$3.50**

THE DEVIL'S DICTIONARY. Ambrose Bierce. Barbed, bitter, brilliant witticisms in the form of a dictionary. Best, most ferocious satire America has produced. 145pp. 5⅜ x 8½. 20487-1 Pa. **$2.50**

HISTORY OF BACTERIOLOGY, William Bulloch. The only comprehensive history of bacteriology from the beginnings through the 19th century. Special emphasis is given to biography-Leeuwenhoek, etc. Brief accounts of 350 bacteriologists form a separate section. No clearer, fuller study, suitable to scientists and general readers, has yet been written. 52 illustrations. 448pp. 5⅝ x 8¼.   23761-3 Pa. $6.50

THE COMPLETE NONSENSE OF EDWARD LEAR, Edward Lear. All nonsense limericks, zany alphabets, Owl and Pussycat, songs, nonsense botany, etc., illustrated by Lear. Total of 321pp. 5⅜ x 8½. (Available in U.S. only)   20167-8 Pa. $4.50

INGENIOUS MATHEMATICAL PROBLEMS AND METHODS, Louis A. Graham. Sophisticated material from Graham Dial, applied and pure; stresses solution methods. Logic, number theory, networks, inversions, etc. 237pp. 5⅜ x 8½.   20545-2 Pa. $4.50

BEST MATHEMATICAL PUZZLES OF SAM LOYD, edited by Martin Gardner. Bizarre, original, whimsical puzzles by America's greatest puzzler. From fabulously rare Cyclopedia, including famous 14-15 puzzles, the Horse of a Different Color, 115 more. Elementary math. 150 illustrations. 167pp. 5⅜ x 8½.   20498-7 Pa. $3.50

THE BASIS OF COMBINATION IN CHESS, J. du Mont. Easy-to-follow, instructive book on elements of combination play, with chapters on each piece and every powerful combination team—two knights, bishop and knight, rook and bishop, etc. 250 diagrams. 218pp. 5⅜ x 8½. (Available in U.S. only)   23644-7 Pa. $4.50

MODERN CHESS STRATEGY, Ludek Pachman. The use of the queen, the active king, exchanges, pawn play, the center, weak squares, etc. Section on rook alone worth price of the book. Stress on the moderns. Often considered the most important book on strategy. 314pp. 5⅜ x 8½.
20290-9 Pa. $5.00

LASKER'S MANUAL OF CHESS, Dr. Emanuel Lasker. Great world champion offers very thorough coverage of all aspects of chess. Combinations, position play, openings, end game, aesthetics of chess, philosophy of struggle, much more. Filled with analyzed games. 390pp. 5⅜ x 8½.
20640-8 Pa. $5.95

500 MASTER GAMES OF CHESS, S. Tartakower, J. du Mont. Vast collection of great chess games from 1798-1938, with much material nowhere else readily available. Fully annotated, arranged by opening for easier study. 664pp. 5⅜ x 8½.   23208-5 Pa. $8.50

A GUIDE TO CHESS ENDINGS, Dr. Max Euwe, David Hooper. One of the finest modern works on chess endings. Thorough analysis of the most frequently encountered endings by former world champion. 331 examples, each with diagram. 248pp. 5⅜ x 8½.   23332-4 Pa. $3.95

THE COMPLETE BOOK OF DOLL MAKING AND COLLECTING, Catherine Christopher. Instructions, patterns for dozens of dolls, from rag doll on up to elaborate, historically accurate figures. Mould faces, sew clothing, make doll houses, etc. Also collecting information. Many illustrations. 288pp. 6 x 9. 22066-4 Pa. $4.95

THE DAGUERREOTYPE IN AMERICA, Beaumont Newhall. Wonderful portraits, 1850's townscapes, landscapes; full text plus 104 photographs. The basic book. Enlarged 1976 edition. 272pp. 8¼ x 11¼. 23322-7 Pa. $7.95

CRAFTSMAN HOMES, Gustav Stickley. 296 architectural drawings, floor plans, and photographs illustrate 40 different kinds of "Mission-style" homes from *The Craftsman* (1901-16), voice of American style of simplicity and organic harmony. Thorough coverage of Craftsman idea in text and picture, now collector's item. 224pp. 8⅛ x 11. 23791-5 Pa. $6.50

PEWTER-WORKING: INSTRUCTIONS AND PROJECTS, Burl N. Osborn. & Gordon O. Wilber. Introduction to pewter-working for amateur craftsman. History and characteristics of pewter; tools, materials, step-by-step instructions. Photos, line drawings, diagrams. Total of 160pp. 7⅞ x 10¾. 23786-9 Pa. $3.50

THE GREAT CHICAGO FIRE, edited by David Lowe. 10 dramatic, eye-witness accounts of the 1871 disaster, including one of the aftermath and rebuilding, plus 70 contemporary photographs and illustrations of the ruins—courthouse, Palmer House, Great Central Depot, etc. Introduction by David Lowe. 87pp. 8¼ x 11. 23771-0 Pa. $4.00

SILHOUETTES: A PICTORIAL ARCHIVE OF VARIED ILLUSTRATIONS, edited by Carol Belanger Grafton. Over 600 silhouettes from the 18th to 20th centuries include profiles and full figures of men and women, children, birds and animals, groups and scenes, nature, ships, an alphabet. Dozens of uses for commercial artists and craftspeople. 144pp. 8⅜ x 11¼. 23781-8 Pa. $4.50

ANIMALS: 1,419 COPYRIGHT-FREE ILLUSTRATIONS OF MAMMALS, BIRDS, FISH, INSECTS, ETC., edited by Jim Harter. Clear wood engravings present, in extremely lifelike poses, over 1,000 species of animals. One of the most extensive copyright-free pictorial sourcebooks of its kind. Captions. Index. 284pp. 9 x 12. 23766-4 Pa. $8.95

INDIAN DESIGNS FROM ANCIENT ECUADOR, Frederick W. Shaffer. 282 original designs by pre-Columbian Indians of Ecuador (500-1500 A.D.). Designs include people, mammals, birds, reptiles, fish, plants, heads, geometric designs. Use as is or alter for advertising, textiles, leathercraft, etc. Introduction. 95pp. 8¾ x 11¼. 23764-8 Pa. $4.50

SZIGETI ON THE VIOLIN, Joseph Szigeti. Genial, loosely structured tour by premier violinist, featuring a pleasant mixture of reminiscenes, insights into great music and musicians, innumerable tips for practicing violinists. 385 musical passages. 256pp. 5⅝ x 8¼. 23763-X Pa. $4.00

TONE POEMS, SERIES II: TILL EULENSPIEGELS LUSTIGE STREICHE, ALSO SPRACH ZARATHUSTRA, AND EIN HELDEN-LEBEN, Richard Strauss. Three important orchestral works, including very popular *Till Eulenspiegel's Marry Pranks,* reproduced in full score from original editions. Study score. 315pp. 9⅜ x 12¼. (Available in U.S. only)
23755-9 Pa. $8.95

TONE POEMS, SERIES I: DON JUAN, TOD UND VERKLARUNG AND DON QUIXOTE, Richard Strauss. Three of the most often performed and recorded works in entire orchestral repertoire, reproduced in full score from original editions. Study score. 286pp. 9⅜ x 12¼. (Available in U.S. only)
23754-0 Pa. $8.95

11 LATE STRING QUARTETS, Franz Joseph Haydn. The form which Haydn defined and "brought to perfection." *(Grove's).* 11 string quartets in complete score, his last and his best. The first in a projected series of the complete Haydn string quartets. Reliable modern Eulenberg edition, otherwise difficult to obtain. 320pp. 8⅜ x 11¼. (Available in U.S. only)
23753-2 Pa. $8.95

FOURTH, FIFTH AND SIXTH SYMPHONIES IN FULL SCORE, Peter Ilyitch Tchaikovsky. Complete orchestral scores of Symphony No. 4 in F Minor, Op. 36; Symphony No. 5 in E Minor, Op. 64; Symphony No. 6 in B Minor, "Pathetique," Op. 74. Bretikopf & Hartel eds. Study score. 480pp. 9⅜ x 12¼.
23861-X Pa. $10.95

THE MARRIAGE OF FIGARO: COMPLETE SCORE, Wolfgang A. Mozart. Finest comic opera ever written. Full score, not to be confused with piano renderings. Peters edition. Study score. 448pp. 9⅜ x 12¼. (Available in U.S. only)
23751-6 Pa. $12.95

"IMAGE" ON THE ART AND EVOLUTION OF THE FILM, edited by Marshall Deutelbaum. Pioneering book brings together for first time 38 groundbreaking articles on early silent films from *Image* and 263 illustrations newly shot from rare prints in the collection of the International Museum of Photography. A landmark work. Index. 256pp. 8¼ x 11.
23777-X Pa. $8.95

AROUND-THE-WORLD COOKY BOOK, Lois Lintner Sumption and Marguerite Lintner Ashbrook. 373 cooky and frosting recipes from 28 countries (America, Austria, China, Russia, Italy, etc.) include Viennese kisses, rice wafers, London strips, lady fingers, hony, sugar spice, maple cookies, etc. Clear instructions. All tested. 38 drawings. 182pp. 5⅜ x 8.
23802-4 Pa. $2.75

THE ART NOUVEAU STYLE, edited by Roberta Waddell. 579 rare photographs, not available elsewhere, of works in jewelry, metalwork, glass, ceramics, textiles, architecture and furniture by 175 artists—Mucha, Seguy, Lalique, Tiffany, Gaudin, Hohlwein, Saarinen, and many others. 288pp. 8⅜ x 11¼.
23515-7 Pa. $8.95

THE CURVES OF LIFE, Theodore A. Cook. Examination of shells, leaves, horns, human body, art, etc., in "*the* classic reference on how the golden ratio applies to spirals and helices in nature . . . . "—Martin Gardner. 426 illustrations. Total of 512pp. 5⅜ x 8½.   23701-X Pa. $6.95

AN ILLUSTRATED FLORA OF THE NORTHERN UNITED STATES AND CANADA, Nathaniel L. Britton, Addison Brown. Encyclopedic work covers 4666 species, ferns on up. Everything. Full botanical information, illustration for each. This earlier edition is preferred by many to more recent revisions. 1913 edition. Over 4000 illustrations, total of 2087pp. 6⅛ x 9¼.   22642-5, 22643-3, 22644-1 Pa., Three-vol. set $28.50

MANUAL OF THE GRASSES OF THE UNITED STATES, A. S. Hitchcock, U.S. Dept. of Agriculture. The basic study of American grasses, both indigenous and escapes, cultivated and wild. Over 1400 species. Full descriptions, information. Over 1100 maps, illustrations. Total of 1051pp. 5⅜ x 8½.   22717-0, 22718-9 Pa., Two-vol. set $17.00

THE CACTACEAE,, Nathaniel L. Britton, John N. Rose. Exhaustive, definitive. Every cactus in the world. Full botanical descriptions. Thorough statement of nomenclatures, habitat, detailed finding keys. The one book needed by every cactus enthusiast. Over 1275 illustrations. Total of 1080pp. 8 x 10¼.   21191-6, 21192-4 Clothbd., Two-vol. set $50.00

AMERICAN MEDICINAL PLANTS, Charles F. Millspaugh. Full descriptions, 180 plants covered: history; physical description; methods of preparation with all chemical constituents extracted; all claimed curative or adverse effects. 180 full-page plates. Classification table. 804pp. 6½ x 9¼.   23034-1 Pa. $13.95

A MODERN HERBAL, Margaret Grieve. Much the fullest, most exact, most useful compilation of herbal material. Gigantic alphabetical encyclopedia, from aconite to zedoary, gives botanical information, medical properties, folklore, economic uses, and much else. Indispensable to serious reader. 161 illustrations. 888pp. 6½ x 9¼. (Available in U.S. only)   22798-7, 22799-5 Pa., Two-vol. set $15.00

THE HERBAL or GENERAL HISTORY OF PLANTS, John Gerard. The 1633 edition revised and enlarged by Thomas Johnson. Containing almost 2850 plant descriptions and 2705 superb illustrations, Gerard's *Herbal* is a monumental work, the book all modern English herbals are derived from, the one herbal every serious enthusiast should have in its entirety. Original editions are worth perhaps $750. 1678pp. 8½ x 12¼.   23147-X Clothbd. $75.00

MANUAL OF THE TREES OF NORTH AMERICA, Charles S. Sargent. The basic survey of every native tree and tree-like shrub, 717 species in all. Extremely full descriptions, information on habitat, growth, locales, economics, etc. Necessary to every serious tree lover. Over 100 finding keys. 783 illustrations. Total of 986pp. 5⅜ x 8½.   20277-1, 20278-X Pa., Two-vol. set $12.00

GREAT NEWS PHOTOS AND THE STORIES BEHIND THEM, John Faber. Dramatic volume of 140 great news photos, 1855 through 1976, and revealing stories behind them, with both historical and technical information. Hindenburg disaster, shooting of Oswald, nomination of Jimmy Carter, etc. 160pp. 8¼ x 11. 23667-6 Pa. $6.00

CRUICKSHANK'S PHOTOGRAPHS OF BIRDS OF AMERICA, Allan D. Cruickshank. Great ornithologist, photographer presents 177 closeups, groupings, panoramas, flightings, etc., of about 150 different birds. Expanded Wings in the Wilderness. Introduction by Helen G. Cruickshank. 191pp. 8¼ x 11. 23497-5 Pa. $7.95

AMERICAN WILDLIFE AND PLANTS, A. C. Martin, et al. Describes food habits of more than 1000 species of mammals, birds, fish. Special treatment of important food plants. Over 300 illustrations. 500pp. 5⅜ x 8½. 20793-5 Pa. $6.50

THE PEOPLE CALLED SHAKERS, Edward D. Andrews. Lifetime of research, definitive study of Shakers: origins, beliefs, practices, dances, social organization, furniture and crafts, impact on 19th-century USA, present heritage. Indispensable to student of American history, collector. 33 illustrations. 351pp. 5⅜ x 8½. 21081-2 Pa. $4.50

OLD NEW YORK IN EARLY PHOTOGRAPHS, Mary Black. New York City as it was in 1853-1901, through 196 wonderful photographs from N.-Y. Historical Society. Great Blizzard, Lincoln's funeral procession, great buildings. 228pp. 9 x 12. 22907-6 Pa. $8.95

MR. LINCOLN'S CAMERA MAN: MATHEW BRADY, Roy Meredith. Over 300 Brady photos reproduced directly from original negatives, photos. Jackson, Webster, Grant, Lee, Carnegie, Barnum; Lincoln; Battle Smoke, Death of Rebel Sniper, Atlanta Just After Capture. Lively commentary. 368pp. 8⅜ x 11¼. 23021-X Pa. $11.95

TRAVELS OF WILLIAM BARTRAM, William Bartram. From 1773-8, Bartram explored Northern Florida, Georgia, Carolinas, and reported on wild life, plants, Indians, early settlers. Basic account for period, entertaining reading. Edited by Mark Van Doren. 13 illustrations. 141pp. 5⅜ x 8½. 20013-2 Pa. $6.00

THE GENTLEMAN AND CABINET MAKER'S DIRECTOR, Thomas Chippendale. Full reprint, 1762 style book, most influential of all time; chairs, tables, sofas, mirrors, cabinets, etc. 200 plates, plus 24 photographs of surviving pieces. 249pp. 9⅞ x 12¾. 21601-2 Pa. $8.95

AMERICAN CARRIAGES, SLEIGHS, SULKIES AND CARTS, edited by Don H. Berkebile. 168 Victorian illustrations from catalogues, trade journals, fully captioned. Useful for artists. Author is Assoc. Curator, Div. of Transportation of Smithsonian Institution. 168pp. 8½ x 9½. 23328-6 Pa. $5.00

SECOND PIATIGORSKY CUP, edited by Isaac Kashdan. One of the greatest tournament books ever produced in the English language. All 90 games of the 1966 tournament, annotated by players, most annotated by both players. Features Petrosian, Spassky, Fischer, Larsen, six others. 228pp. 5⅜ x 8½. 23572-6 Pa. $3.50

ENCYCLOPEDIA OF CARD TRICKS, revised and edited by Jean Hugard. How to perform over 600 card tricks, devised by the world's greatest magicians: impromptus, spelling tricks, key cards, using special packs, much, much more. Additional chapter on card technique. 66 illustrations. 402pp. 5⅜ x 8½. (Available in U.S. only) 21252-1 Pa. $5.95

MAGIC: STAGE ILLUSIONS, SPECIAL EFFECTS AND TRICK PHO- TOGRAPHY, Albert A. Hopkins, Henry R. Evans. One of the great classics; fullest, most authorative explanation of vanishing lady, levitations, scores of other great stage effects. Also small magic, automata, stunts. 446 illus- trations. 556pp. 5⅜ x 8½. 23344-8 Pa. $6.95

THE SECRETS OF HOUDINI, J. C. Cannell. Classic study of Houdini's incredible magic, exposing closely-kept professional secrets and revealing, in general terms, the whole art of stage magic. 67 illustrations. 279pp. 5⅜ x 8½. 22913-0 Pa. $4.00

HOFFMANN'S MODERN MAGIC, Professor Hoffmann. One of the best, and best-known, magicians' manuals of the past century. Hundreds of tricks from card tricks and simple sleight of hand to elaborate illusions involving construction of complicated machinery. 332 illustrations. 563pp. 5⅜ x 8½. 23623-4 Pa. $6.95

THOMAS NAST'S CHRISTMAS DRAWINGS, Thomas Nast. Almost all Christmas drawings by creator of image of Santa Claus as we know it, and one of America's foremost illustrators and political cartoonists. 66 illustrations. 3 illustrations in color on covers. 96pp. 8⅜ x 11¼. 23660-9 Pa. $3.50

FRENCH COUNTRY COOKING FOR AMERICANS, Louis Diat. 500 easy-to-make, authentic provincial recipes compiled by former head chef at New York's Fitz-Carlton Hotel: onion soup, lamb stew, potato pie, more. 309pp. 5⅜ x 8½. 23665-X Pa. $3.95

SAUCES, FRENCH AND FAMOUS, Louis Diat. Complete book gives over 200 specific recipes: bechamel, Bordelaise, hollandaise, Cumberland, apri- cot, etc. Author was one of this century's finest chefs, originator of vichyssoise and many other dishes. Index. 156pp. 5⅜ x 8. 23663-3 Pa. $2.75

TOLL HOUSE TRIED AND TRUE RECIPES, Ruth Graves Wakefield. Authentic recipes from the famous Mass. restaurant: popovers, veal and ham loaf, Toll House baked beans, chocolate cake crumb pudding, much more. Many helpful hints. Nearly 700 recipes. Index. 376pp. 5⅜ x 8½. 23560-2 Pa. $4.95

ILLUSTRATED GUIDE TO SHAKER FURNITURE, Robert Meader. Director, Shaker Museum, Old Chatham, presents up-to-date coverage of all furniture and appurtenances, with much on local styles not available elsewhere. 235 photos. 146pp. 9 x 12. 22819-3 Pa. $6.95

COOKING WITH BEER, Carole Fahy. Beer has as superb an effect on food as wine, and at fraction of cost. Over 250 recipes for appetizers, soups, main dishes, desserts, breads, etc. Index. 144pp. 5⅜ x 8½. (Available in U.S. only) 23661-7 Pa. $3.00

STEWS AND RAGOUTS, Kay Shaw Nelson. This international cookbook offers wide range of 108 recipes perfect for everyday, special occasions, meals-in-themselves, main dishes. Economical, nutritious, easy-to-prepare: goulash, Irish stew, boeuf bourguignon, etc. Index. 134pp. 5⅜ x 8½. 23662-5 Pa. $3.95

DELICIOUS MAIN COURSE DISHES, Marian Tracy. Main courses are the most important part of any meal. These 200 nutritious, economical recipes from around the world make every meal a delight. "I . . . have found it so useful in my own household,"—N.Y. Times. Index. 219pp. 5⅜ x 8½. 23664-1 Pa. $3.95

FIVE ACRES AND INDEPENDENCE, Maurice G. Kains. Great back-to-the-land classic explains basics of self-sufficient farming: economics, plants, crops, animals, orchards, soils, land selection, host of other necessary things. Do not confuse with skimpy faddist literature; Kains was one of America's greatest agriculturalists. 95 illustrations. 397pp. 5⅜ x 8½. 20974-1 Pa. $4.95

A PRACTICAL GUIDE FOR THE BEGINNING FARMER, Herbert Jacobs. Basic, extremely useful first book for anyone thinking about moving to the country and starting a farm. Simpler than Kains, with greater emphasis on country living in general. 246pp. 5⅜ x 8½. 23675-7 Pa. $3.95

PAPERMAKING, Dard Hunter. Definitive book on the subject by the foremost authority in the field. Chapters dealing with every aspect of history of craft in every part of the world. Over 320 illustrations. 2nd, revised and enlarged (1947) edition. 672pp. 5⅜ x 8½. 23619-6 Pa. $8.95

THE ART DECO STYLE, edited by Theodore Menten. Furniture, jewelry, metalwork, ceramics, fabrics, lighting fixtures, interior decors, exteriors, graphics from pure French sources. Best sampling around. Over 400 photographs. 183pp. 8⅜ x 11¼. 22824-X Pa. $6.95

ACKERMANN'S COSTUME PLATES, Rudolph Ackermann. Selection of 96 plates from the Repository of Arts, best published source of costume for English fashion during the early 19th century. 12 plates also in color. Captions, glossary and introduction by editor Stella Blum. Total of 120pp. 8⅜ x 11¼. 23690-0 Pa. $5.00

THE ANATOMY OF THE HORSE, George Stubbs. Often considered the great masterpiece of animal anatomy. Full reproduction of 1766 edition, plus prospectus; original text and modernized text. 36 plates. Introduction by Eleanor Garvey. 121pp. 11 x 14¾. 23402-9 Pa. $8.95

BRIDGMAN'S LIFE DRAWING, George B. Bridgman. More than 500 illustrative drawings and text teach you to abstract the body into its major masses, use light and shade, proportion; as well as specific areas of anatomy, of which Bridgman is master. 192pp. 6½ x 9¼. (Available in U.S. only) 22710-3 Pa. $4.50

ART NOUVEAU DESIGNS IN COLOR, Alphonse Mucha, Maurice Verneuil, Georges Auriol. Full-color reproduction of *Combinaisons ornementales* (c. 1900) by Art Nouveau masters. Floral, animal, geometric, interlacings, swashes—borders, frames, spots—all incredibly beautiful. 60 plates, hundreds of designs. 9⅜ x 8-1/16. 22885-1 Pa. $4.50

FULL-COLOR FLORAL DESIGNS IN THE ART NOUVEAU STYLE, E. A. Seguy. 166 motifs, on 40 plates, from *Les fleurs et leurs applications decoratives* (1902): borders, circular designs, repeats, allovers, "spots." All in authentic Art Nouveau colors. 48pp. 9⅜ x 12¼.
23439-8 Pa. $6.00

A DIDEROT PICTORIAL ENCYCLOPEDIA OF TRADES AND IN-DUSTRY, edited by Charles C. Gillispie. 485 most interesting plates from the great French Encyclopedia of the 18th century show hundreds of working figures, artifacts, process, land and cityscapes; glassmaking, paper-making, metal extraction, construction, weaving, making furniture, clothing, wigs, dozens. of other activities. Plates fully explained. 920pp. 9 x 12.
22284-5, 22285-3 Clothbd., Two-vol. set $50.00

HANDBOOK OF EARLY ADVERTISING ART, Clarence P. Hornung. Largest collection of copyright-free early and antique advertising art ever compiled. Over 6,000 illustrations, from Franklin's time to the 1890's for special effects, novelty. Valuable source, almost inexhaustible.
Pictorial Volume. Agriculture, the zodiac, animals, autos, birds, Christmas, fire engines, flowers, trees, musical instruments, ships, games and sports, much more. Arranged by subject matter and use. 237 plates. 288pp. 9 x 12.
20122-8 Clothbd. $15.00

Typographical Volume. Roman and Gothic faces ranging from 10 point to 300 point, "Barnum," German and Old English faces, script, logotypes, scrolls and flourishes, 1115 ornamental initials, 67 complete alphabets, more. 310 plates. 320pp. 9 x 12. 20123-6 Clothbd. $15.00

CALLIGRAPHY (CALLIGRAPHIA LATINA), J. G. Schwandner. High point of 18th-century ornamental calligraphy. Very ornate initials, scrolls, borders, cherubs, birds, lettered examples. 172pp. 9 x 13.
20475-8 Pa. $7.95

GEOMETRY, RELATIVITY AND THE FOURTH DIMENSION, Rudolf Rucker. Exposition of fourth dimension, means of visualization, concepts of relativity as Flatland characters continue adventures. Popular, easily followed yet accurate, profound. 141 illustrations. 133pp. 5⅜ x 8½.
23400-2 Pa. $2.75

THE ORIGIN OF LIFE, A. I. Oparin. Modern classic in biochemistry, the first rigorous examination of possible evolution of life from nitrocarbon compounds. Non-technical, easily followed. Total of 295pp. 5⅜ x 8½.
60213-3 Pa. $5.95

PLANETS, STARS AND GALAXIES, A. E. Fanning. Comprehensive introductory survey: the sun, solar system, stars, galaxies, universe, cosmology; quasars, radio stars, etc. 24pp. of photographs. 189pp. 5⅜ x 8½. (Available in U.S. only)
21680-2 Pa. $3.75

THE THIRTEEN BOOKS OF EUCLID'S ELEMENTS, translated with introduction and commentary by Sir Thomas L. Heath. Definitive edition. Textual and linguistic notes, mathematical analysis, 2500 years of critical commentary. Do not confuse with abridged school editions. Total of 1414pp. 5⅜ x 8½.   60088-2, 60089-0, 60090-4 Pa., Three-vol. set $19.50

Prices subject to change without notice.

Available at your book dealer or write for free catalogue to Dept. GI, Dover Publications, Inc., 31 East 2nd St. Mineola., N.Y. 11501. Dover publishes more than 175 books each year on science, elementary and advanced mathematics, biology, music, art, literary history, social sciences and other areas.